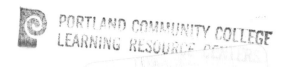 

ENGINE DIAGNOSTICS AND TUNE-UP

ENGINE DIAGNOSTICS AND TUNE-UP

JACK ERJAVEC
Columbus Technical Institute

PRENTICE-HALL, INC., Englewood Cliffs, NJ 07632

Library of Congress Cataloging in Publication Data

Erjavec, Jack.
Engine diagnostics and tune-up.

Includes index.
1. Automobiles—Motors—Maintenance and
repair.   I. Title.
TL210.E74   1986   629.2'5'0288   85-6454
ISBN 0-13-277823-8

Editorial/production supervision and
  interior design: Ellen Denning
Manufacturing buyer: Rhett Conklin

Printed in the United States of America

10 9 8 7 6 5 4 3 2 1

ISBN 0-13-277823-8 01

PRENTICE-HALL INTERNATIONAL (UK) LIMITED, *London*
PRENTICE-HALL OF AUSTRALIA PTY. LIMITED, *Sydney*
PRENTICE-HALL CANADA INC., *Toronto*
PRENTICE-HALL HISPANOAMERICANA, S.A., *Mexico*
PRENTICE-HALL OF INDIA PRIVATE LIMITED, *New Delhi*
PRENTICE-HALL OF JAPAN, INC., *Tokyo*
PRENTICE-HALL OF SOUTHEAST ASIA PTE. LTD., *Singapore*
EDITORA PRENTICE-HALL DO BRASIL, LTDA., *Rio de Janeiro*
WHITEHALL BOOKS LIMITED, *Wellington, New Zealand*

To my family:
ROSE, MEGAN, MOIRA, and ERIN

CONTENTS

PREFACE

The technology of the automobile is advancing at a rate never before experienced in the history of the automobile. While the consumers and the environment benefit from these advances, automotive technicians struggle with the new systems in an attempt to keep these complex automobiles running well and efficiently. Simple diagnostics are no longer the rule. Detailed testing and intelligent interpretation of the results are necessary for effective troubleshooting of these complex systems.

Many of these advances have focused on the engine and its systems. But what had been distinct systems are now integrated into one complex system. Only through an understanding of the system and its components can problems be diagnosed quickly and correctly. It is important to understand how each component functions and how it affects the operation of the total system.

The main topic of this text is the diagnosis of engine problems. Every attempt is made to allow the reader to develop diagnostic skills for the new technology. The development of these skills requires a functional understanding of the systems and of the appropriate test equipment. The text takes an applied theoretical approach, beginning with some very basic facts and ending with a detailed look at the more complex systems.

The aftermarkets' and manufacturers' service manuals contain the step-by-step procedures for testing the components of particular systems. The text does not attempt to duplicate this information. In fact, the reader is often told to go to these manuals for the exact procedures. The text does attempt to provide the information needed to gain an understanding of the systems. This understanding will assist a technician in determining when to conduct the test procedures given in service manuals and will permit a clearer understanding of the test results.

The text provides not only a functional understanding of the advanced systems, but also defines the reasons for the refinements or changes brought into the systems by the new technology. This prepares a person, through a knowledge of the basics, to understand the changes yet to come in the future.

Testing sequences and instruments, covered in the text, are approached from the most basic to the most complex. Proper testing techniques and result interpretation are vital to effective troubleshooting. An understanding of what a test actually tests and how certain failures affect the test results helps the technician not only to troubleshoot accurately but also helps in deciding when it is appropriate to perform a particular test.

The entire progression of the text is from the basic to the complex. The first six chapters are concerned with the basic applied theories of the engine and its systems: induction, ignition, and emission controls. Readers who have a good solid understanding of these basics may elect to start the text at Chapter 7 and use the preceding chapters as reference.

Personal safety and the safety of others should be the concern of all. Therefore, it is recommended that every reader study the contents of Chapter 7, which deals with that topic.

Chapter 8 covers the proper procedures for conducting a maintenance tune-up. This discussion includes the replacement of typical tune-up items for engines equipped with electronic and point-type ignitions, carburetion and fuel injection, and emission control devices.

The next seven chapters cover the essentials of troubleshooting. This section begins with an examination of logical approaches to diagnosis (Chapter 9). Common diagnostic equipment, such as compression testers, vacuum gauges, exhaust analyzers, and engine analyzers, are covered in these chapters. Two chapters, Chapters 14 and 15, cover the use and interpretation of a "scope."

The last three chapters discuss the operation and testing of the most common and latest electronic engine controls, and fuel injection and diesel systems. Every attempt is made in these chapters to point out the similarities of the different systems used by the manufacturers.

I would like to acknowledge and give a word of thanks to those who helped and supported me in the writing of this text. To my wife, Rose, who not only put up with the time I spent working on this project but also helped in the editing and typing of the manuscript. To my friend, Mike Bell, who spent hours on the word processor for me, as did Deb Hausch. To the automotive faculty, former and present students, and the administration of Columbus Technical Institute, who helped and supported me in this project, often not knowing they were doing so. And to Herb Ellinger, who, during the review of the manuscript, gave me many helpful suggestions and comments.

Many organizations and individuals have provided reference material and illustrations for use in this text. The author wishes to express his gratitude to the following for their cooperation and contributions:

Bear Automotive

Champion Spark Plug Company

Chrysler Corporation

Countdown, Inc.

Ford Motor Company

General Motors Corporation

    Buick Motor Division

    Cadillac Motor Car Division

    Chevrolet Motor Division

    Delco-Remy

    Oldsmobile Motor Division

    Pontiac Motor Division

    AC/Delco Parts Division

Robert Bosch Company

Sun Electric Corporation

Toyota Motors

JACK ERJAVEC

1

INTRODUCTION

Figure 1–1

How often have you heard someone say, "It probably just needs a tune-up!", when their car is hard to start or is not running right. To the typical owner of a poor-running car, a tune-up is what is needed to make the car run well again. Some car owners blame the hard starting or poor performance on themselves for not having a tune-up performed on their car recently. It is commonly thought that to avoid hard starting and poor performance, you must have a tune-up done periodically.

Tune-ups are perceived by most car owners and drivers as an operation that both prevents and cures engine running problems. Therefore, cars are brought into service departments for a tune-up either because they are running poorly or because the owner feels that it is time for a tune-up. It is time for a tune-up because it is not running quite right or it is merely time for its semiannual or annual tune-up.

TYPICAL TUNE-UP

Along with the perception of what a tune-up can do for a car, there is also a common perception of what is included in a tune-up. Car owners learn what is included in a tune-up through advertisements in the newspapers or on the radio and television. Service facilities offer many tune-up specials and with the announcement of

1

these specials they include a description of what the tune-up includes. Figure 1-2 shows those items that are included in a typical tune-up. A typical description of this service includes: "checking the charging, starting, and fuel systems; adjusting the timing, carburetor, and choke; installing new spark plugs; checking the air filter and PCV valve; and road testing the car." This description and others like it define to car owners what a tune-up is. The installation of new spark plugs and the inspection of certain parts together with some adjustments will prevent and cure their engine problems, or so they feel.

There are many things that can cause an engine to run poorly and the items included in the typical tune-up are only a few of them. For an engine to run well, everything must happen at the right time and place. There are many individual systems that must work together to make the engine run well. If there is a fault in one of these systems, the engine will run poorly or not run at all. The typical tune-up does not include all of these systems and therefore cannot and will not cure or prevent all engine problems.

The items included in the typical tune-up are those items that are the most

Figure 1-2   (Courtesy of Countdown, Inc.)

familiar to the public and service facilities. These are items that have become rather familiar terms and common service items, because in the past they were the cause of a majority of problems. Because of federally regulated standards for exhaust emission quality, fuel consumption, and safety, systems have been introduced that are not so familiar to car owners or to service facilities. These new systems work together with the old and can be the cause of driveability problems. The driveability of a car is its ability to start properly, run smoothly, and have immediate throttle response. A maintenance tune-up may improve driveability, but it may not if there is a fault in a system not part of the typical tune-up.

MAINTENANCE TUNE-UP

Each manufacturer of automobiles today has a list of suggested services and replacement of parts with a suggested service interval. Many of these items play a role in how the engine runs. As you can see in Fig. 1-3, these items should be replaced and/or checked on a routine basis. These items should be part of a tune-up. These services and the frequency with which they should be completed are suggested by the car makers to prevent running problems. For car owners who have a tune-up done to avoid hard starting or poor performance problems, a tune-up that includes the checking, adjusting, and replacement of the items recommended by the car manufacturer is precisely what they want. This sort of tune-up is best referred to as a scheduled maintenance tune-up. The typical tune-up as advertised by service facilities is a scheduled maintenance tune-up and assumes that either the car has no running problem, or that if it does, the problem is caused by those items normally included in the service.

MAINTENANCE SCHEDULE "B"	Service Interval — Time in Months or Miles (or kilometers) in thousands, whichever occurs first, unless otherwise specified.					
MONTHS	10	20	30	40	50	60
MILES (000)	10	20	30	40	50	60
KILOMETERS (000)	16	32	48	64	80	96
MAINTENANCE OPERATION:						
Change Engine Oil (1)	B	B	B	B	B	B
Replace Engine Oil Filter (1)	B	B	B	B	B	B
Replace Spark Plugs (1, 2)			B			B
Check* Coolant Protection and Condition (3)	ANNUALLY					
Check* Cooling System, Hoses and Clamps; replace coolant** (3)			B			B
Check* Accessory Drive Belt Condition and Tension	(B)		B			B
Replace Engine Timing Belt *						B*
Replace Carburetor Air Cleaner Element (4)			B			B
Check* Idle Speeds (2)	B	(B)				B
Replace Crankcase Emissions Filter (4, 5)		(B)				B
Choke System Linkage: Clean and verify freedom of movement			B			B
Replace EGR Valve and Clean EGR Manifold Passages — 1.3L Only (7)						
Inspect Exhaust System Heat Shields (6)			B			B
Drain and Refill Automatic Transaxle Fluid — Severe or Continuous Service Only (1)		B		B		B
Check Automatic Transaxle T.V. Linkage	B	B				

Figure 1-3 (Courtesy of Ford Motor Company)

DIAGNOSTIC TUNE-UP

Although some problems can be corrected by a maintenance tune-up, other engine performance problems must be carefully and completely diagnosed. When a car owner desires a tune-up because a car does not run right and the normal items of a tune-up do not cure the problem, a diagnostic tune-up must be performed. Simply said, the problems must be corrected and then a maintenance tune-up performed.

There are many methods for solving a problem. Unfortunately, one of these is to guess! The other methods require a certain amount of knowledge, skill, and training. The first step in problem solving is simply identifying the problem. You need to know what the problem is before you can correct it. Imagine being told to go out into the field in Fig. 1–4 to look for something. Not knowing what this "something" is, your only chance of finding it is if you happen to get lucky and make a good guess. All engine running problems are caused by something, and to correct the problem you need to know what that something is or to be a good guesser.

Your chances of finding something in a field are far greater if you know some details about the thing you are looking for. To solve an engine problem you need to know the details of the problem. To get these details you must have an understanding of how all the systems of that engine work. You must be able to test, to determine which system has the problem, and you must have an understanding of that system, in order to determine the cause of the problem within that system. To gain an understanding of how the engine and its systems work, you need to start at the basics and build your understanding from there.

An engine produces the power necessary to move a car through a series of reactions between air, fuel, and heat (Fig. 1–5). The right amount of air mixed with the right amount of fuel will quickly explode when the right amount of heat is present. The exploding of the fuel-and-air mixture is caused by the rapid expansion of the mixture due to the heat. How much the mixture will expand depends on the amount of air, fuel, and heat. How quickly the mixture explodes or expands depends on how quickly the heat is introduced. If heat is gradually introduced, the mixture will gradually expand and there will be no apparent explosion. However,

Figure 1–4   How many somethings are in this field?

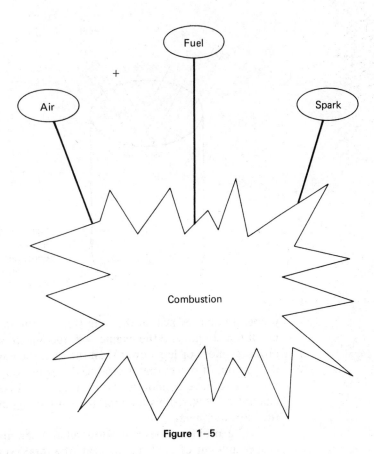

Figure 1-5

if high heat is introduced quickly, there will be a violent explosion due to the sudden expansion of the mixture.

The power or force produced by an explosion has been demonstrated many times in the movies. We have seen doors blown down, cars blown up, and cities destroyed by explosions. All of these explosions are caused by a rapid expansion of a fuel-and-air mixture. The power produced by an engine is the result of power produced by a series of minor explosions. These explosions take place in a sealed container which has a bottom that moves downward, as shown in Fig. 1-6, as the expansion of the fuel-and-air mixture occurs. This container is designed to allow the bottom to move downward and yet remain sealed. The bottom moves with the same force or with the same amount of power produced by the expansion.

As shown in Fig. 1-7, if the container were not sealed, the air-and-fuel mixture, while expanding, could leak out and the bottom would not have as much force on it. Power would be wasted. Ideally, most of the power produced by the expansion of the fuel and air mixture is used to move the car. Most automobile engines are made up of several containers for this expansion or explosion of fuel-and-air mixtures. Most have four, five, six, or eight containers producing power to move the car. In engines with more than one container, the explosions in the different containers do not occur at the same time. The expansion of the fuel-and-air mixture occurs first in one container, then another, and another. The even staggering of these explosions has the effect of one continuous explosion and a continuous power supply, allowing the car to move smoothly down the road. The power produced by each container is added to the power produced by the other containers; therefore, there is more power available to move the car.

If one container does not produce as much power as it should, total engine output would be affected. For each container to produce the amount of power that it should, it must receive the right amount of fuel and air and be introduced to the

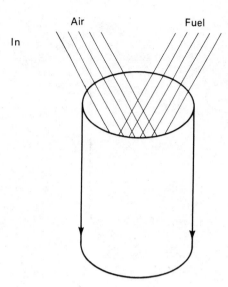

Figure 1-6 After the container is filled with air and fuel, it is sealed and then heat is added to the container. This causes an expansion of the mixture and forces the bottom to move downward as the expansion occurs.

proper amount of heat at the right time. If the same amount of power is produced by all the containers, the engine will operate smoothly. However, if one container produces more or less power than the rest, the engine will not run smoothly. Since the amount of power that is produced depends on the amount of fuel, air, and heat, in order to have a smooth-running engine, all containers must produce the same amount of power; thus all containers in an engine must receive equal amounts of fuel, air, and heat.

If a problem does not allow all the containers of an engine to receive the proper amount of fuel, air, or heat, the total power output of that engine will be affected, as each container is affected. Since they are all producing the same amount of power, the engine will operate smoothly but will produce less power. Engines are designed with equal-size containers and systems to provide equal amounts of fuel, air, and heat to all containers. Some parts of these systems are unique to each container and contribute to the explosion in that particular container. These parts are not common to all containers and will have only an indirect effect on the total

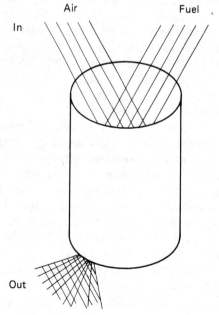

Figure 1-7 If the bottom of the container does not seal well while it is moving downward, the mixture will leak out from the areas where the bottom is not sealing. There will be less force on the bottom to cause it to move than if the bottom were well sealed.

engine. However, there are other parts in these systems that affect all the containers, because they have a purpose that is common to all the containers of that engine. These parts have a direct effect on the entire engine.

The purpose of these systems, whether they are common to all containers or are not common, is to provide proper amounts of fuel, air, and heat at the correct time. When these systems are working properly and all the containers are well sealed, the engine will smoothly produce the right amount of power and there will be little waste of power. A problem in one of these systems will either cause the engine not to run smoothly or allow the engine to run smoothly but be wasteful and not produce the proper amount of power, depending on whether or not the fault is common to all containers.

An understanding of which parts of the engine's systems are common to all the containers and which affect the individual containers is one of the keys to solving an engine running problem. Figure 1-8 shows some typical common and noncommon engine components.

ENGINE DIAGNOSTICS

While diagnosing an engine running problem, you should look for the cause of the problem only in the parts of the system that can cause the symptoms. The details needed to solve the problem will be found through an understanding of the systems and the precise testing of the parts of those systems.

APPROACH OF TEXT

The approach of this text will be to build an understanding of the basic operation of the individual engine systems and of the integration of these systems, which allows for efficient operation of the engine. An understanding of how these systems work and how to test these systems is the key to being able to perform a good

Figure 1-8 Some of these components will affect all the containers or cylinders of an engine, while others will affect only one.

diagnostic tune-up. An understanding of the engine's systems will also aid in performing a good maintenance tune-up and the methods of performing such a tune-up will be discussed prior to the explanation of detailed testing. The very basic facts explained in this chapter will be used to explain even the most complex of systems. In later chapters, methods for testing the sealed containers and the fuel, air, and heat systems will be discussed.

REVIEW QUESTIONS

1. When should a maintenance tune-up be performed?
2. What is the major difference between a maintenance tune-up and a diagnostic tune-up?
3. To cause an explosion, what three things must be present?
4. When high heat is quickly introduced to a mixture of fuel and air, a violent explosion takes place. Why?
5. What causes an engine not to run smoothly?
6. What four conditions must be met for an engine to produce the amount of power it was designed to produce?
7. If the engine has a poor-running problem, what three systems should be tested for faults?

2

ENGINE OPERATION FUNDAMENTALS

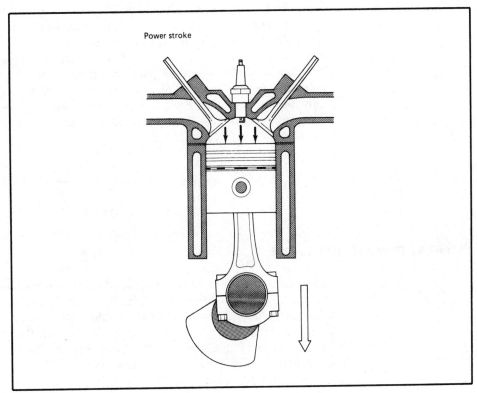

Power stroke

Figure 2-1

The operation of an engine depends on having a sealed container and the right amount of fuel, air, and heat. When high heat is introduced into the fuel-and-air mixture, a rapid expansion of that mixture occurs, causing the bottom of the container to be pushed down. The force on the bottom of the container is used to power the car down the road. This is a very simple explanation of how an engine operates. Although it is very simple and does not include technical terms and an explanation of how the systems for the fuel, air, and heat work, this explanation will help to define the purpose of all the parts of an engine and its related systems. There has to be a way to allow fuel, air, and heat into the sealed container. There must also be a way to allow only the right amount of fuel, air, and heat into the container. Somehow the power from the movement of the bottom of the container must be able to move the car. These statements outline the need to explain further how an engine operates.

COMBUSTION PROCESS

When the fuel-and-air mixture is shocked by the introduction of a high temperature, *combustion* is taking place. Combustion is the process of burning. In the sealed container of an engine, the fuel-and-air mixture is rapidly heated or burned. Total combustion takes place when all the conditions for combustion are correct, that is, when there is a correct amount of fuel mixed with the correct amount of air and this mixture is shocked by a correct amount of heat. The result of total combustion is a complete burning of the fuel present in the sealed container. In the combustion process, the fuel-and-air mixture goes through a chemical change and becomes a harmless gas and water, as shown in Fig. 2-2. The resulting gas should not be confused with fuel or "gasoline." The air around us, which is the same air that an engine uses in the combusion process, is made up of several gases. Air is the invisible mixture of gases (mostly nitrogen and oxygen) that surrounds the earth and is often referred to as the atmosphere of the earth.

The engine of a car does not operate in ideal conditions and it is presently impossible for an engine to receive the ideal amounts of fuel, air, and heat. Therefore, combustion in a car's engine is not total. The results from the incomplete combustion process are some poisonous gases and unburned fuel mixed with the water and harmless gases. This is the reason for the federal regulations on exhaust emissions. The inability to operate an engine with and under ideal conditions has caused automobile manufacturers to add systems that reduce the amount of harmful gases released by the engine.

$$HC + O_2 \xrightarrow{\text{Heat}} H_2O + CO_2$$

Figure 2-2

INTERNAL COMBUSTION ENGINE

The engine of a car is called an *internal combustion engine* simply because combustion takes place within the engine itself, in the sealed containers. These containers are the cylinders of the engine. Most automobiles have four, five, six, or eight cylinders. The *cylinder block* or *engine block* is the unit made of iron or aluminum which contains the cylinders of an engine. The engine block also contains those parts needed to convert the power produced by the combustion process into a usable form of power to move the car. The downward force of the bottom of the cylinder is changed to a rotating power and this power turns the tires of the car, allowing it to move. The crankshaft, which is mounted at the bottom of the cylinder block, is responsible for this change in motion.

The top of the cylinder is referred to as the *combustion chamber,* as this is where the combustion takes place. The combustion chamber or the top of the cylinder is sealed by the cylinder head. For combustion to occur again in that cylinder, the bottom of that cylinder must be pushed back up to the top of the cylinder.

This is accomplished by the momentum of the bottom moving down and reaching its most downward position and having some power left to continue moving. As it is being pushed down, it is rotating the crankshaft and as it continues to move, the bottom reverses its direction and moves back toward the top of the cylinder through the rotation of the crankshaft. In multiple-cylinder engines, the other cylinders of the engine contribute to the continuous rotation of the crankshaft. Keeping the cylinder sealed while its bottom is moving up and down is accomplished by the use of a piston and piston rings.

A *piston* is a short cylinder fitted into the engine's cylinder and connected to the crankshaft by a connecting rod. The up-and-down or reciprocating motion of the piston due to the expansion of the fuel and mixture is changed to a rotary motion by the crankshaft (Fig. 2–3). The piston top serves as the bottom of the cylinder. To seal the piston to the walls of the cylinder, piston rings are used. *Piston rings* are flat metal seals that expand from the outside of the piston to the walls of the cylinder. The walls of the engine's cylinder must be specially machined so that the piston rings can form a good seal. How the piston is fitted to the cylinder block is shown in Fig. 2–4. If a good seal is not present while the piston is moving up and down in the cylinder, fuel and air can leak out and power can be lost as the mixture is expanding and driving the piston down. If some of the force is lost past the piston rings, there will be less power on the piston top driving it down. Therefore, there would be less power available to move the car and fuel would be wasted.

After the combustion process is complete in a cylinder, not only does the piston need to be returned to the top, but a new supply of fuel-and-air mixture must also be present. The results of combustion, the *exhaust,* must first be let out of the cylinder to make room for the new supply of fuel and air. To achieve these things, *valves* are used. These valves are normally located in the cylinder head. An intake valve is used to allow fuel and air to enter the cylinder. It is open only long enough to allow the cylinder to be filled with fuel and air. It must be tightly closed during and after combustion to prevent the expanding mixture from leaking out and reducing the power output of the engine. When the combustion process is complete, an exhaust valve opens and allows the exhaust gases to flow out of the cylinder. This valve must also be tightly sealed during combustion. The positioning of the valves in the cylinder head is shown in Fig. 2–5.

With the removal of the exhaust from the cylinder and a new supply of fuel and air, all that is needed for combustion to occur again is heat. The heat for combustion is supplied by a *spark plug*. A spark plug is a plug that allows electricity to flow into the combustion chamber in the form of a spark. The amount of heat generated by the spark is determined by the amount of electricity that flows from it. Figure 2–6 shows the typical placement of the spark plug in the combustion chamber.

To ensure that the proper amount of electricity is flowing from the spark plug,

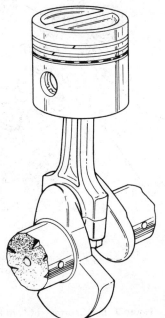

Figure 2–3 A piston connected to the crankshaft by a connecting rod. The design of the crankshaft changes the reciprocating motion of the piston to a rotating motion of the crankshaft.

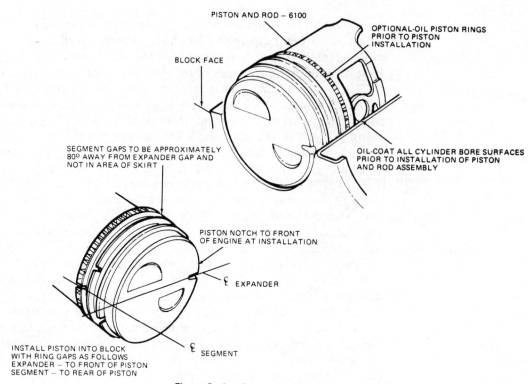

PISTON AND ROD – 6100

OPTIONAL-OIL PISTON RINGS PRIOR TO PISTON INSTALLATION

BLOCK FACE

SEGMENT GAPS TO BE APPROXIMATELY 80° AWAY FROM EXPANDER GAP AND NOT IN AREA OF SKIRT

OIL-COAT ALL CYLINDER BORE SURFACES PRIOR TO INSTALLATION OF PISTON AND ROD ASSEMBLY

PISTON NOTCH TO FRONT OF ENGINE AT INSTALLATION

₵ EXPANDER

₵ SEGMENT

INSTALL PISTON INTO BLOCK WITH RING GAPS AS FOLLOWS EXPANDER – TO FRONT OF PISTON SEGMENT – TO REAR OF PISTON

Figure 2-4 (Courtesy of Ford Motor Company)

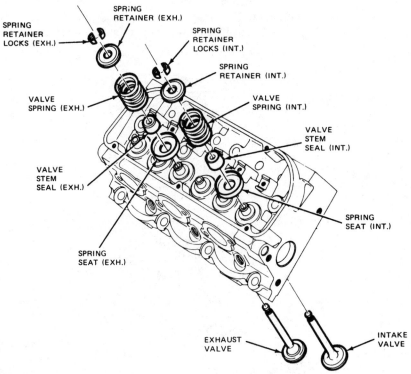

SPRING RETAINER (EXH.)

SPRING RETAINER LOCKS (EXH.)

SPRING RETAINER LOCKS (INT.)

SPRING RETAINER (INT.)

VALVE SPRING (EXH.)

VALVE SPRING (INT.)

VALVE STEM SEAL (INT.)

VALVE STEM SEAL (EXH.)

SPRING SEAT (INT.)

SPRING SEAT (EXH.)

EXHAUST VALVE

INTAKE VALVE

Figure 2-5 (Courtesy of Ford Motor Company)

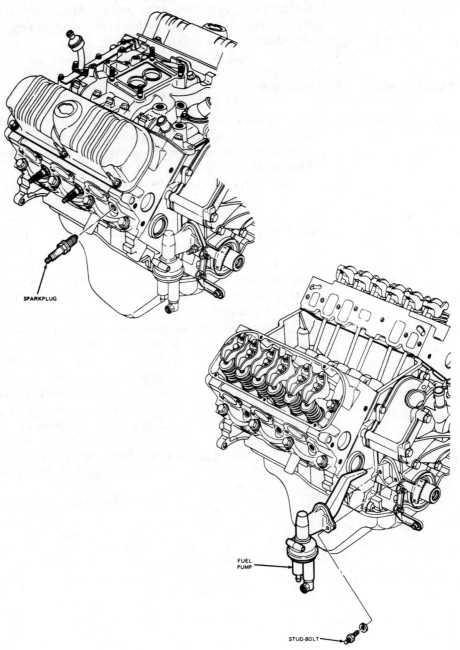

SPARKPLUG

FUEL
PUMP

STUD-BOLT

Figure 2-6   (Courtesy of Ford Motor Company)

an engine is equipped with an *ignition system*. This system is responsible for the spark of the spark plugs and for the timing of this spark.

Let's review how an engine operates using the components identified thus far. The intake valve opens and allows a mixture of fuel and air to fill the cylinder. When filled, the valve closes and the ignition system allows electricity to flow into the combustion chamber in the form of a spark, causing a sudden increase of the temperature present in the chamber. This heat shock causes combustion to take place and the pressure produced by the rapid expansion of the mixture pushes the piston down. The downward motion of the piston causes the crankshaft to rotate. As the piston returns to the top of the cylinder, the exhaust valve opens to allow the burnt gases to escape and make room for the next input of fuel-and-air mixture needed for the next combustion process. This summary suggests that there are four major events that must take place in order for an engine to run. These four events are referred to as the four *strokes*.

FOUR-STROKE-CYCLE ENGINES

Most automotive engines are four-stroke internal combustion engines. The foregoing summary of how an engine operates, together with a detailed discussion of what takes place during each one of these events, should allow for a good understanding of how an engine really does run.

The term "stroke" refers to the movement of the piston. A four-stroke engine has four piston movements, two from the bottom to the top and two from the top to the bottom. When the piston is at the top of a cylinder it is said to be at *top dead center* or TDC. When the piston is at the bottom it is at *bottom dead center* or BDC. These positions are shown in Fig. 2-7. Therefore, a piston stroke is its movement from TDC to BDC or from BDC to TDC. The piston's movement, up and down and then up or from TDC to BDC back to TDC, causes the crankshaft to turn one complete revolution or turn. Therefore, the four strokes of an engine require two complete crankshaft revolutions.

For the engine to run, the first event must be filling the cylinder with the fuel-and-air mixture. With the piston moving from the top to the bottom, the volume of the cylinder increases. This increase in volume creates a low pressure in the cylinder. Low pressure in the cylinder is the result of increasing the volume and not letting additional air into the cylinder. This low pressure is called a *vacuum*. With a vacuum formed in the cylinder and the piston continuing to move downward, the intake valve opens. *Atmospheric pressure,* the pressure of the air around us, is higher than the low pressure in the cylinder and rushes into the cylinder past the intake valve. The rushing air from the outside brings with it fuel particles provided by the fuel system and enters the cylinder (Fig. 2-8). When the piston reaches BDC, the intake valve begins to close.

The opening of the intake valve while the piston moves downward is called

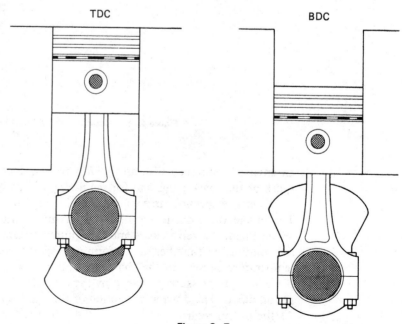

TDC   BDC

Figure 2-7

Intake stroke

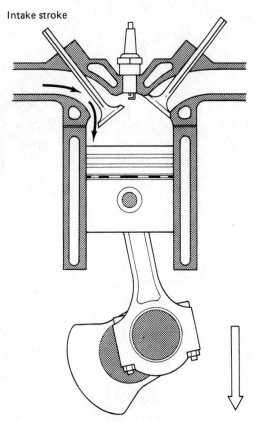

Figure 2–8 With the intake valve open and the piston moving downward, the air-and-fuel mixture enters the cylinder.

the *intake stroke*. Now the cylinder is filled with a fuel and air mixture and the piston is ready to move upward. The intake valve is closed, as is the exhaust valve. The cylinder is sealed and is at its largest volume. The rotation of the crankshaft causes the piston to move upward and this decreases the volume of the cylinder. This reduction in volume causes an increase in the pressure within the cylinder. This is the process of *compression*. Figure 2–9 shows how the movement of the piston upward compresses the mixture into a volume that may be one-eighth of its original volume. With the piston at TDC, the fuel-and-air mixture in the cylinder is highly pressurized. The movement of the piston from BDC to TDC with the valves closed is called the *compression stroke*. The compressed mixture will allow for a more violent explosion to occur in the cylinder when it is shocked by heat.

When the piston is about at TDC, the ignition system supplies electricity to the spark plug. This sets fire to or ignites the compressed fuel-and-air mixture. Combustion takes place, causing a rapid expansion of the mixture. This rapid expansion creates an extremely high pressure on the piston top, forcing it downward. The movement of the piston downward caused by the power produced by the combustion of the compressed fuel-and-air mixture is called the *power stroke* and is shown in Fig. 2–10.

The piston moves downward to BDC. At this time the exhaust valve opens and the piston moves upward toward TDC (Fig. 2–11). The movement of the piston upward pushes the exhaust out of the cylinder. By the time the piston is at TDC, the cylinder is empty and ready for another intake stroke. The *exhaust stroke* is the movement of the piston from BDC to TDC with the exhaust valve open.

These four strokes—intake, compression, power, and exhaust—are continuously repeated as the engine is running. In multiple-cylinder engines, these four

Compression stroke

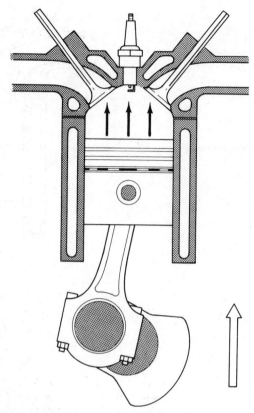

Figure 2-9 After the intake stroke, the intake valve closes and the piston begins to move upward, compressing the mixture.

events are evenly staggered among the cylinders. While one cylinder is on the power stroke, another may be on the intake or exhaust. This staggering of strokes allows the engine to provide for continuous power, simply by having one cylinder always on the power stroke.

THE ENGINE'S VALVE TRAIN

Important to the operation of the engine is not only the movements of the piston up and down the cylinder but the precise movements of the valves. In order for the strokes of the piston to accomplish those things that are necessary for combustion, the valves must open and close at the right times. It is best to think of a valve as a plug that seals the cylinder when closed. The valves are opened by cams on the engine's camshaft only when the pistons are in the proper position. A cam is like a wheel with a lobe or projection on it. As the cam rotates, the lobe comes around and through a variety of ways causes the valve to open (Fig. 2-12).

The camshaft is turned by the crankshaft at half of the crankshaft speed. Therefore, the camshaft turns one complete turn as the crankshaft rotates twice. The camshaft is driven by the crankshaft through gears, and a belt or chain. Because the camshaft rotates at half the speed of the crankshaft and because the exhaust and intake valves are opened only once each during the four strokes, there is a cam for each exhaust valve and a cam for each intake valve on the camshaft. As the lobe of the cam rotates, it allows the valves to close and open. The valves are held

Power stroke

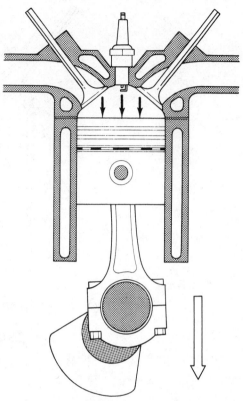

Figure 2–10 With the valves closed and the mixture compressed, ignition occurs and the resultant explosion forces the piston down.

in the cylinder head by heavy springs (Fig. 2–13). These springs allow the valves to close completely when the cam permits their closing.

The camshaft can be located either in the cylinder block or in the cylinder head. If the camshaft is in the block, it is positioned above the crankshaft and the valves are opened through a lifter, push rod, and rocker arm. As the cam lobe comes around, it pushes up on the lifter, which pushes the push rod up and moves one end of the rocker arm up while the other end pushes the valve down to open it. As the cam rotates and allows the valve to close, the valve spring maintains the contact between the valve and the rocker arm and therefore tends to keep the push rod and lifter in contact with the rotating cam. Engines with the camshaft in the engine block and the valves in the cylinder head are referred to as *overhead valve* (OHV) *engines* (Fig. 2–14).

OHC or overhead camshaft engines have the camshaft mounted in the cylinder head (Fig. 2–15). Positioned here, there is no need for push rods. As the camshaft rotates, the cams ride directly above the valves. The lobes will open the valves by either directly depressing the valve or by depressing the valve through the use of a cam follower, which is quite similar to a rocker arm. The closing of the valves is still the responsibility of the valve springs.

Some automobile manufacturers install two camshafts in the cylinder head, one for the exhaust and one for the intake. The advantage of doing so is more control over the opening and closing of the valves. These engines are referred to as *dual overhead camshaft* (DOHC) *engines.* Engines with a single camshaft in the cylinder head are sometimes called *single overhead camshaft* (SOHC) *engines.*

Exhaust stroke

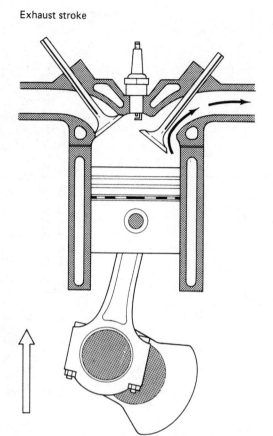

Figure 2–11 The exhaust valve opens and allows the exhaust gases to leave
the cylinder after the power stroke.

TWO-STROKE-CYCLE ENGINES

So far we have discussed the basics of a four-stroke engine; however, some small
engines use a two-stroke cycle. Although these are not commonly used in automobiles, they have been and are still classified as internal combustion engines and
therefore worthy of some discussion.

The cycle starts with the piston at TDC starting the power stroke. As the piston
moves downward near BDC, the exhaust opens to allow the burned gases out. The
intake valve opens shortly afterward and allows the fuel-and-air mixture to enter
the cylinder. The movement of the incoming mixture helps push the exhaust out of
the cylinder. This increases the chances of the cylinder being filled with fuel and
air. About the time the piston reaches BDC, both valves close and the piston begins
the compression stroke. Ignition occurs when the piston approaches TDC. A two-
stroke engine has a power stroke every crankshaft revolution. The two strokes are
called the *compression stroke* and the *power stroke*. Notice in Fig. 2–16 that the
intake and exhaust strokes both occur near the end of the power stroke. The camshafts of these engines rotate at the same speed as the crankshaft. Two-stroke engines are not used much in automobiles and by design they need oil to be mixed
with the fuel to lubricate them properly.

There are many types of internal combustion engines used in automobiles. So
far, we have looked at two- and four-stroke engines and OHV, SOHC, and DOHC
engines. There are two other designs of engines that operate differently than the
engines discussed previously.

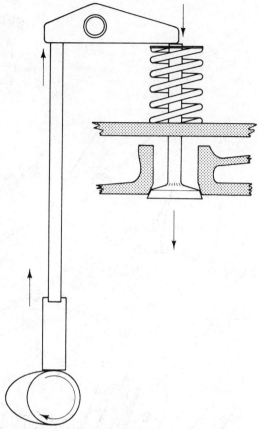

Figure 2–12   Notice that as the cam lobe (lower left) rotates, it raises or lowers the push rod, which in turn opens and closes the valve.

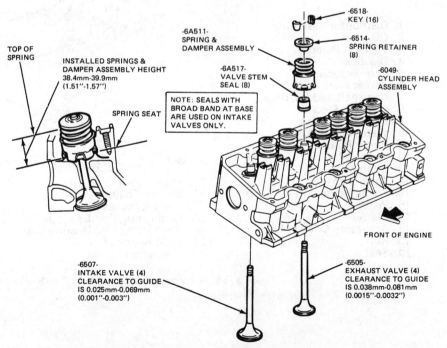

Figure 2–13   (Courtesy of Ford Motor Company)

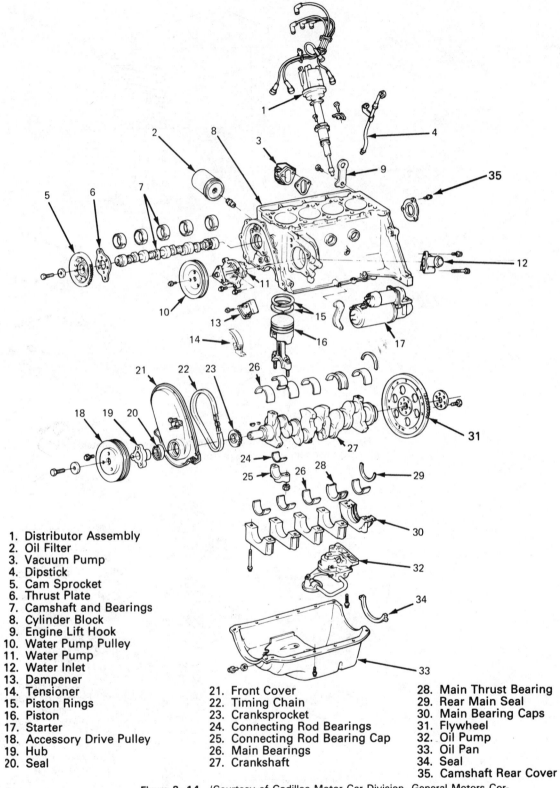

1. Distributor Assembly
2. Oil Filter
3. Vacuum Pump
4. Dipstick
5. Cam Sprocket
6. Thrust Plate
7. Camshaft and Bearings
8. Cylinder Block
9. Engine Lift Hook
10. Water Pump Pulley
11. Water Pump
12. Water Inlet
13. Dampener
14. Tensioner
15. Piston Rings
16. Piston
17. Starter
18. Accessory Drive Pulley
19. Hub
20. Seal

21. Front Cover
22. Timing Chain
23. Cranksprocket
24. Connecting Rod Bearings
25. Connecting Rod Bearing Cap
26. Main Bearings
27. Crankshaft

28. Main Thrust Bearing
29. Rear Main Seal
30. Main Bearing Caps
31. Flywheel
32. Oil Pump
33. Oil Pan
34. Seal
35. Camshaft Rear Cover

Figure 2-14 (Courtesy of Cadillac Motor Car Division, General Motors Corporation)

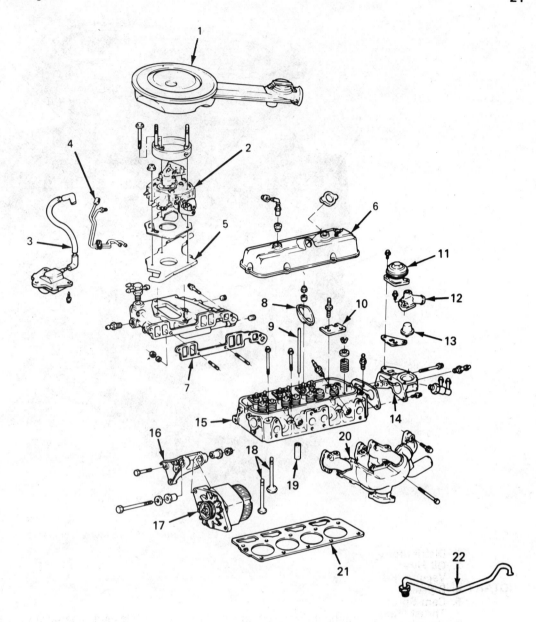

1. Air Cleaner
2. TBI Unit
3. Coil and Coil Wire
4. Fuel Line
5. E.F.E. Grid
6. Rocker Arm Cover
7. Intake Manifold & Gasket
8. Rocker Arm
9. Push Rod
10. Push Rod Guide
11. E.G.R. Valve
12. Thermostat Outlet
13. Thermostat
14. Adapter
15. Cylinder Head
16. Generator Bracket
17. Generator
18. Valves
19. Lifter
20. Exhaust Manifold
21. Cylinder Head Gasket
22. Pulsair Pipe

Figure 2–14 (*continued*)

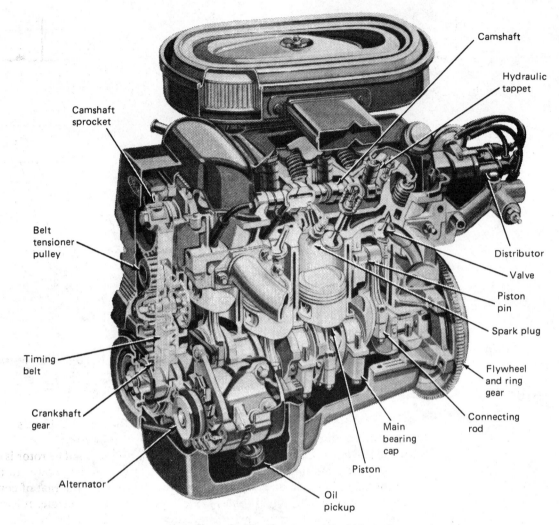

Figure 2–15   (Courtesy of Ford Motor Company)

ROTARY ENGINES

Throughout the development of the automobile many engine designs have been tried. One design of the internal combustion engine that is used today and may gain popularity in the future is the *rotary engine.* This engine operates as a four-stroke engine but uses different parts to accomplish its power. It uses continuously rotating components to complete the cycle as opposed to the up-and-down movement in a conventional engine. The basis of operation is a continuously rotating combustion chamber. The design of the engine consists of a three-lobe rotor that rotates with a noncircular motion within a specially designed and shaped housing. This eccentric motion is caused by the placement of the shaft on which the rotor pivots or turns. This shaft is not in the center of the rotor and thereby causes the eccentric motion. As the rotor turns within the housing, a changing volume occurs. This changing volume is similar to the changing volume resulting from a piston moving up and down the cylinder. Intake and exhaust valves are not used, but rather ports are positioned in the housing to allow for the filling and emptying of the combustion chamber as the rotor turns. The spark plug is also properly positioned in the housing so that ignition can also occur at the right time. Having these placed in an exact position allows for the production of maximum combustion chamber pressure, be-

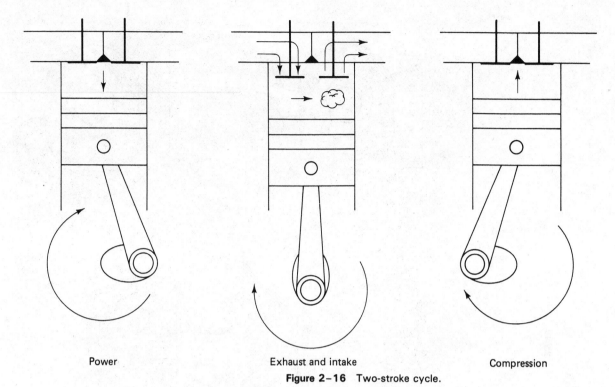

Power   Exhaust and intake   Compression

Figure 2–16 Two-stroke cycle.

fore and after combustion, forcing the rotor to turn. Figure 2–17 displays how the four strokes are accomplished by the movement of the rotor.

The power that turns the rotor is used to move the car. The rotor is connected to a set of gears that are connected to a shaft that transmits power to the wheels of the car. Although the design of this engine is different from that of conventional piston-type engines, similar things are still needed for it to operate. It is still a four-cycle engine, and therefore it needs precise control of both the fuel-and-air system and the ignition system. The basics of operation still apply to this rotating-combustion-chamber four-stroke-cycle internal combustion engine.

All the engines discussed so far used electricity at a spark plug for ignition. A diesel is a two- or four-stroke internal combustion engine that uses no spark plug. This type of engine is gaining popularity for use in automobiles because it is extremely fuel efficient. Rather than use electricity to ignite the air-and-fuel mixture, it uses the heat of compression.

DIESEL ENGINES

As a gas is compressed, not only does the pressure increase but so does its temperature. In *diesel engines,* on the compression stroke, the air is compressed much more than in a spark-ignited engine. When the compression stroke is nearly complete, fuel is squirted under high pressure into the compressed air (Fig. 2–18). The extreme heat that was produced by the pressure causes the fuel to ignite. The heat of compression causes the fuel to ignite instead of the heat released by a spark plug. The basic construction of a diesel engine is similar to that of a spark-ignited engine. However, because of the extreme pressures produced by compression, the components of this type of engine tend to be heavier and stronger. Diesel engines can also operate as two-stroke engines but require the addition of a means to deliver the air under some pressure on the intake stroke. Because the fuel is squirted or injected

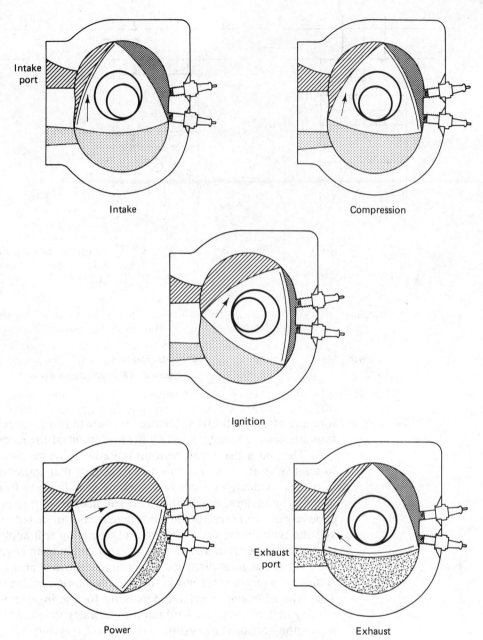

Intake port

Intake   Compression

Ignition

Power   Exhaust

Exhaust port

Figure 2-17 Follow the marked side of the rotor around the inside of the engine's housing as the rotor completes its four strokes. Some rotary engines are equipped with two spark plugs, as is the one used for this drawing. The two spark plugs allow for more complete combustion.

after the incoming air is compressed, a special fuel delivery system is used. The fuel used in a diesel is also different from the fuel used in a spark-ignited engine. Other than the operational differences from a spark-ignited engine, all basics of the operation of an internal combustion engine apply to a diesel engine.

ENGINE CLASSIFICATION

The purpose of all engines is to supply power to be used. It can be said that the power that an engine produces is dependent on the amount of air it uses. The amount of air that an engine can use is dependent on the size of its cylinders, the number

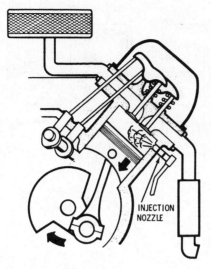

Figure 2-18 (Courtesy of Oldsmobile Division, General Motors Corporation)

of cylinders, and how well the engine allows the air to flow in and out. Engine designers must consider these factors as they design an engine for a particular application. Typically, as more power is needed, more or bigger cylinders are used. Cylinders can be made bigger by increasing the movement of the piston as well as increasing the diameter of the cylinder. The total volume of a cylinder is its displacement (Fig. 2-19). An engine's size or displacement is determined by finding the displacement of one cylinder and multiplying this by the number of cylinders.

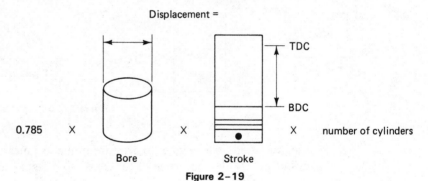

Figure 2-19

Engines are classified not only by how they operate (i.e., diesel, SOHC, and OHV) but also by their size and the number of cylinders. The size of an engine is the *displacement* of that engine. Displacement of an engine is measured either in cubic inches or in liters. These are both measurements of volume. The number of cylinders usually also determines another design characteristic, the arrangement of cylinders. Four-, five-, and six-cylinder engines often use an in-line cylinder arrangement. An *in-line engine* is one that has its cylinders arranged in a row, one after another. These are usually located in a single cylinder block with a common crankshaft.

Most eight-cylinder engines are made as two in-line four-cylinder engines set apart at a 90-degree angle and using a common crankshaft. The shape of this design is that of a "V" and therefore, these engines are called *V-8's*. This engine design uses two cylinder heads and has a common fuel and ignition system for all cylinders. This design allows for the increased displacement without a drastic increase in the bulk or external size of an engine. Six-cylinder engines are also commonly built in this design because of the compactness of this design.

Figure 2-20 displays the many cylinder arrangements that are and have been

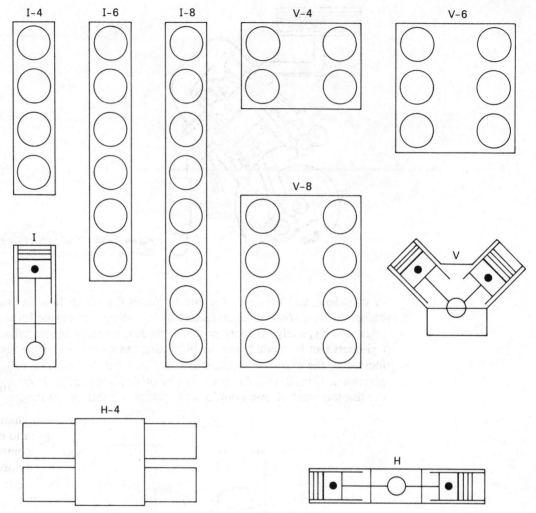

Figure 2–20

used in automobile engines. There are in-line engines of four, five, six, and eight cylinders. Vee designs are used for four-, six-, eight-, and even twelve-cylinder engines. Another design used for four- and six-cylinder engines is called the *horizontally opposed design*. These engines have half of the cylinders positioned 180 degrees from the other half. The centerlines of the cylinders are usually in a horizontal plane—thus the name.

In spite of the cylinder arrangement, the cylinders are usually numbered in sequence. This numbering refers to their arrangement, not to the order in which they occur in the power stroke. The numbering of cylinders is for reference only and it is important to remember that the cylinders do not "fire" in that order. Included in the specifications of any particular engine is the *firing order*. This refers to the order in which the cylinders fire or are on the power stroke. To ensure that the engine produces power smoothly, the firing of cylinders is staggered (Fig. 2–21). An understanding of the firing order is important when doing diagnostic work.

COMPRESSION RATIO

After the intake stroke, the mixture is compressed. With the piston at BDC, the rotation of the crankshaft pushes the piston up, thereby decreasing the size of the cylinder. The air particles are now being pushed closer together and this results in

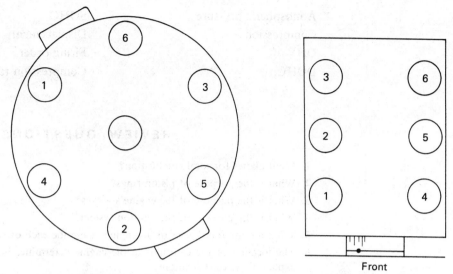

Figure 2–21 The numbering sequence on the left represents the firing order and the arrangement of the spark plug wires on the engine's distributor cap. The numbering sequence on the right represents the arrangement of the cylinders on the same engine.

a higher pressure. The change in volume of the cylinder when the piston is at BDC to the volume present when the piston is at TDC determines the *compression ratio* of the engine. The compression ratio of a cylinder determines the amount of pressure formed by compression. The volume of the combustion chamber when the piston is at TDC compared to the volume of the combustion chamber when the piston is at BDC is the compression ratio. If the compression ratio of an engine is 8:1, this means that the volume of the cylinder at BDC will be compressed into an area one-eighth of that volume. Most modern engines have a compression ratio of 8:1 or 9:1. Exceptions to this are diesel engines, which require much more compression pressure and therefore much higher compression ratios, perhaps as high as 25:1.

SUMMARY

Changes in the volume and pressure of air are not only important in the operation of an engine, but these changes also affect how the engine's supportive systems, such as the ignition and fuel systems, meet the needs of the engine. For an engine to operate, the right amount of air must be mixed with the right amount of fuel in a sealed cylinder and be shocked by the right amount of heat at the right time. In the next chapter we will discuss how the right amount of fuel and air are delivered to the cylinder.

TERMS TO KNOW

Combustion	Ignition system
Internal combustion engine	Stroke
Cylinder block	TDC
Combustion chamber	BDC
Spark plug	Vacuum

Atmospheric pressure   SOHC
Compression   Displacement
OHV   Firing order
DOHC   Compression ratio

REVIEW QUESTIONS

1. What allows for total combustion?
2. What is the purpose of piston rings?
3. What is the purpose of the engine's valves?
4. What is the purpose of the ignition system?
5. What are the four cycles of an engine? Describe each of them.
6. The location of the camshaft in the engine determines how it is classified. Name and explain these classifications.
7. What are the two strokes of a two-cycle engine?
8. What is the major difference between a rotary and a conventional engine?
9. What causes combustion in a diesel engine?
10. How does the engine's ability to breathe affect its power output?

3

ENGINE AIR/FUEL FUNDAMENTALS

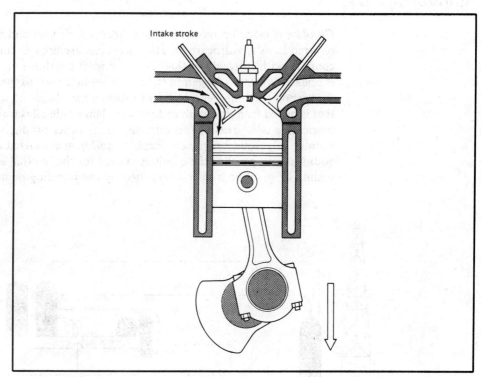

Intake stroke

Figure 3–1

One of the most effective ways to put out a fire is by suffocating it. A fire cannot continue to burn without air. The fuel for the fire needs air to burn. Similarly, the fuel used in engines needs air to burn or combust. For total combustion to take place, there must be the right amount of fuel mixed with the right amount of air. The proper ratio of air and fuel needed by an engine depends on the conditions in which the engine must operate.

INDUCTION SYSTEM

The engine must be able to start whether it is cold or hot. It must be able to idle, run at part throttle and full throttle, and be able to accelerate. All of these conditions require a different *air/fuel ratio*. These conditions cause the engine to run at a variety of speeds and loads. Each variation demands a different volume of air and a corresponding difference in fuel volume. The purpose of the *induction system* of the engine is to provide for these different ratios.

The induction system draws in air from the atmosphere, mixes it with fuel, and then distributes this mixture to the cylinders. The ideal air/fuel ratio for an engine operating at a constant speed and load is about 15:1, or 15 parts of air to 1 part of fuel. When there is less than 15 parts of air mixed with 1 part of fuel, the mixture is called *rich*. A rich mixture burns cooler and therefore longer because it lacks oxygen or air. A *lean* mixture is composed of more than 15 parts of air mixed with 1 part of fuel and tends to burn hotter and quicker. The variations in speed and load require different mixture ratios, richer or leaner, because these conditions are better overcome by the engine if burn times are controlled.

AUTOMOTIVE FUELS

Gasoline is made up mostly of the elements hydrogen and carbon and is therefore referred to as a hydrocarbon. Hydrocarbons are used as fuels because they readily combine with oxygen and this is exactly what combustion is. In combustion, heat is released as the hydrogen in the fuel combines with oxygen and forms water and the carbon combines with oxygen to form a gas. Gasoline is derived from crude oil that is pulled from the earth and refined. This crude oil is transformed into gasoline, lubricating oils, greases, fuel oils, and many other products (Fig. 3-2). During the refining of gasoline, many chemicals or *additives* are mixed into the gasoline. These additives give gasoline the qualities needed for the internal combustion engine. The quality of gasoline is also determined by the blending or mixing of particular hy-

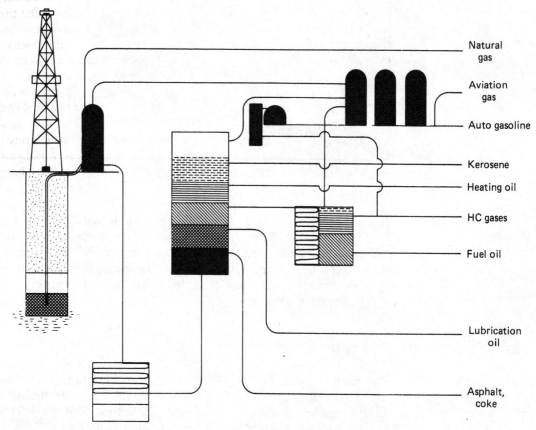

Natural gas

Aviation gas

Auto gasoline

Kerosene

Heating oil

HC gases

Fuel oil

Lubrication oil

Asphalt, coke

Figure 3-2

drocarbons present in crude oil. Therefore, gasoline is not simply produced. Many characteristics of the fuel are designed and controlled to ensure that it meets the requirements of an automobile's engine.

One characteristic of gasoline is its *volatility,* or how easily it can become a vapor. When water boils, it is becoming a vapor. A liquid that boils or becomes a vapor at low temperatures has a *high volatility. Low volatility* means that the liquid will not vaporize until there are high temperatures. The various hydrocarbons found in crude oil have different volatilities and therefore, through careful blending of these hydrocarbons, gasoline can be designed to have a particular volatility. Gasolines with the wrong volatility will hinder the overall performance of an engine. Typically, the blends used in gasoline will vary from climate to climate. Warmer climates require a lower volatility than colder climates because it is undesirable to allow the gasoline to vaporize before it is sent to the combustion chamber. High-volatile fuels are needed for cold climates because this aids combustion in a cold engine.

The major difference between a gasoline engine and a diesel engine is the method of ignition. Gasoline engines use the heat of spark plugs for combustion, whereas a diesel uses the heat of compression. Diesel fuel and gasoline are blends of hydrocarbons that are designed for their application. It is undesireable for gasoline to ignite by the heat of compression. A gasoline's ability to resist ignition by the heat of compression is referred to as its *octane rating.* Higher compression ratios require higher octane simply because the heat generated by compression is greater. In the normal combustion process of a gasoline engine, the compressed air-and-fuel mixture in the combustion chamber burns rapidly and evenly as the heat produced at the spark plug moves to heat all of the mixture. This even burning allows for an even amount of pressure to be present on the piston top, pushing it down.

If the gasoline present in some part of the combustion chamber started to expand due to the heat of compression and not by the spark, the pressure on the piston top would not be even and this could cause severe damage to the piston. This does occur sometimes and this is called *detonation.* When this does occur, this uneven pressure causes a sound that is like a hammer tapping a piece of metal. This sound is commonly called *pinging* or *knocking.* Normally, the heat of the spark at the spark plug starts the burning of the mixture in the combustion chamber. From the spark plug, a wall of flame or intense heat moves rapidly in all directions through the compressed mixture until all the mixture is burned. At times, as shown in Fig. 3-3, the end of gas or the last part of compressed air-and-fuel mixture ignites before the flame front or wall arrives; this is detonation. The ability of a gasoline to resist detonation is determined by its octane rating or *antiknock value.* Chemical additives

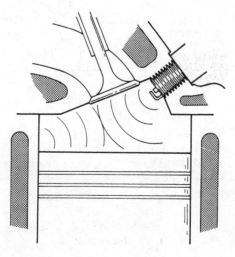

Figure 3-3 One flame front in the cylinder caused by the heat generated by the spark plug colliding with another flame front caused by some heat buildup causing detonation.

such as lead (tetraethylead) are added to gasoline to increase the octane rating. Low-octane fuel is not the only cause of detonation. Causes of detonation will be discussed further in later chapters.

Diesel fuels ignite by the heat of compression and therefore it can be assumed that they have low octane ratings. However, diesel fuels must still control when combustion takes place. Diesel fuels must burn quickly during combustion. The fuel is sprayed into compressed air and combustion should take place at that instant. Diesel fuels are not rated by octane but rather by *cetane,* the ease with which it ignites. A high cetane number means that the fuel will ignite easily or will ignite with a low temperature. There is no direct relationship between an octane rating and a cetane rating, even though there seems to be.

There are many other additives used in gasoline and diesel fuels to aid in the efficient operation of an engine. These chemicals are added to prevent dirt build-up in the combustion chamber, to separate water from the fuel, to prevent ice formation, and for many other reasons. Currently, there are two major types of gasoline available, *leaded* and *unleaded.* Lead was added to gasoline to increase octane ratings, but lead is a poison and does not get burned in the combustion process. The lead is therefore expelled from the chamber with the exhaust and pollutes our air. The use of unleaded gasoline in newer cars has been required since 1975, when the fear of lead poisoning and the use of catalytic converters became dominant. A *catalytic converter* is used to convert poisonous gases in the exhaust of an engine to harmless gases, and the use of leaded gasoline hinders the effectiveness of the catalyst. The removal of lead from gasoline lowers the octane rating, so automobile manufacturers lowered the compression ratios of their engines to allow for the use of low-octane gasoline. Oil refining companies have developed other chemicals that increase octane ratings of gasoline, and this has allowed newer engines to have higher compression ratios. Car manufacturers have also redesigned the combustion chambers of engines to allow the use of lower-octane fuels.

AIR REQUIREMENTS

Having the right amount of air available to combine or mix with the fuel is as important as using the correct fuel. Air is supplied to the cylinders by the difference of pressure between the vacuum produced on the intake stroke of the piston and atmospheric pressure. Ideally, the cylinder becomes totally filled with air during full throttle. It is the role of the induction system to fill the cylinders rapidly and completely on each intake stroke. As engine speed increases, this task becomes more difficult. A lack of air in the mixture will not allow the engine to operate efficiently.

On the intake stroke when the intake valve opens, air rushes into the cylinder. Not only is it important to get as much air in the cylinder as possible, but this air must also be clean. If dirt enters the cylinder, it can cause damage to the pistons, valves, cylinder head, or the finely machined cylinder walls. To provide for this clean air, automobiles have an air filtration system to prevent even the smallest of dirt particles from entering the cylinder. Through the *air filtration system* or *air cleaner,* air is pulled from the atmosphere into the induction system to be mixed with fuel and then distributed to the individual cylinders.

Air cleaner assemblies used on cars today serve purposes other than cleaning the air. These assemblies are used to reduce the noise caused by the air rushing into the engine. They are also used to help the incoming air into the induction system. Although these two functions should not be slighted, the main purpose of the air cleaner is to provide clean air for the engine.

The typical air cleaner assembly (Fig. 3-4) is composed of a container that is sealed completely with the exception of a large hole to let air in and another to let

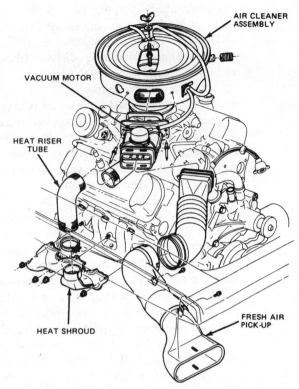

Figure 3-4 (Courtesy of Ford Motor Company)

air out into the engine. Inside the container is an air filter element, which fits snugly inside the container. Air is filtered in the air cleaner assembly in two ways; larger particles of dirt are removed with the movement of air toward the air filter and then smaller particles are removed as the air moves through the filter. Currently, air filters are made of a dry paper element (Fig. 3-5). This paper is folded to form pleats, which increases the filtering area of the filter. These paper pleats are supported by a wire mesh and molded into a plastic or rubber gasket that seals the top and bottom of the filter to the inside of the air cleaner assembly. Sealing the top and bottom of the filter prevents unwanted dirt from entering the engine. All air that enters the

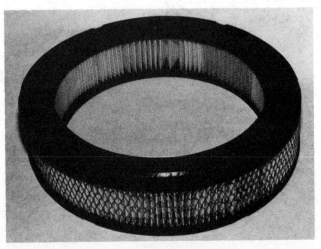

Figure 3-5

engine should go through the air filter. Within time, the air filter gets dirty and tends to restrict the amount of air that can enter, causing the engine to lose power and to waste gasoline. The replacement of an air filter is one of the required services of a maintenance tune-up. A dirty air filter can cause a rich or lack-of-air condition, and regardless of how well the other parts of an engine are performing will decrease the efficiency of the engine.

The opening used to allow air into the air cleaner assembly usually has a tube or duct constructed over it to aid in directing the air. This duct, or *snorkle* (Fig. 3-6), is positioned in a place where there is good under-the-hood airflow. While a car is moving, the air under the hood is moving in many directions. The air entering from the front of the car serves to cool the engine as well as serving the combustion process. The air cleaner duct attempts to grab undisturbed air from under the hood. Attached to the duct is another tube that directs air from around the exhaust of the engine (Fig. 3-6). The purpose of this system is to allow warm air to enter the engine when it is cold, which improves the performance of a cold engine by aiding in the vaporization of the fuel. When the engine is warmed up, this system is no longer needed and the duct draws in air from the outside. This switching of the source of air for the duct is controlled by a thermostatic switch (Fig. 3-6). This switch operates a door located at the end of the duct, which allows air to enter only from the exhaust area or from the outside. Warm air entering a warm cylinder may allow the fuel to vaporize too quickly and affect the performance of the engine. The proper functioning of this thermostatically controlled air cleaner duct contributes to the overall efficiency and driveability of the engine.

The other opening of the air cleaner assembly is fastened to a device that regulates the amount of air that enters the engine, thereby controlling engine speed. This device can be either a carburetor or a throttle body. Both of these govern the amount of air that enters the engine and work with other devices to allow for the proper amount of gasoline to mix with that amount of air for a proper air-and-fuel mixture.

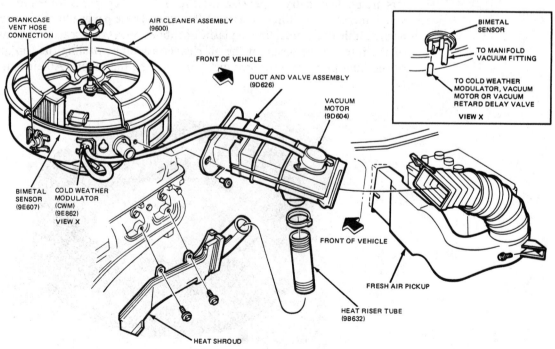

Figure 3-6 (Courtesy of Ford Motor Company)

BASIC CARBURETION

A *carburetor* (Fig. 3–7) is a metering device that is used to mix air and fuel in the proper ratios to accommodate the engine's needs for operation under a variety of conditions. The formation of the air-and-fuel mixture occurs when the fuel is divided into fine particles or atomized and mixed with the air. Atomization is accomplished by adding air to the liquid gasoline as it moves through the carburetor passages and then this air/fuel mixture is sprayed into a stream of moving air which is rushing into the engine. The movement of the air is caused by the pressure difference caused by the low pressure or vacuum created by the piston's intake stroke and by the higher pressure of the atmosphere. A carburetor works on the principle of *pressure differential:* When two different pressures are exposed to each other, the greater of the two will rush toward the other, attempting to balance the pressures. The more pressure differential that is present, the faster the air will flow through the carburetor and therefore the more fuel will be drawn out of the passages of the carburetor. The carburetor has two basic functions: to regulate engine speed by controlling the amount of fuel and air the engine receives, and to deliver to the engine the correct ratio of fuel-and-air mixture.

Not only is a pressure differential formed on the intake stroke but one is also formed by the carburetor itself. In the throat or bore of the carburetor is a *venturi,* which is a streamlined restriction in the bore of the carburetor. This restriction causes an increase in air velocity and lowers the pressure of the air passing through it. A carburetor controls engine speed and air/fuel ratios by controlling air velocity. Pressure differentials are thereby controlled by the carburetor and its venturi. Notice in Fig. 3–8 that as the air enters the venturi, the center column of air has no direct contact with the venturi. The surrounding air is directed toward the center by the venturi and this tends to push the center column of air through the venturi faster. Beyond the restriction, the throat or bore of the carburetor increases in size. Although the air velocity has increased, the volume of air entering the larger area is less than the amount before the venturi. There is now a low pressure after the venturi and this further increases the speed of the air as it passes through the venturi rushing to equalize the pressure.

An understanding of the effects of pressure differential and the venturi will aid in understanding how a carburetor works. One other occurrence with air must

Figure 3–7

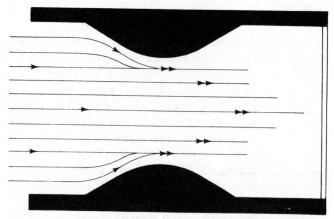

Figure 3-8 Venturi action.

be explained before going into detail as to how the carburetor works. If air passes rapidly through a tube or cylinder that has a small hole in it, air will be drawn through the hole to combine with the air already flowing in the tube. If a small hose were connected to the hole in the cylinder and to a container of liquid, the liquid would be drawn out of the container into the passing air (Fig. 3-9). The greater the speed of the air, the more liquid is drawn from the container.

If another hose (Fig. 3-10) is connected to the first hose to form a "Y" connection and its other end exposed only to the atmosphere, a mixture of air and liquid will be drawn off into the airstream moving through the cylinder. This occurrence helps explain the atomization of gasoline by the carburetor. Air velocity dictates the amount of fuel that is drawn in and mixed with the air and therefore dictates how an engine will run. However, at engine idle speeds, the air velocity is very low because the engine needs little air. This low air velocity draws little fuel and is not enough to allow the engine to run. Other operating conditions also affect air velocity and therefore affect the functioning of a carburetor. There are systems built into the carburetor that allow for operation under these conditions. With these additional systems, a carburetor is quite complex, but its operation relies simply on differences of pressure.

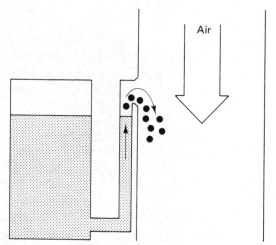

Figure 3-9 Liquid is being drawn out of its reservoir by the flow of air past the small hole that leads to the liquid reservoir.

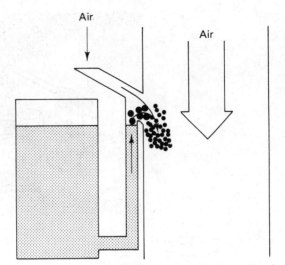

Figure 3–10 Air mixes with the liquid before entering the main column of air.

FLOAT SYSTEM

Regardless of operating conditions, the correct amount of fuel is necessary for an engine to operate efficiently. To do this the carburetor must have a constant supply of fuel available to meet the demands of the engine. Carburetors have a float system which has the purpose of storing gasoline in the carburetor and maintaining the amount of fuel to a specified level. The gasoline is stored in a small container called the *float bowl* and the level of fuel is controlled by a float assembly. The level of fuel in the bowl lowers or raises the float, which in turn opens or closes a fuel inlet valve, allowing gasoline to enter when the level is low and shutting off the fuel supply to the bowl when it is at the proper level. There are many types and models of carburetors, and each type requires a different amount of stored gasoline. To ensure that the carburetor has the right amount of fuel available, each carburetor has a particular level at which the float should be positioned and to which it should be adjusted. Too high or too low a fuel level can cause running and/or starting problems.

The float circuit includes the float bowl and a float attached to a *needle valve,* as shown in Fig. 3–11. The needle valve is located at the fuel inlet of the carburetor and moves in and out of a small hole called a *seat,* controlling the inlet of fuel. The needle valve is tapered on one end and when the point of the taper only is in the seat, the valve is open. As the diameter of the taper increases, it closes off fuel flow through the seat. When the needle valve is completely in the seat, fuel flow into the bowl stops.

If the level of fuel in the float bowl is too high, too much fuel will be released from the fuel nozzle, resulting in a rich mixture. A lean mixture will result from too low a fuel level. If the fuel enters the float bowl faster than it is discharged, the fuel level will rise, causing the float to rise. This causes the needle valve to drop into its seat, shutting off the inlet of fuel. When the fuel level drops, the float moves down, causing the needle valve to lift off its seat, allowing fuel to enter the float bowl. All carburetors have a vent or an inlet for air into the float bowl to allow atmospheric pressure in to help maintain a balance between the pressure of the air entering the carburetor and the pressure on the fuel in the float bowl. A pressure differential between these two can also cause a rich or lean mixture.

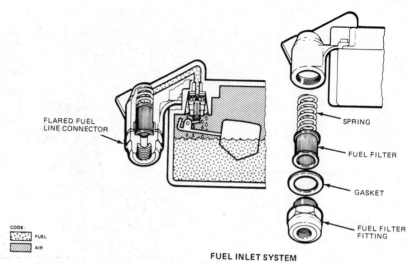

FUEL INLET SYSTEM

Figure 3-11   (Courtesy of Ford Motor Company)

IDLE CIRCUIT

At idle and low speeds, air velocity in the carburetor is extremely low. Therefore, the venturi effect does not allow for enough fuel to be discharged or drawn off by the flow of air. There is, however, a pressure differential between the atmosphere and the cylinders. The vacuum produced on the intake stroke of the piston is high because little air is entering the system to fill the void or equalize the pressure. The throttle plates determine how much air and fuel can be drawn into the engine. These plates are located at the bottom of the carburetor and function as doors, allowing the air-and-fuel mixture to enter the intake manifold. The wider the plates are open, the more mixture that can enter. This controls engine speed. At idle speeds the throttle plates are open only slightly, thus confining the vacuum of the engine to the area of the carburetor below the throttle plates. In this part of the carburetor is a small hole called the *idle discharge hole,* which allows fuel to be drawn off from the float bowl to allow the engine to idle. The discharge hole will allow fuel to enter the cylinders as long as the throttle plates are nearly closed and there is vacuum present below the carburetor. To aid in the atomization of the fuel, there is a small hole, called an *air bleed,* which allows air from the atmosphere to enter and mix with the fuel in the discharge passageways. Carburetors have adjustment screws that regulate the amount of air and/or fuel that can be discharged through these holes (Fig. 3-12).

As engine speed slowly increases due to the opening of the throttle plates, more air flows into the engine and there is a need for more fuel to maintain the proper mixture. This extra fuel is provided by the *secondary idle discharge holes.* These holes are located in the carburetor just above the regular idle discharge holes, as shown in Fig. 3-13. The secondary holes become uncovered as the throttle plate opens and the vacuum from the cylinders draws fuel out from the holes. During this time the regular idle discharge hole is also supplying fuel to the engine; the secondary holes simply add more fuel to the flow.

MAIN METERING CIRCUIT

Increasing the opening of the throttle plates causes the engine to speed up by allowing more air to flow into the engine. This increase in airflow and the effects of the venturi cause an increase in air velocity, which in turn creates a low pressure in

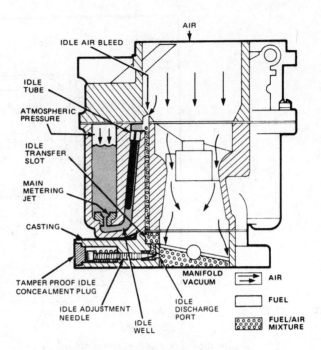

Figure 3-12 (Courtesy of Ford Motor Company)

the lower part of the venturi. This flow of air and the consequent lower pressure draws fuel from the *main metering circuit*. The amount of fuel that flows from the main metering system is determined by the flow of air passing through the venturi and by the size of the main jet. The *main jet* is simply a plug with a hole drilled in it and it is usually located in the bottom of the float bowl. Increasing the size of the hole in the jet will allow more fuel to flow into the moving column of air through the venturi. At this time, both the main metering system and the idle circuits will discharge fuel. However, the amount of fuel discharged by the idle circuits has decreased because of the decrease in vacuum due to the wider opening of the throttle plates (see Fig. 3-14).

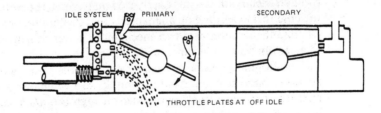

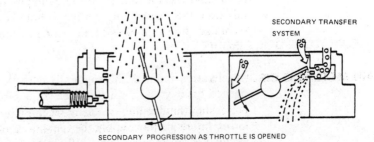

Figure 3-13 (Courtesy of Ford Motor Company)

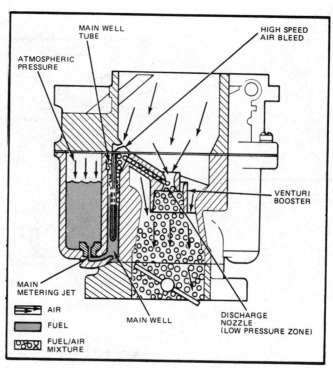

Figure 3-14 (Courtesy of Ford Motor Company)

POWER CIRCUIT

Whenever the throttle is opened enough to cause a large decrease in engine vacuum, the engine needs more fuel than the amount delivered by the main metering circuit. This additional fuel is supplied by the power circuit. Fuel from the typical power circuit is released by the presence of low vacuum. At idle or low speeds, high vacuum is routed to the power circuit, and this prevents the extra fuel from flowing. Low vacuum causes the power valve to open and allow additional fuel to flow. Power valves are calibrated to open only when a particular vacuum drop is reached. If the power valve remains open or opens at too high a vacuum, the engine will not perform efficiently due to this overly rich mixture. The power circuit is open only when the engine is operating at high speeds and heavy loads.

Some carburetors use a mechanical power system. As the throttle is opened to a certain point, the throttle linkage pushes on a power valve which allows more fuel to flow. There are also some carburetors that do not use a separate power system. These carburetors use a step-up rod or metering rod in the main metering system. To allow more fuel to flow through the main metering circuit, these rods lift out of the center of the main jets as the throttle is opened (Fig. 3-15). This action increases the size of the hole in the main jet and therefore allows more fuel to flow through it.

ACCELERATOR PUMP CIRCUIT

Acceleration requires the engine to increase speed rapidly. The throttle plates are opened quickly and a much larger volume of air enters the carburetor and the engine. The cylinders momentarily receive a very lean mixture because the main metering system cannot supply the fuel necessary to maintain the proper air/fuel ratio as quickly as air is entering. When the cylinders receive this overly lean mixture, the engine will hesitate and not increase speed until the main metering system can

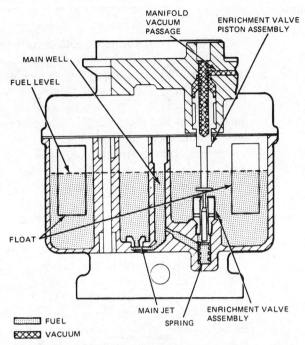

Figure 3-15 (Courtesy of Ford Motor Company)

match the increased airflow with fuel. To allow for smooth acceleration, carburetors are equipped with an *accelerator pump,* which squirts a predetermined amount of fuel into the air as the throttle plates open. Because the main metering circuit will eventually catch up to the airflow, the accelerator pump will squirt fuel out for only a brief time. If the amount of fuel released and the timing of the squirt are correct, the engine will not hesitate upon acceleration. The accelerator pump is activated mechanically by the movement of the *throttle plates* (Fig. 3-16). The fuel for the accelerator pump is normally supplied through the float bowl.

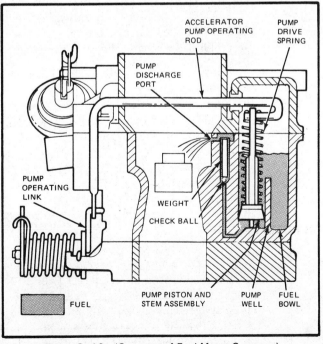

Figure 3-16 (Courtesy of Ford Motor Company)

CHOKE SYSTEM

To start an engine equipped with a carburetor, the first procedure is to open the throttle plates, in other words, to press down on the "gas pedal." This activates the accelerator pump, which squirts fuel into the engine to allow for easier starting. When the engine is cold, this action also activates the choke system. The choke valve or plate closes off outside air entering the carburetor. This is necessary to provide an extra-rich fuel mixture for starting in cold weather. A rich mixture is required because fuel vapor has a tendency to condense into droplets of raw gasoline when it comes in contact with the cold walls of the bore of the carburetor. This can stop much of the fuel needed to start the engine from reaching the cylinders by causing these fuel droplets to collect on the walls of the carburetor and the passageway into the cylinder. To allow for the amount of fuel needed, the choke plate closes and thereby prevents the low pressure formed by the engine from being balanced by the atmosphere. The vacuum produced by the engine is retained in the carburetor body, and this low pressure draws fuel out of the idle and main metering circuits. Fuel from both of these sources is enough to get the engine started and to allow it to run until the engine warms up.

To prevent engine stalling while the engine is warming up and receiving this overly rich mixture, the choke assembly includes a *fast-idle circuit*. This is designed to raise engine speed while the choke plate is closed or partially closed. When the choke plates are fully open or when the engine is warm, the idle returns to normal. This increased idle speed is accomplished through the use of a *fast-idle cam,* which is connected by a rod to the choke plate pivot rod. As you can see in Fig. 3–17, this cam has a series of steps of varying heights which contact an adjustment screw that is attached to the throttle plate. When the choke is fully closed, the fast-idle cam rests with its highest step contacting the screw on the throttle shaft. This allows the engine to run at its specified maximum idle speed. As the engine warms up and the choke begins to open, the fast-idle cam rotates and allows the engine to slow down.

When the engine starts, the choke must open a sufficient amount to allow more air to enter the engine. This is accomplished by a *choke piston* or *choke pull-off* (Fig. 3–18). High manifold vacuum is applied to the piston or pull-off as soon as the engine starts and pulls the choke partially open. Air rushing into the carburetor, against the offset choke plate, assists in opening the choke.

A *choke unloader* is also used to provide a way to open the choke if the engine becomes flooded during starting or if the driver seeks to accelerate quickly. This is accomplished mechanically by linkage that causes the choke to open when the throttle plates are fully opened.

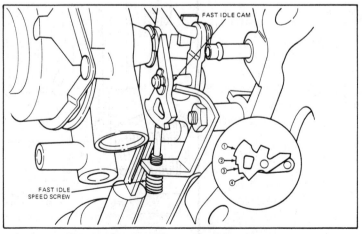

Figure 3–17 (Courtesy of Ford Motor Company)

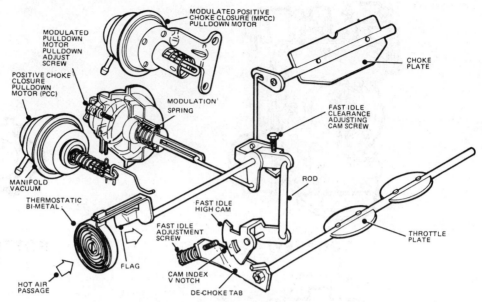

Figure 3-18 (Courtesy of Ford Motor Company)

What has just been described are the six basic circuits of a carburetor. Actually, there are seven circuits; however, one circuit plays a role with most of the rest and its duties were described together with those of the other circuits. This system is the *throttle system,* which is responsible for changes of speed and is connected to the gas pedal. The other six circuits are the float system, idle circuit, main metering circuit, power circuit, accelerator pump circuit, and choke system. All carburetors have these seven circuits. There are many types and appearances of carburetors and many ways to design these circuits, but all will have some means to accomplish these tasks that are necessary to operate an engine.

The bore or throat of a carburetor is often referred to as a *barrel.* The size of the barrel determines how much air can flow through the carburetor. Naturally, a large-displacement engine that requires a lot of air to fill its cylinders will require a larger-barreled carburetor. A large carburetor barrel has less of a venturi effect, so to achieve the same results as with a large barrel, manufacturers produce carburetors that are made of more than one barrel. Two barrels of equal size can flow about the same amount of air as a barrel twice its size, and without any loss of the venturi effect. Figure 3-19 shows the carburetors currently used: one-, two-, and four-barrel carburetors.

Four-barrel and some two-barrel carburetors have primary and secondary barrels. With this arrangement, half the number of barrels are used for normal driving. These primary barrels supply the engine with its air/fuel requirement until the throttle plates are opened wide or the manifold vacuum is very low. At this time, the secondary barrels assist the primaries in meeting the air/fuel requirements of the engine. The activation of the secondaries is accomplished either by a linkage or by vacuum, which causes the throttle plates of the secondary to open. During low-speed operation, only the primary throttle plates are open. However, during full-throttle conditions, the throttle plates of both the primary and secondary barrels are fully opened.

The number of barrels is a general way to classify or identify carburetors. Most carburetors have an identification tag or a model number to allow an automotive technician to find the correct adjustment specifications and replacement parts. Carburetors have also been classified in the past by the direction of the air flow in and out. Today's cars that are equipped with a carburetor use *down-draft carburetors,* which means that the air enters the top and flows down into the engine.

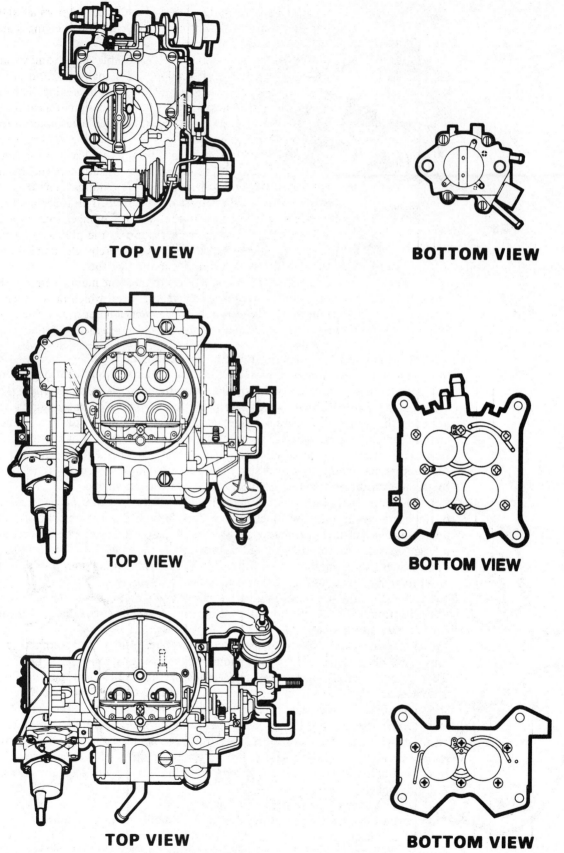

TOP VIEW

BOTTOM VIEW

TOP VIEW

BOTTOM VIEW

TOP VIEW

BOTTOM VIEW

Figure 3-19 (Courtesy of Ford Motor Company)

Older domestic and foreign cars used up-draft and side-draft carburetors as well as down-drafts. Further classification of carburetors results from design changes made to be more suitable for a particular application.

Two current variations of the basic design are the variable venturi and variable jet carburetors. *The variable venturi carburetor* (Fig. 3–20) changes venturi size as the fuel and air requirements change due to speed and load of an engine. The *variable jet carburetor* does the same with the main jet. The basic design of a carburetor has a fixed diameter for the venturi, the size of which has been determined to be a good compromise for all engine operating conditions. By being able to increase and decrease the size of a venturi, a higher air velocity through the carburetor can be maintained. This results in very accurate metering and atomization of the air-and-fuel mixture. The changing of the size of the venturis also requires a change in the amount of fuel to needed to maintain the proper air/fuel ratio. This is accomplished by having a tapered metering rod attached to the movable plunger that regulates the size of the venturi. This rod fits into a main jet and as the plunger moves, it causes the rod to fit looser or tighter into the main jet, thereby changing the size of the main jet and allowing for a change in the amount of fuel.

This same basic design allows manufacturers to have the main jet be variable. Variable jet carburetors (Fig. 3–21) also have a metering rod which passes through a jet, allowing the delivery of fuel to be altered. Rather than depend on the movement of a venturi plunger, these carburetors have the rods attached to a diaphragm. This diaphragm is activated by vacuum and as it moves, it moves the rod in and out of the jet.

For the sake of improving engine efficiency, manufacturers have added many circuit-assist devices to the basic carburetor, as well as making other changes in the individual circuits. These changes further classify the types of carburetors. It is al-

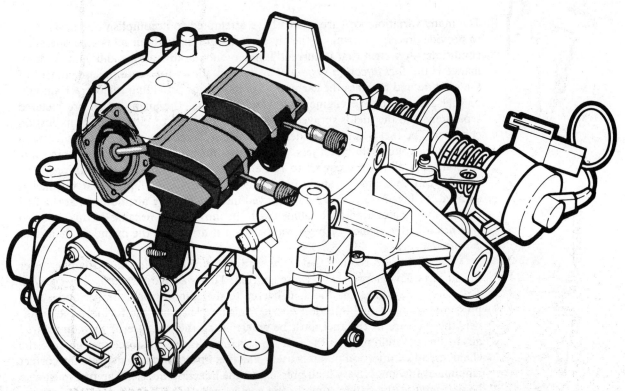

Figure 3–20 VV2700 variable venturi carburetor. (Courtesy of Ford Motor Company)

Figure 3-21 Varijet carburetor.

ways wise to identify and secure accurate specifications from shop manuals before getting too involved with individual carburetors. Throughout time, changes have been made continuously in an attempt to make a carburetor efficient in terms of engine power and fuel economy. It is rather ridiculous to expect any one person to know the details of each type of carburetor; the necessary information is readily available in a number of shop manuals. Understand the basics and use the material in the manuals when working with a carburetor.

FUEL INJECTION SYSTEMS

The many variations of carburetors have attempted to accomplish one basic task: to provide precise metering of the air/fuel mixture throughout all engine operating conditions. A system that is currently undergoing modifications and gaining dominance is the *fuel injection system*. This system allows for more precise control of the mixture and replaces the carburetor in its role of providing the correct amount of fuel and air for the engine. With fuel injection, the air and fuel are metered separately, and the fuel is sprayed into the moving air in or close to the combustion chamber. The fuel is sprayed under pressure and therefore a venturi is not required to draw fuel out of the main jets. This allows the air/fuel mixture to remain quite constant and allows the engine to run efficiently at all operating conditions. The mixture can be changed simply by changing the amount of time that the fuel is sprayed. In Fig. 3-22 fuel is being sprayed into the moving air; notice that the fuel becomes more atomized as it collides with the moving air. Because of this, injected fuel has a better chance of being vapor when it arrives at the cylinders than does carbureted fuel.

Diesel engines require fuel to be sprayed into the combustion chamber after the air has been compressed to cause ignition. This is accomplished by a *mechanical fuel injection system*. The fuel is delivered to the combustion chamber under high pressure through a fuel injection nozzle. This nozzle sprays the fuel into the chamber; the pressure of the fuel must be greater than the pressure of the compressed air or the air would prevent the fuel from being released. To force the fuel into the chamber, a fuel injection pump is used. The fuel injection pump delivers the correct amount of fuel under very high pressure to the injection nozzle. The fuel must also be delivered at the correct time in the engine cycle. Fuel injection nozzles are designed to spray fuel into the chamber in a particular pattern to allow for proper ignition (Fig. 3-23).

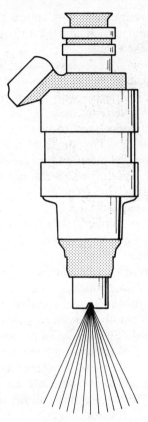

Figure 3-22 Spray of fuel from injection nozzle into the air.

The sequence in which the nozzles spray fuel into the cylinders determines the firing order of the diesel engine. The injector pump controls this by distributing high-pressure fuel to one injector nozzle at a time. The pump is driven by the engine and therefore must be timed to allow fuel to be delivered at the correct time. The injection nozzle will allow fuel to be sprayed only when high pressure is being applied to it. A spring inside the nozzle must be overcome by the fuel pressure before it can open. The opening pressure of the nozzle can usually be adjusted by changing the spring tension either by an adjustment screw or by the use of shims. This opening pressure for an injection nozzle is often referred to as the *pop-off pressure*. This

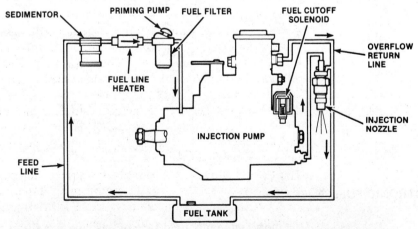

Figure 3-23 The flow of fuel for a typical diesel engine. Notice that the injector pump is responsible for both the delivery of fuel to the injector and the pressure at which it is delivered. This pressure is important for the proper spraying of fuel into the cylinders. (Courtesy of Ford Motor Company)

type of fuel injection is called "mechanical" because it uses a mechanically operated and controlled pump, which controls the opening of the nozzles and thereby controls the firing of the engine.

Mechanical fuel injection systems are also used on gasoline engines, but since the ignition is not caused by the injection of fuel, they operate differently. Gasoline engines are typically injected with fuel before the combustion chamber or intake valve. The vacuum of each cylinder then draws the fuel into the chamber. Because of this, the fuel does not need to be delivered at the extremely high pressures required in diesel engines. The timing of the injection of fuel is also not important with gasoline engine fuel injection because it is not the injecting of fuel that dictates when ignition will occur. The fuel can be continuously injected into the airstream and each intake stroke of a cylinder will draw in the air-and-fuel mixture. For gasoline engines the mechanical injection system most commonly used is the *Bosch Continuous Injection System* (CIS) (Fig. 3–24). It continuously injects fuel before the intake valve during the operation of the engine. Because the injection is continuous and therefore need not be timed, it is not driven by the engine. The amount of fuel injected depends on the amount of air at the throttle plate. Above the throttle plate is an airflow sensor which measures the volume of air present, and this sensor controls a fuel distributor which regulates the amount of fuel delivered to each cylinder. This ensures a proper air-and-fuel mixture for all cylinders under all engine operating conditions. Changes in engine vacuum affect the demand for air in an engine and therefore by monitoring the airflow to control the fuel flow, the mixture can be precisely controlled. Although this system is not mechanically driven, it is a mechanical system because all parts are mechanically operated or controlled by levers and plungers. There are some additional systems added to this basic system which are electrically controlled, such as the *cold-start system,* which measures the temperature and allows a separate injector nozzle to release fuel when the engine is cold.

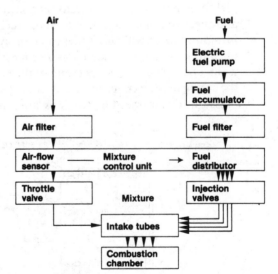

Figure 3-24   (Courtesy of Robert Bosch Company)

ELECTRONIC FUEL INJECTION

The most commonly used type of gasoline fuel injection is electronically operated and controlled. These *electronic fuel injection systems* operate much the same as the gasoline mechanical type but use electricity to control certain devices, which in turn control the mixture. Most electronic fuel injection systems are not continuous and inject fuel only at a particular time. Engine fuel requirements are determined by airflow or by engine vacuum. Sensors for airflow and vacuum send electrical signals to a microcomputer, which determines the fuel requirements for the engine. The use of electronics in fuel injection gives the system more precise control of the mixture because the speed of electricity is much quicker than something mechanical responding to changes in airflow or vacuum. Once the computer has determined the needs of the engine, it sends an electrical signal to the injection nozzle, which then opens. The amount of time the injector is open determines the amount of fuel that will be sprayed into the air. The longer the injector is open, the more fuel is delivered. The time that the injectors are open is called the *injector pulse width*. When the injector nozzles are located before the intake valve, as shown in Fig. 3–25, the system is referred to as a *port injection system*. Fuel is injected into the intake valve outside the combustion chamber. Each cylinder has its own injection nozzle and through the use of a computer, the mixture for each cylinder has the potential of being precisely controlled.

Another type of electronic fuel injection is the *single-point injection system*. This system does not have individual injectors for each cylinder but rather, one or two injectors mounted in a throttle plate assembly located where a carburetor would be (Fig. 3–26). This throttle assembly is similar to the lower half of a carburetor in appearance. The injection nozzles spray fuel into the engine past the throttle plates and the vacuum of the engine distributes the fuel-and-air mixture to the cylinders. Other than the difference in the location of the injectors, this system operates similarly to the port type.

Electronic fuel injection systems have many sensors that monitor a variety of conditions to allow for precise control of the fuel. Through the use of electronics and automotive applications of computers, *electronic fuel injection* (EFI) is widely used today in the automotive industry and it is believed that it will eventually replace the carburetor. In a later chapter these types of fuel injection are discussed in more detail.

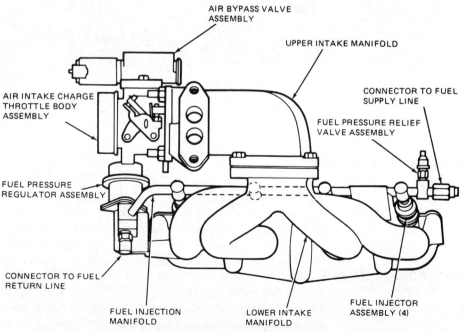

Figure 3–25 (Courtesy of Ford Motor Company)

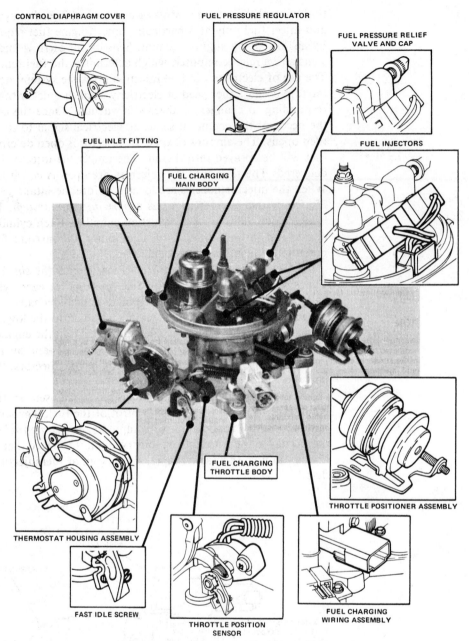

Figure 3-26 (Courtesy of Ford Motor Company)

FUEL DELIVERY SYSTEMS

No matter whether the engine is equipped with fuel injection or a carburetor, fuel must be delivered to the air/fuel system. This is accomplished by a *fuel pump*. Fuel pumps, like fuel injection systems, are either mechanical or electrical (Fig. 3-27). The purpose of either type is to deliver fuel from the gas tank to the carburetor or fuel injection system. A mechanical fuel pump is driven by a cam on the camshaft of the engine. As the camshaft and fuel pump cam rotate, a rocker arm positioned on the cam moves up and down. This rocker arm is positioned on the other end against a diaphragm which moves inversely to the movement of the rocker arm on the pump cam. The movement of this diaphragm against the pump spring creates a vacuum and a high pressure as it moves up and down. The vacuum formed draws

Figure 3-27 Typical electric pump (left); mechanical pump (right).

gasoline from the gas tank, while the pressure formed pushes the fuel to the carburetor's float bowl or the fuel injection system. The pump continues to work all the while the engine is working. The fuel is delivered to the carburetor at a relatively low pressure and the needle valve in the float system controls how much fuel will enter the float bowl. When no fuel is needed, the fuel being delivered by the pump does not continue to flow but rather pressure builds up in the system until the needle valve allows fuel to enter. Fuel will not be drawn out of the carburetor back to the gas tank when the pump creates a vacuum because there is a check valve on the outlet side of the pump which only allows fuel to flow out. Similarly, on the inlet side of the pump there is a check valve which allows fuel to enter only through the opening. Fuel pumps must deliver a specific amount of fuel and must deliver it under pressure to allow the float bowl to maintain the proper level. There are specifications for both volume output and pressure. The pumps are not typically adjustable, and if test results are not within specifications, a pump should be replaced.

Electric fuel pumps can be used on engines with carburetion but are used more commonly with fuel injection. An electric pump is not driven by the engine but uses electricity. It can be mounted in or at the gas tank and pushes the gasoline to the engine. Some electric fuel pumps use a diaphragm assembly like that of a mechanical system but use an electric solenoid to move the diaphragm. In place of a diaphragm, others use a metal bellows, which creates a vacuum when stretched out and pressure when it is collapsed. The basic functioning is that of a mechanical pump.

Another type of fuel pump, which is more common today, especially with fuel injection, is the *impeller-type electric fuel pump.* This is mounted in the bottom of the gas tank and is designed to operate while submerged in gasoline. The pump is made up of an electric motor and an impeller blade. As the motor turns, it spins the impeller blades at a high speed. Because the impeller blades are submerged in gasoline, the gasoline enters the blades without the creation of a vacuum. The gasoline passing through the blades becomes pressurized and is pushed out of the gas tank.

The gasoline in the gas tank also serves the purpose of cooling and lubricating the motor of the pump as the gasoline flows by the pump. This type of pump has no check valves and therefore, when the engine is turned off, there is no pressure in the system. The other types of fuel pumps, electric and mechanical, hold pressure within the system until fuel is allowed to flow. By relieving the pressure when the engine is not running, pressure is taken off the float's needle valve and the fuel injection system. This tends to extend their life and prevents fuel from seeping into the engine while it is not running. Regardless of the type of fuel pump, it is important to have the pump operate within the desired specifications because if the pressure or volume is too great or too low, driveability problems can occur.

Fuel pumps have the purpose of supplying the induction system with fuel. They deliver the fuel from the gas tank through fuel lines to the induction system (Fig. 3–28). These lines are designed to allow for the travel of fuel without losing any fuel. They are also designed to prevent a decrease in the amount of fuel volume and pressure. A damaged fuel line can cause the engine to be starved for fuel, causing the engine to run lean and inefficiently. These lines are usually made from a rust-resistant steel alloy that resists crushing. Flexible rubber hoses are used to connect ends of the steel lines and where considerable movement of the fuel line is normal. This helps prevent the possibility of the line cracking or becoming crimped. The length of these lines is usually kept to a minimum because with age, the rubber tends to crack. Longer lengths flex more, increasing the chance for a fuel leak. Low fuel volume or pressure delivered to the engine can be caused by damaged fuel lines as well as by the fuel pump. These lines should be inspected routinely.

The *gas tank* serves as the storage for fuel and is usually made of rust-resistant steel or stamped steel with a rust-resistant coating. Most gas tanks have baffles inside to reduce the movement of gas as the car moves. The fuel is put in the tank through a filler tube, which is covered by a removable cap that seals the gas tank from dirt and prevents gasoline vapors from entering the atmosphere. Years ago, gas tank caps were not sealed and had a hole in them to allow atmospheric pressure in the tank. This cap is referred to as a *vented gas cap*. However, gasoline vapor is a regulated pollutant and the hole in the vented cap allowed gasoline vapors to escape into the atmosphere. The concern for cleaner air caused automobile manufacturers to use *nonvented gas caps* to prevent the escape of these vapors. However, the gas tank must still be vented to allow atmospheric pressure in. Without the presence of atmospheric pressure, the vacuum produced by the fuel pump would not be able to draw fuel out.

With a nonvented gas cap, a *vent line* is used to allow atmospheric pressure in. The vent line is usually a tube leading from the top of the tank to a cannister which absorbs the gasoline vapors. Somewhere in the fuel line that connects the gas tank vent line to the cannister is a liquid-vapor separator, which allows only gasoline vapor to be delivered to the cannister. The liquid gasoline that may be present is returned to the gas tank. When the engine is running, the vapors trapped in the cannister are drawn into the engine by engine vacuum, which pulls air through the cannister, purging it of gasoline vapors.

Nonvented gas caps are fitted with a special valve that prevents a buildup of pressure or a vacuum in the gas tank. This valve will open to allow pressure to escape and to allow atmospheric pressure in. Too much pressure in the tank could

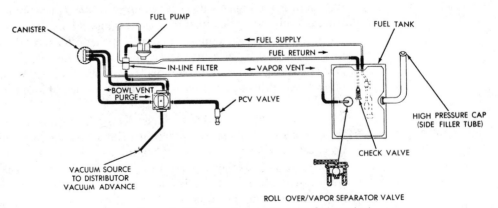

Figure 3–28 (Courtesy of Chrysler Corporation)

cause it to rupture, whereas a vacuum could cause fuel starvation of the engine or cause the tank to collapse. The system used with the nonvented gas cap is called the *evaporative emission control system*. An example of this system is shown in Fig. 3–29.

The purpose of the fuel pump, fuel lines, and gas tank is to supply the engine with the fuel it needs. This fuel must be clean when it arrives at the engine. Not only do we not want dirt to enter into the engine, but even small particles of dirt can plug a passage in the carburetor or fuel injection nozzle. Most automobiles have two *fuel filters* to remove these dirt particles. There is usually one located in or at the gas tank to clean the fuel of the larger particles of dirt as the gasoline leaves the tank. The other is located in the engine compartment and cleans the fuel of the smaller particles. The fuel filter in the engine compartment is replaceable and can be one of the four types shown in Fig. 3–30, depending on the application. *Sediment-bowl-type fuel filters* use a glass or metal bowl with a fiber, paper, ceramic, sintered metal, or fine metal screen to filter the fuel. As the fuel flows through the bowl, the larger particles of dirt fall to the bottom of the bowl and the smaller particles get trapped in the filter element. This type of filter is sometimes mounted on the fuel pump.

Screw-on housing-type fuel filters are mounted in the fuel line or can be part of the fuel pump. These filters have a disposable filter element or are replaced as a complete unit. These filters have the appearance of a can and are screwed onto a housing which has fuel flowing through it.

An *in-line fuel filter* is positioned in the fuel line between the fuel pump and the carburetor. This type of filter allows for the full flow of fuel because it directly filters the fuel as it moves to the engine, without causing the fuel to be delayed for filtering. In-line filters are enclosed in a glass, plastic, or metal housing. The entire assembly of this type of filter is disposable. Like all fuel filters, these should be replaced routinely.

Two types of fuel filters are used right at the fuel inlet of the carburetor. One type has a disposable element and the other is totally disposable. Because these filters are located at the end of the fuel lines, they tend to be very efficient in cleaning the fuel before the fuel enters the carburetor. Being efficient, they tend to collect more dirt and therefore will plug up sooner. When a fuel filter becomes plugged up, it can decrease fuel flow or stop fuel flow totally. When a fuel filter is suspected

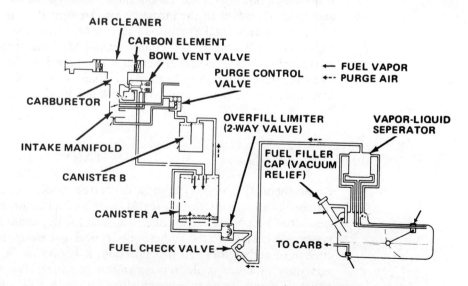

DODGE CHALLENGER/PLYMOUTH SAPPORO

Figure 3–29 (Courtesy of Chrysler Corporation)

Figure 3-30   Four different types of fuel filters.

of being partly or completely plugged, the fuel delivery system should be tested for proper delivery volume.

THE ENGINE'S MANIFOLDS

So far in this chapter we have examined the fuel system and its major parts. We have taken the fuel from the gas tank, cleaned it, and sent it to the induction system. We have seen what the fuel injection system and carburetor do to meet the demands of the engine for air and fuel. There are two more important features of an engine which aid in its "breathing." These are the passageways for the air and fuel to enter the cylinders and the passageways for the exhaust to be released to make room for the next charge of mixture. These passageways are referred to as *manifolds*. There is an intake manifold and an exhaust manifold. The *intake manifold* delivers the air-and-fuel mixture to the individual cylinders of carbureted and throttle-body fuel-injected engines and delivers air to the cylinders of diesel and other injected engines. It not only must deliver the air, but must also keep the air-and-fuel mixture mixed and retain the speed of the incoming air. To keep the fuel vaporized as it moves into the cylinders, intake manifolds receive heat from the exhaust manifold or the engine's cooling system. The *exhaust manifold* connects the exhaust valves to an exhaust pipe, which releases the exhaust from the car. The exhaust manifold must allow for the exhaust gases from each cylinder to exit the chambers in the manner best suited for the engine's valve timing. A quick exit of the gases is usually desired.

SUMMARY

For an engine to operate efficiently, the correct amount of fuel must be mixed with the correct amount of air in a sealed container, and must be shocked by the right amount of heat at the correct time. The heat for ignition is supplied by the ignition system, which is operated by both mechanical and electrical means. From the discussion of carburetors and fuel injection, it is easy to see that electricity plays a very important role in the way an engine performs. The next chapter covers the basics of electricity and the many electrical components that govern or influence how an engine runs.

TERMS TO KNOW

Induction system	Carburetor
Rich mixture	Pressure differential
Lean mixture	Venturi
Volatility	Pop-off pressure
Octane rating	Fuel injection
Detonation	Injector pulse width
Cetane rating	TBI

REVIEW QUESTIONS

1. What is the purpose of the induction system?
2. What is the ideal air/fuel ratio?
3. What determines the needed air/fuel ratio of an engine?
4. What is a hydrocarbon? Give an example of one.
5. What causes "pinging"?
6. What are the three purposes of an air cleaner assembly?
7. What two major things determine the rate of airflow through a carburetor?
8. Explain the action of a venturi.
9. What is the major purpose of the float system?
10. What is the purpose of the idle discharge hole?
11. What is the purpose of the main jet?
12. Why is the power circuit of a carburetor needed?
13. Why is an accelerator pump needed?
14. What is a choke unloader? Why is it needed?
15. Why is a venturi not needed with fuel injection?
16. What is the major difference between diesel and gasoline fuel injection?
17. What are the advantages of fuel injection over carburetion?
18. Why should rubber fuel lines be kept to a minimum length?

4

ELECTRICITY FUNDAMENTALS

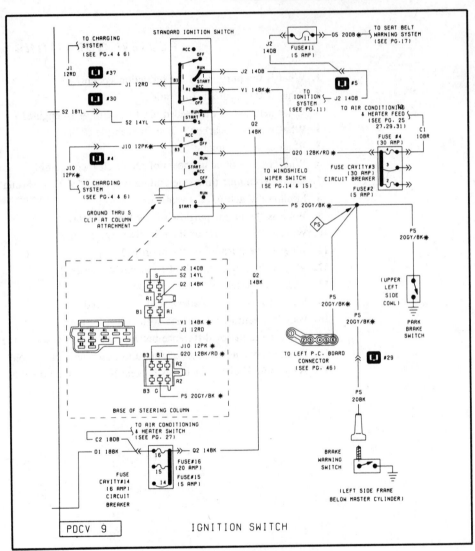

Figure 4–1 (Courtesy of Chrysler Corporation)

The automobiles of today are designed to be safe, comfortable, reliable, and efficient. Most of the accessories and components that contribute to these qualities are operated and controlled by electricity. For safety there are the lights, windshield wipers, and window defoggers. The radio, heater fan, power seats and windows, and interior lighting are some of the comfort items that are operated by electricity. The advances made by the automotive industry utilizing computers have provided increased reliability and efficiency. Computers are totally electric. Many of these items could be operated and have been operated by means other than electricity. Electricity is chosen to operate these components because it can be used without depending heavily on the mechanical parts of an engine. Vacuum is also used to operate some accessories, but this tends to deplete some of the vacuum needed for induction. Also, vacuum signals are much slower than electrical signals. This is another reason for the preferred use of electricity.

Electricity is used for the ignition of gasoline engines, the heat needed for combustion is supplied by electricity, and the process of starting the engine is also handled by electricity. To start an engine, the crankshaft must turn at a speed fast enough to allow the pistons to form a vacuum which allows the fuel-and-air mixture to be drawn in.

The turning of the crankshaft is accomplished by an electric motor called the *starter*. A *battery* supplies the power needed by the starter and the power for the ignition system when the engine is starting. A battery is a storage place for electricity and also supplies the power needed for operation of the car. Because there is a constant drain on the battery as the engine runs, a *battery-charging system* is installed on the engine to keep power available at the battery. The charging system is driven by the engine and as the engine runs, it recharges the battery. The electricity used or the actual drain the ignition system has on the battery is low. It is the drain of the starter and accessories such as lights that use a lot of power from the battery for operation. If these systems drain down the battery and the charging system cannot adequately recharge the battery, the consequent loss of power could affect the operation of the ignition system. The ignition system takes the power supplied by the battery and multiplies it to provide for the spark across the gap of the spark plug to generate the heat needed for combustion. Before examining how the ignition system multiplies the power available from the battery, the basics of electricity must be discussed.

Electricity cannot be heard, felt, or seen, but the effects of electricity can be heard, felt, and seen. Electricity is all around us. If we could hear it, that would be all we could hear. Our bodies function through electricity or electrical pulses. Everytime we think or do something, there is an electrical exchange of information between our brain and our foot, hand, tongue, or whatever. We do not feel the electric pulses; we just walk or talk. Our eyes cannot see electricity because it is faster than we can even imagine. What causes electricity is the movement of something we cannot see, even with a microscope. In fact, electricity is the movement of electrons. Electrons are part of an atom and we cannot even see atoms. To understand what electricity is, we must discuss atoms and electrons.

Everything is made up of atoms and all atoms have electrons. The atoms of one element are different from those of another. Everything is made up of an element or a combination of elements. *Matter* is the name given to everything that exists. It is composed of a combination of elements or is a group of atoms of the same element. *Elements* are the building blocks for all matter and each element has its own unique characteristics. All atoms of an element are similar, but atoms of different elements are different. There are approximately 103 known elements and everything that exists is made up of them. The smallest part of an element is an *atom* and the atom has all the features and characteristics of the element. The characteristics and features of an atom which make it different from an atom of a different element are the things that must be examined to provide an understanding of electricity.

An atom is made up of electrons, protons, and neutrons (Fig. 4–2). The *protons* and *neutrons* are in the center of the atom and make up the *nucleus* of the atom. The *electrons* orbit the nucleus of the atom in much the same way that the planets in our solar system orbit around the sun. The electrons orbit in a set or fixed

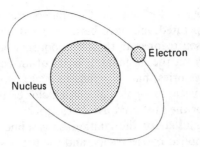

Figure 4–2

path, as do the planets. The only difference between the atoms of one element and those of another is the number of protons, neutrons, and electrons. The number of electrons also dictates the space occupied by the orbiting electrons.

Parts of the atom have *electrical charges*. The electrons have a negative charge and the protons have a positive charge. The neutrons, located with the protons in the nucleus, have a neutral charge. Negative charges are labeled with a minus (−) sign and positive charges are marked with a plus (+) sign, as shown in Fig. 4–3. These symbols are used to help define drawn electrical circuits. A *balanced* or *normal atom* has an equal number of protons and electrons; in other words, it has the same number of positive charges as negative charges. The electrons stay in orbit around the protons because they are attracted to them. Unlike (negative versus positive) charges attract each other, whereas like (positive versus positive) charges repel each other. The electrons never leave their orbit to go to the proton because of the presence of the other electrons in the orbit, which repel them and keep them in motion in the orbit. When an atom has more electrons than protons or more protons than electrons, it is called an *unbalanced atom*. If an atom has more protons than electrons, it has a positive charge because of this inbalance. An atom will have a negative charge when the number of electrons is greater than the number of protons. A negatively or positively charged atom is referred to as an *ion*. Ions try to become balanced and will exchange or share electrons to do so. This exchange of electrons and the movement of the electrons among atoms is *electricity*.

Figure 4–4 shows the paths that the electrons follow while orbiting around the protons; these are called *shells*. There are a fixed number of electrons that will be able to orbit within a given shell. If there are too many electrons in the first or closest shell, the extra electrons will move to the next shell. Atoms with many electrons have many shells for the electrons to orbit in; there can be as many as seven shells of electrons.

The outermost shell, called the *valence ring,* can combine with other atoms to exchange or share electrons. In a discussion of electricity, we are concerned only with the electrons present in the valence ring. If a valence ring has three or fewer electrons, it has room for more and the electrons present are held loosely in the shell. These electrons are loose in the shell because with few electrons, the repelling effect has less strength to hold them together. These loose electrons are called *free electrons* because it takes very little to move them into the shell of another atom.

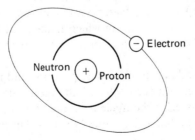

Figure 4–3

2
8
18
+ 1
___
29

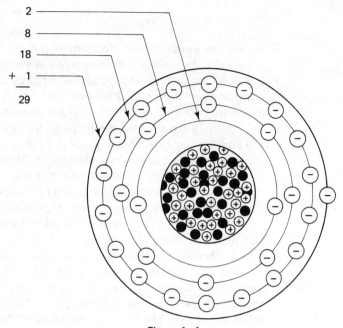

Figure 4-4

Therefore, an atom with three or fewer electrons in the valence ring can give off an electron as easily as it can accept a free electron from another atom. The movement of electrons from atom to atom is called *electrical current* (Fig. 4-5). When the valence ring of an atom has five or more electrons, it is nearly full. The electrons in this valence ring cannot easily be moved and are called *bound electrons*. It is hard for a free or drifting electron to join a shell of bound electrons.

Materials that have free electrons are good *conductors* of electricity, because electricity is the movement of electrons, and materials with three or fewer electrons in the valence ring allow for easy movement of electrons. Some metals, such as copper, silver, and aluminum, are good conductors of electricity and are commonly used to allow electrical current to flow. The path of the movement of electrons or of electrical current is set by the placement of a conductor.

Materials with many bound electrons are good electrical *insulators* because they tend not to allow electrons in or out of the valence ring. Plastic, rubber, and glass are common materials used for insulating electricity. Insulators are used on the outside of a conductor to prevent the flowing electrons from straying and not following the path set by the conductor.

Electronic devices such as diodes and transistors are composed of materials that are neither good conductors nor good insulators. The atoms that make up these materials have four electrons in their valence ring. These materials are called *semiconductors*.

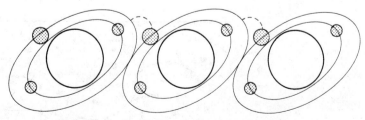

Figure 4-5 An electron moving from atom to atom.

CURRENT

A random movement of electrons is not electrical *current* flow; The controlled movement of electrons from one point to another is. If a few electrons move, there is a small amount of current, and when large amounts of electrons flow, there is a large amount of current. Electrical current or electricity is the directed movement of electrons from atom to atom within a conductor.

To help explain what electricity is, let's look at the piece of copper wire in Fig. 4–6 and put some electrical current through it. Copper is a good conductor of electricity because it has only one electron in its valence ring. Now let's grab a proton from somewhere other than the wire and put it at one end of the wire. This end of the wire is now positively charged. The electron in the valence ring of the atom closest to the proton jumps over to the proton because the proton needs an electron to be balanced and the electron is attracted to the proton. Now the atom that just lost the electron attracts an electron from the atom closest to it and is again balanced. Then that atom attracts an electron, and this sequence continues all through the wire until the last atom at the other end of the wire is reached. This end of the wire is now electrically charged negative. The movement of electrons will continue through the wire as long as the negative end remains negative and the positive end remains positive. The force of attraction of the electrons to protons is the same as the force of the electrons pushing away from each other, and this keeps the electrons flowing in one direction as long as the charges are present at both ends of the wire.

To understand electricity, it is important to remember that it is the electrons that move and that they move toward something positive. The number of electrons that flow determines the amount of current. Obviously, the larger a conductor is, the more current flow it will allow. Is it the attraction of the protons that causes the electrons to move, or is it the electrons that seek something positive that causes current flow? This question points out a continuing battle over which way current flows. Does it flow from negative to positive, or does it flow from positive to negative? It does not matter. We use electrical current in the same way regardless of which way it flows. To have current flow, there must be a conductor with a positive charge on one end and a negative charge on the other end. To avoid unnecessary confusion, let's say that current flows from positive to negative (Fig. 4–7), which is the way most automobile industry people like to think. In this book, all explanations of electricity will be based on the idea that current flows from positive to negative and that current is the result of electrons moving to something positive.

To maintain the opposing charges at either end of a conductor, an electrical power source must be used. This power source must have in storage, or must be producing, positive and negative charges. If the power source runs out of either positive or negative particles, current will cease to flow. Current will also stop if the conductor is cut. This separates the two different charges, current would no longer have a path to follow, and the circuit would be incomplete.

Current can be likened to water flowing in a pipe. The amount of water that flows through the pipe can be measured. We merely measure the amount of water that flows through a certain point in the pipe over a particular period of time. With

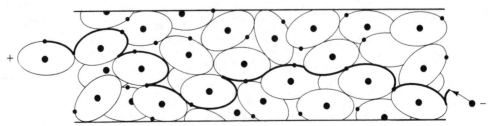

Figure 4–6   The flow of electrons through a conductor.

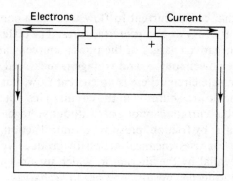

Figure 4–7 Electricity is the flow of electrons, and current is the result of that movement of electrons.

current, we measure the number of electrons that flow through a particular point in the circuit in a second. To give you an idea of just how small and fast electrons are, we do not have a unit of measurement for the amount of current flow until 6.25 billion billion electrons pass a point in 1 second! One ampere of current is equal to 6,250,000,000,000,000,000 electrons flowing through a point in a circuit in 1 second. An *ampere* or "amp" is the unit of measurement for the rate of electrical current flow. Referring back to water, it is easier to measure amounts in gallons than it is to count the drops it takes to fill a 1-gallon container.

VOLTAGE

The power source needed to cause current flow does not become part of the path for current flow; it simply causes it. The supplying of opposite charges keeps the current going and tends to push the current through the conductor. The power source contains the opposite electrical charges and keeps them separated. The power source contains the potential for current flow. Until a conductor connects the positive with the negative, current never occurs. The electrons are pushed toward the protons by this potential. This potential or electrical pressure is called *voltage* (Fig. 4–8). Volt-

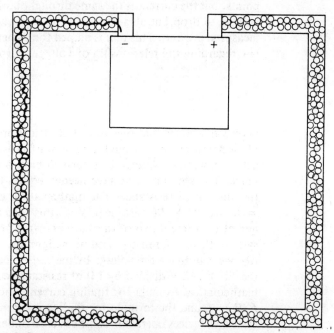

Figure 4–8 In an open circuit, electrons move to fill the conductor but current does not flow. There is a potential to cause current flow at the wire ends but because there is not a complete path for the flow of electrons, only a potential or voltage is present at the wire ends.

age is the force that causes current to flow through a conductor. Voltage is sometimes referred to as EMF, or *electromotive force*. The difference in strength between the positive and negative charges of the power source is its potential difference, or the voltage of the power source, and voltage is measured in *volts*.

In an incomplete circuit there is no current flow, but there is the potential for current flow if the power source is in the circuit. This potential or voltage can exist without current, but current cannot exist without voltage. Voltage can be produced in a number of ways: by friction, pressure, chemical action, and magnetic induction. The battery of a car uses chemical action to produce voltage. The battery is recharged with potential by the alternator, which uses magnetic induction to supply the battery with the needed positive and negative charges. An understanding of how magnetic induction and chemical reactions produce voltage will come with a better understanding of what electricity is. So far we know that current is the directed movement of electrons through a conductor and that voltage is the pressure that causes current to flow.

RESISTANCE

The size of a conductor determines the amount of current that can flow through the circuit. If a conductor is small, fewer electrons can travel through it within a given period of time. Therefore, the size of the conductor limits or opposes current flow. This opposition to current flow is called *electrical resistance*. There is some resistance in all conductors; there may be a great amount or a small amount. The amount of resistance in a circuit limits the amount of current that can flow in the circuit. Resistances within a circuit also decrease the amount of voltage in the circuit. Some electrical pressure is lost while pushing current through some resistance. While trying to maintain a constant rate of current flow through a resistor, the voltage loses some of its force and the amount of voltage decreases. Measuring voltage drops throughout a circuit is a way of determining the amount of resistance that a circuit has. The voltage throughout a circuit may be different at different points, but the current is the same throughout the circuit. Resistance not only causes voltage to drop but also limits the amount of current that can flow. Resistance is measured in *ohms*. This unit is named for a scientist who calculated a very important law regarding the relationship of voltage, current, and resistance.

OHM'S LAW

Ohm's law states that one volt of electrical pressure is needed to push one ampere of electrical current through one ohm of resistance. With Ohm's law you can calculate how much voltage is needed to have a certain flow through a certain resistance. The amount of voltage needed equals the amount of current multiplied by the amount of resistance. The mathematical expression for this law is: E(voltage) $= I$(current) $\times R$(resistance). With this formula one can figure out the values of any of the three if two of them are known. If on a 12-volt (V) automobile electrical system, there is a resistor used in the ignition system that has a resistance of 1 ohm (Ω), we can divide the voltage by the resistance to find the current that will flow in the circuit: 12 V divided by 1 Ω of resistance equals 12 amperes (A) of current. The mathematical formula for finding current when voltage and resistance is known is $E/R = I$, and the formula for finding resistance is $E/I = R$. Explanations of how electrical devices operate become easier to understand if they can be related to Ohm's law. It will be helpful to insert some numbers in the formulas of Ohm's law and practice the calculations, as shown in Fig. 4-9. When working on a car, you will

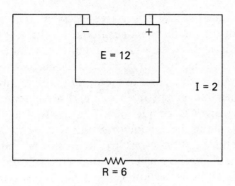

Figure 4–9 Six times 2 equals 12 (R × I = E).

seldom, if ever, need to calculate the values for a circuit. You do need to know the relationships of the values to one another. For example, on a car with a 12-V power source, as resistance increases in a circuit it can be expected that the current will decrease. Remember that resistance limits the amount of current that will flow in a circuit. With a constant voltage at the power source, we can expect current to decrease with an increase in resistance and current to increase with a decrease in resistance.

All conductors have some resistance. Electrical devices use conductors to perform work or to allow the devices to perform work. The amount of resistance a conductor has depends on a number of factors. Naturally, a good conductor has very little resistance. All electrical devices are called *loads* because they use electrical power to perform work.

Power is the term used to represent the amount of work done in a particular period of time. Electrical power or wattage is the amount of voltage used to push a certain amount of current through some resistance, or a load. Wattage is measured in *watts* and represents the voltage dropped across some resistance multiplied by the current that is able to flow through it (Fig. 4–10). Therefore, if we have a load that drops 10 V and has a current flow through it of 4 A, the power used at that load was 40 watts (W). Again the resistance in a circuit dictates the amount of power used. Any electrical resistance or load will use some electrical power. Most of these loads in a car's electrical system are there by design, to perform a particular function. Undesired resistances are the leading cause of electrical problems. These tend to decrease the amount of voltage and current available for the desirable loads or resistances, causing them not to function properly or not to function at all.

There are five basic factors that determine the resistance of a conductor: the condition, the length, the cross-sectional area, the atomic structure, and the temperature of the conductor. Usually, the higher the temperature, the higher the re-

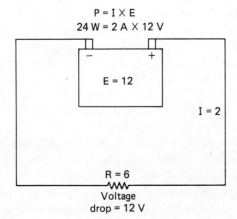

Figure 4–10 Watt's law.

sistance of a conductor. The fewer free electrons that a material has, the more resistance a conductor made of that material will have. "Cross-sectional area" refers to the thinness or thickness of a conductor; the thinner a conductor is, the more resistance it has. In a long wire or conductor the current has farther to travel. When electrons are traveling from atom to atom, they collide with each other, which tends to slow down their movement from one end of the conductor to the other. In a longer wire, electrons will collide more frequently, and this causes an increase in resistance. Points along the conductor can have a higher resistance than the rest of the conductor. This can be the result of corrosion at the connections or simply a loose connection. A partial cut in the conductor will have the same effect on the resistance as a small cross-sectional area would. The condition of the conductor determines how well electrons can flow through it. Looseness, cuts or frays, and corrosion cause the resistance of a conductor to increase.

To have current flow, there must be a complete circuit. Within the circuit there must be a load to limit the current flow or the conductor will burn up. The current flows from the battery through the conductor to the load and through it, then back to the battery. The battery serves as the power source for the circuit. The voltage present at the battery will push as much current through the wire as it can. If there is no load in the system, the circuit resistance is only that of the conductor, which is very low. The amount of current flow through the wire will be very high and the wattage given off as heat will be so great that it will burn the wire. Electrical energy, like all forms of energy, cannot be used up or destroyed but will change to another form of energy as it is used to perform work. As in the case of the burning wire, electrical energy changes to heat energy and causes the wire to burn. Not only would the conductor of a circuit with no load burn up, it would also be useless. The only work performed would be the burning of the wire. If a light bulb were inserted somewhere in the conductor, the conductor would not burn up. The resistance of the load or light bulb would limit the current (Fig. 4-11). Inside the bulb is a fine conductor called a *filament* made of a special metal which glows as it is heated up from the current flow. This glowing is the result of electrical power slowly burning away the filament. The heat energy given off by this burning allows the bulb to light. When a bulb burns out, the filament is burned away, causing an incomplete circuit, so that the light no longer glows.

SIMPLE SERIES CIRCUIT

The simple light circuit in Fig. 4-11 is an example of a *series circuit*. In a series circuit, current has only one path to take. A series circuit is continuous; it starts at the positive terminal of the battery (the side of the battery where the voltage is present and is pushing the current through) and ends at the negative or ground side of the battery (this side has no voltage present). The direction of current flow is always from a point of higher potential or voltage to a point of lower potential. This is why electrical devices are often marked with a positive (+) sign on one terminal and a negative (−) sign on the other, positive being a higher potential and negative being lower. Between the two terminals of the battery is a conductor leading to a load and a conductor from the load back to the battery. The load of a circuit is turned off and on by a switch, also wired in series with the load (Fig. 4-12). The switch opens or closes the circuit. A switch is not considered a load because it does not perform work; it merely controls the work of the load.

A series circuit is a continuous path for current from the positive side of a battery through a load and to the negative side of the battery. Throughout the circuit, amperage is the same and Ohm's law can be used to calculate any of the values within the circuit. The amount of current in a series circuit depends entirely on the

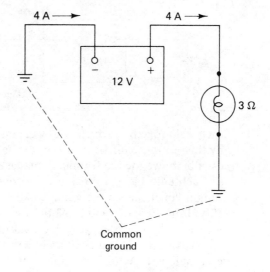

Common
ground

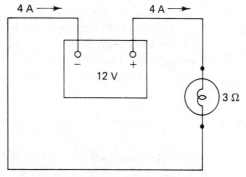

Figure 4-11

total resistance in that circuit. A series circuit may have more than one load. The total resistance of the circuit would simply be the sum of the resistances. By adding together all the individual resistances within the circuit, we can determine the current that will flow in that circuit. Although these resistances may vary in amount from one another, the current through all of them will be the same.

At each load or resistance, there will be a voltage drop, and the sum of all the voltage drops will equal the source voltage or the voltage of the battery. In Fig. 4–13 we have a simple series circuit with two light bulbs connected to a 12-V battery. Each bulb has a resistance of 3 Ω, the current is the same throughout the circuit, and the total of the voltage drops is 12 V. Using Ohm's law, we can see what happens at each load. To find the total resistance of the circuit, we add the resistance of

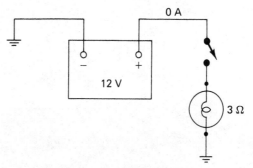

Figure 4-12 Incomplete circuit.

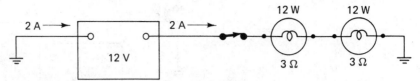

Figure 4-13

each bulb together, which is 6. Current is determined by dividing the voltage (12) by the resistance (6). The current of the circuit is 2 A. To find the voltage drop for each bulb, we use the formula: voltage equals current times resistance, or 3 × 2 = 6. Each bulb has a voltage drop across it of 6 V. The total of the voltage drops should equal the source voltage and it does, as we have 12 V at the source and 12 V dropped in the circuit. Notice that all of the voltage was dropped in the circuit, which leaves zero voltage going back to the negative side of the battery. All the voltage available for a circuit will be used in that circuit. The wattage given off by each bulb is 12 W (6 V multiplied by 2 A). The voltage applied to the first bulb in this series circuit was 12 V, and it used 6 of those volts. This left 6 V for the second, which is what it needed and what it used.

Any series circuit, regardless of the number or types of loads it has, will always have the same characteristics: The current will be the same throughout the circuit, the total resistance will be the sum of all the resistances, and the sum of the voltage drops will always equal the voltage of the source.

PARALLEL CIRCUITS

Parallel circuits are designed to allow current to travel in more than one path. This allows a single power source to operate more than one device independent of another (Fig. 4-14). In a two-bulb series circuit it is impossible to have one bulb work and not the other. A series circuit is a continuous circuit and an *open* or break any where in the circuit will cause current to stop flowing. In a parallel circuit, current has more than one path to follow, and therefore if one of the paths is open, it will not affect the others.

A parallel circuit has a common path from the battery and back to the battery. The additional paths branch out to additional paths from the common point and then join together again at the common point back to the battery. Each path or *leg* of the parallel circuit can operate one or more devices independent of the other legs. This allows for the ability to have a switch in each leg to control just that leg. The current flow in one leg does not affect the current flow in the others but does affect the total circuit amperage or the amperage at the common point. The amperage of a parallel circuit is the sum of the amperages of all the legs of that circuit. The voltage applied to each leg or *branch* of a parallel circuit is the same. In Fig. 4-15

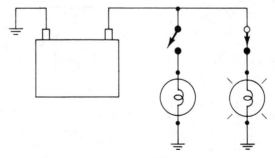

Figure 4-14

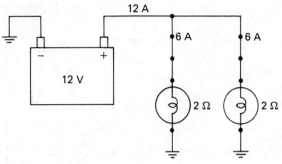

Figure 4–15

we have a two-branch parallel circuit with a 2-Ω light bulb in each hooked to a 12-V battery; the amperage of each leg is 6 A (current equals voltage divided by resistance). The total amperage for the circuit is the sum of the amperages of all the legs and is therefore 12 A. If we calculate the total resistance of the circuit, we find that it is 1 Ω (resistance equals voltage divided by amperage). In a parallel circuit the total resistance is always smaller than the smallest resistor in the circuit. This is why, although the resistance of each bulb is 2 Ω, we have a total resistance of only 1 Ω. This happens because current has more than one path to follow and there is an equal amount of voltage applied to each leg to push the current through the resistors. When paths are added to a parallel circuit, additional conductive paths for current flow have been added.

The three characteristics of parallel circuits that set them apart from series circuits are (1) the total amperage of the circuit is the sum of the amperages of all branches; (2) the voltage applied to each branch is the same; and (3) the total resistance of the circuit is less than the smallest resistor in the circuit. It is important to understand the difference between the characteristics of a series circuit and those of a parallel circuit. This understanding will allow you to diagnose engine and electrical problems more quickly and easily.

In parallel circuits, Ohm's law can be used directly only to calculate totals within each branch. Circuit totals can be determined with Ohm's law only after the circuit's total resistance has been determined. In a series circuit, to find total resistance, the resistances are added to find the total. In a parallel circuit, there are two formulas that can be used to calculate the total resistance. When only two resistances are in parallel, the *product-over-sum formula* can be used:

$$\text{total } R = \frac{R_1 \times R_2}{R_1 + R_2}$$

Using the example in Fig. 4–16 with two bulbs in parallel, each having 2 Ω of resistance, we find that 2 times 2 equals 4 and 2 plus 2 equals 4, and dividing the product (4) by the sum (4), we have a total resistance of 1 Ω. This formula can be used for parallel circuits with more than two branches, but you must calculate their values in pairs until one total resistance has been determined.

Another formula that can be used when there are more than two resistors in a parallel circuit, like the one shown in Fig. 4–17, is

$$\text{total } R = \frac{1}{\dfrac{1}{R_1} + \dfrac{1}{R_2} + \dfrac{1}{R_3} + \ldots + \dfrac{1}{R_n}}$$

Each branch of a parallel circuit should be treated as a series circuit; the total resistance in a branch is the sum of the resistances (Fig. 4–18). This is true in all

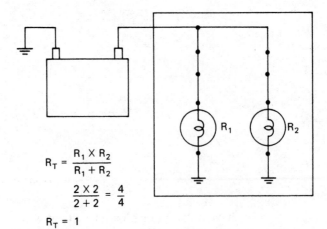

$$R_T = \frac{R_1 \times R_2}{R_1 + R_2}$$

$$\frac{2 \times 2}{2 + 2} = \frac{4}{4}$$

$$R_T = 1$$

Figure 4–16

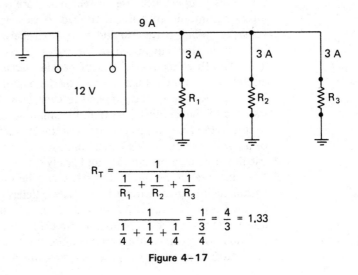

$$R_T = \frac{1}{\dfrac{1}{R_1} + \dfrac{1}{R_2} + \dfrac{1}{R_3}}$$

$$\frac{1}{\dfrac{1}{4} + \dfrac{1}{4} + \dfrac{1}{4}} = \frac{1}{\dfrac{3}{4}} = \frac{4}{3} = 1.33$$

Figure 4–17

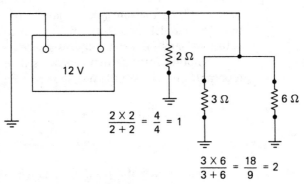

$$\frac{2 \times 2}{2 + 2} = \frac{4}{4} = 1$$

$$\frac{3 \times 6}{3 + 6} = \frac{18}{9} = 2$$

Figure 4–18

cases except when there is a parallel circuit in a branch of a parallel circuit. In this case, the total resistance in the branch should be calculated first using the appropriate formula. This type of circuit is a combination circuit and it is best to attempt to reduce the complexities of the circuit to a simple series or parallel circuit when attempting to calculate its values.

MAGNETISM

Magnetism is a form of energy that can cause current flow in a conductor and is caused by the movement of electrons just like electricity. It is also similar to electricity because it can not be seen. But magnetism is not electricity. Current does not flow in a magnet as a result of the movement of electrons; rather, *lines of force* surround the magnet (Fig. 4–19). When a conductor passes through these lines of force, voltage is produced in the conductor. These lines of force move in one direction, setting up a polarity. The positives and negatives used in electricity define a circuit's polarity. A magnet's polarity is similar but has north and south poles instead. The magnet's lines of force or *magnetic field* result from a concentration of magnetic energy at the poles, and the attraction between the two poles forms parallel loops around all sides of the magnet. It is believed that the lines of force are caused by the alignment of the atoms inside the object that is a magnet. Not all materials can become magnets. Rubber, plastic, aluminum, and wood cannot be magnetized, whereas soft iron ore can exist as a magnet in nature. Other materials can become magnets through heat or electricity.

A magnetic field is formed around a conductor when it has current flowing in it. The number and strength of the lines depends on the current that is flowing. The higher the rate of current, the stronger the lines of force. Soft-iron ore is easily magnetized and is the metal most often used for magnets. When a current-carrying conductor is wrapped around soft iron, the magnetic field formed becomes quite strong. When current in the conductor is stopped, the strength decreases. This type of magnet is called an *electromagnet* (Fig. 4–20) and its strength is controlled by the current flowing in the conductor that surrounds it. Current does not flow in the iron; therefore, the magnet is nonelectrical. The effects of magnetism are used in a number of systems in a car. One such system is the ignition system, which is discussed in great detail later in this text.

Figure 4–21 shows that the opposite poles of a magnet (*unlike poles*) attract each other and tend to pull together to form a larger magnet, whereas *like poles* repel each other and tend to push each other away. These principles of attraction and repelling of the poles are used to make electric motors work (Fig. 4–22). By controlling the current going to two separate electromagnets, the strength of the

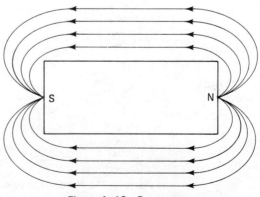

Figure 4–19 Bar magnet.

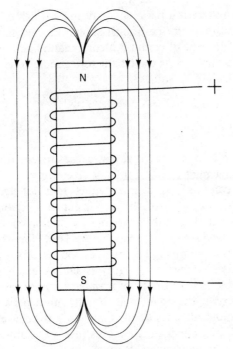

Figure 4-20

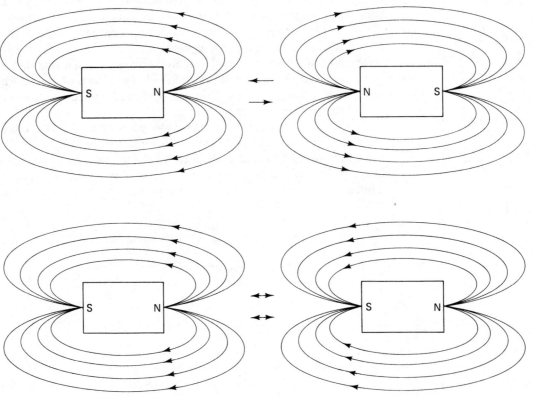

Figure 4-21 Repelling of poles (top); attraction of poles (bottom).

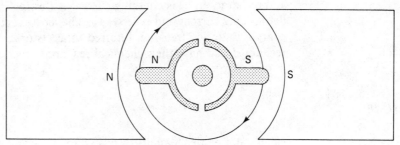

Figure 4-22 Simple motor.

attracting and repelling of the poles is controlled. Transforming the action of the poles to a rotary motion enables a motor to turn.

Perhaps the most important motor in the car is the starter motor. The starter motor, as do all electrical devices on the car, receives its power to operate from the battery. A battery converts chemical energy to electrical energy to supply the heavy demands for current and voltage. The recharging of the battery is the responsibility of the charging system, which uses principles of magnetism to generate the amount of electricity to allow the battery to meet the demands of the car. When a conductor passes through a magnetic field material, voltage is caused or induced in the conductor. The process of producing electricity through the use of magnetism is called *electromagnetic induction* (Fig. 4-23). Voltage is induced whenever a conductor passes through a magnetic field. In an *alternator,* the main component of the car's charging system, a magnetic field is rotated past a number of conductors. As the field rotates, the stationary conductors continuously pass through the field. The amount of voltage induced depends on the speed of the rotating field, the strength of the field, and the number of conductors that pass through the field. As any or all of these three conditions are increased, the output of the alternator increases.

When current flow leaves the battery, chemical reactions take place within the battery. These reactions reduce the amount of power available from the battery, and the battery becomes discharged. The alternator sends current back to the battery

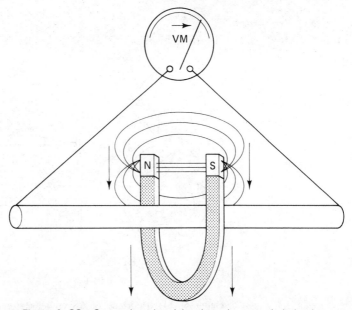

Figure 4-23 Generating electricity through magnetic induction.

and recharges it. This charging current is flowing in an opposite direction from the discharging current and reverses the chemical reactions that took place during discharge. Batteries are often referred to as "storage" batteries because they can be recharged and have their chemical reactions reversed to allow for continuous use.

SWITCHES

Nearly all electrical circuits have a switch to turn the circuit's load on and off. With a switch in the "on" position, the circuit is closed. The circuit becomes incomplete or open when the switch is in the "off" position. Some switches, like the ignition switch, control more than one circuit at a time. There are many types of switches used and there are many devices used to act as a switch in a circuit. A *relay* is a switch that uses the current of another circuit to open or close the circuit to which it is connected. This is an electromagnetic switch. The current from the other circuit is connected to an electromagnet in the relay and when there is current flowing in that circuit, the magnet closes the contacts of the relay, which allows the relay's circuit to be complete. All switches have *contacts,* which are opened or closed. Contacts are conductors that, when they come together, complete the circuit and allow for current flow. The contacts of switches may be manually opened and closed, like the ignition switch, or may be opened and closed through some other means, such as electromagnetism. A relay is used to control a circuit so that a particular device is switched on only when another circuit is in operation.

Another type of electromagnetic switch is the *solenoid.* This type of switch uses electrical energy to perform a mechanical task. The starter motor is not directly hooked up or engaged by the engine until it is activated. Then a solenoid pushes the starter drive gear into the engine's flywheel to allow the starter to spin the crankshaft. The starter solenoid (Fig. 4–24) also connects the battery directly to the starter to allow the starter motor to spin. Solenoids are also used for other purposes, such as to operate power door locks. Switches are used in a car in a variety of ways: Some are operated directly by the driver; some, such as the ignition points, are manually controlled by a mechanical function of the car; and some are controlled entirely by the operation of the engine or other systems on the car.

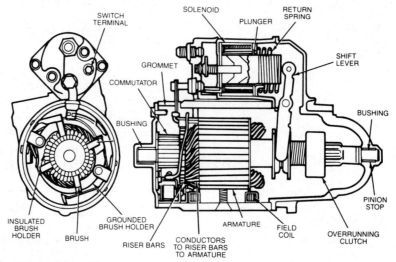

Figure 4–24 (Courtesy of Chevrolet Motor Division, General Motors Corporation)

SEMICONDUCTORS

Semiconductors are switches that are totally electrical, with no contacts that open or close. These switches function on principles of electronics and will operate only when the electrical conditions of the circuit they are in are suitable or are within a particular range. These electronic switches or devices are referred to as *solid-state* components. The two most commonly used solid-state devices are *diodes* and *transistors*. These allow current to flow in a circuit only when the voltage and/or current of that circuit is at a predetermined level. A diode allows current to flow in only one direction (Fig. 4–25). In circuits that are connected to another circuit, it is necessary that current flow only in the desired direction. Diodes open the circuit for current flowing in the wrong direction and close it for current of the correct direction. Diodes are also used in alternators to change alternating current (ac) into direct current (dc). Ac is produced by the alternator as the field passes by the conductors. The field enters into the conductor with its north pole and leaves with its south pole. This change of direction causes the current induced to flow in two directions. *Direct current* is the continuous flow of current in one direction. *Alternating current* allows for current flow in two directions within a conductor. The automobile uses direct current and it is the job of the diodes to allow only that part of the ac which is in the desired direction to leave the alternator and charge the battery. Many refer to the diode as a "one-way" valve.

Figure 4–26 shows a "switching" transistor that allows current to flow from it only when the correct amount of voltage and current are going to it. There is no need to go into how a transistor works. It is important to know, however, that transistors allow for current flow only when the conditions are correct. If there is not enough voltage present at one part of the transistor, it will not allow current to flow. The use of transistors is growing as the technology of electronic and computer applications advances. Currently, and in the near future, automobile technicians do not replace or diagnose individual transistors. Most transistors in a car are contained in a little box or *module,* and entire modules are replaced when one or more transistors are bad. The overall function of the modules should be of concern, and those that influence the way an engine runs are discussed in later chapters.

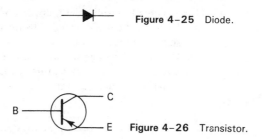

Figure 4–25 Diode.

Figure 4–26 Transistor.

SUMMARY

In this chapter, electricity, voltage, current, and resistance have been discussed. Particular electrical components have also been discussed. In the next chapter, these electrical principles will be applied to the engine's ignition system. The ignition system is a very important part of the engine because it provides the heat needed for combustion. Principles of electricity and magnetism are used to provide this heat.

TERMS TO KNOW

Electricity	Watts
Valence ring	Series circuit
Current	Parallel circuit
Conductor	Open
Insulator	Magnetism
Semiconductor	Electromagnetic induction
Amperage	Relay
Voltage	Solenoid
EMF	Diode
Resistance	Transistor
Electrical load	

REVIEW QUESTIONS

1. What three things make up an atom?
2. Of what materials are good conductors composed?
3. To keep current flowing, what must a power supply do?
4. Current is measured in what units of measurement?
5. How does resistance affect the amount of current in a circuit?
6. In what units is electrical power measured?
7. What are the five basic factors that determine the resistance of a conductor?
8. What would happen if there were no loads in an electrical circuit?
9. What are the three characteristics of a series circuit?
10. What are the three characteristics of a parallel circuit?
11. What causes a magnetic field to be formed?
12. What happens when a conductor is passed through a magnetic field?
13. State the basic operation of an electrical motor.
14. What are the three factors that determine the amount of current produced by an alternator?
15. What are the purposes of the diodes in an alternator?
16. Using Ohm's law, solve the following problems:
    (a). How much current would flow through a circuit that is connected to a 12-V power source and has a total resistance of 4 Ω?
    (b). How much voltage would be dropped by a resistor that has a value of 3 Ω, when 3 A of current is flowing through it?
    (c). What is the total resistance of the following circuit?

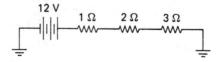

    (d). What is the total circuit current in the following circuit?

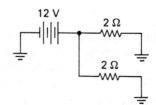

(e). What is the total resistance of the following circuit?

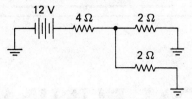

5

IGNITION FUNDAMENTALS

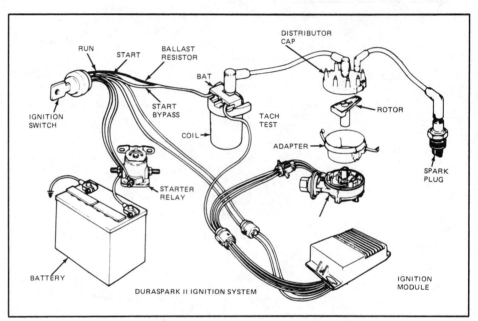

Figure 5-1   (Courtesy of Ford Motor Company)

The purpose of the ignition system is to provide a spark that will ignite the air-and-fuel mixture. The ignition system increases the battery voltage to thousands of volts, which jump across the gap of a spark plug. This spark ignites the mixture and the continued burning of the mixture causes the pressure of the mixture to rise quickly. This high pressure is the force that pushes the piston down during the power stroke. The amount of voltage present at the gap of the spark plug depends on the needs of the engine and of the ignition circuit. Unlike electrical systems in which the voltage is constant, the ignition system will supply the amount of voltage necessary to cause current flow.

In a simple light circuit, if a conductor is cut, current will cease to flow. There is not enough voltage present in the circuit to push the current across the cut in the wire to complete the circuit. The gap of a spark plug is like a cut in the wire; for current to flow, the circuit must be complete. The ignition system supplies enough voltage to push the current across the open spot, thereby completing the circuit and allowing current to flow. The gap of the spark plug offers a large amount of resistance to the circuit and the ignition system supplies the voltage necessary to overcome the resistance of that gap.

There are two basic types of ignition used on automobiles: the *electronic* and the *mechanical* or *breaker-point* types. To understand the ignition system, breaker-point ignition will be discussed first in detail and then the evolution of the electronic types will follow. The two types accomplish the same task, but the electronic types do the job better and quicker.

BASIC IGNITION CIRCUITS

All ignition systems can be divided into two circuits: primary and secondary. The *primary circuit* is that part of the ignition system which uses battery voltage to control the sequence of events necessary to provide the system with the high voltage that is required at the proper time. The *secondary circuit* provides for the high voltage to each cylinder of the engine. The secondary always has a much higher voltage than that of the primary or battery circuit. The primary provides the conditions necessary to allow the secondary to work.

There are many parts of the ignition system, each with a unique task to perform. All these parts must perform well for the ignition system to provide the amount of heat needed for combustion at the correct time. Before discussing how the ignition system functions, the parts of the system will be briefly described and defined.

The battery and the alternator are the voltage sources for the primary circuit. They supply voltage to the circuit whenever the ignition switch is closed. The ignition switch is simply the "on" and "off" switch for the primary. Primary wires connect the battery to the ignition circuit. Part of the primary wiring includes a *ballast resistor* or *resistance wire* which ensures that the correct amount of current flows in the primary. This ballast resistor is bypassed during the operation of the starter motor because battery voltage tends to decrease during its operation. The bypassing of the ballast resistor allows full available battery voltage to the ignition system when starting the engine.

The primary wires connect to the ignition coil, which has the task of stepping up or increasing the battery voltage to thousands of volts. The ignition coil is part of both the primary and secondary circuits. Both circuits of the ignition coil connect to the distributor of the engine, but at different points. Figure 5–2 displays a simple ignition circuit with its connections to the ignition coil.

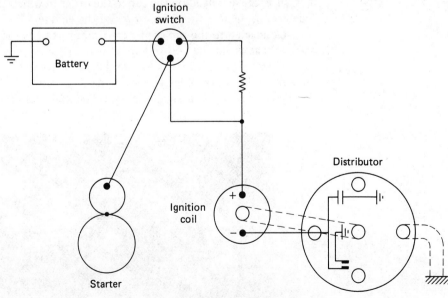

Figure 5–2

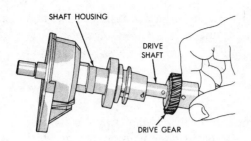

SHAFT HOUSING

DRIVE SHAFT

DRIVE GEAR

Figure 5-3 (Courtesy of Chrysler Corporation)

The *distributor* is driven by the camshaft of the engine and is the component used to ensure that the spark plug firings occur at the correct time. By moving the distributor in relation to the camshaft, the ignition timing is changed. As the camshaft rotates, the distributor shaft rotates through a gear connected on the shaft driven by the camshaft (Fig. 5-3). The turning of the distributor shaft controls the timing of the primary and secondary ignition circuits. The primary wires from the coil are connected to the breaker points and condenser, which are located in the distributor. The *breaker points* open and close through the rotation of the distributor shaft. A cam, which rotates on the shaft, causes the points to open and close. On the distributor cam there are typically as many lobes as there are cylinders in the engine. This allows the points to open and close once for each cylinder. The opening and closing of the points interrupts and allows current to flow in the primary circuit. Hooked in parallel to the points inside the distributor is the condenser (Fig. 5-4). The main purpose of the *condenser* is to allow for the immediate interruption of primary current flow when the points open. When the points suddenly open, there is a tendency for the voltage present to cause an arc or a spark across the air gap of the open points. The condenser reduces the amount of arcing, thereby allowing a more precise stopping of current flow in the primary.

The secondary of the coil is connected by a high-tension lead to the *distributor cap*. The high-tension lead or coil wire delivers high voltage to the distributor, where it is distributed to the individual cylinders. Also connected to the distributor cap are the high-tension wires that connect to the individual spark plugs. These spark plug wires are arranged on the cap in the order in which the spark plugs should fire. This arrangement is the *firing order* of the engine.

Connected to the top of the distributor shaft is the *rotor,* which connects the high voltage from the coil to the individual spark plug wires one at a time. The rotor rides below the distributor cap and as it rotates it connects the secondary of the coil to the spark plugs. Between the rotor tip and the spark plug terminals in the distributor cap is a small air gap. The high voltage from the coil must jump

Figure 5-4 Typical distributor with points and condenser in place.

across this gap to continue its path through the spark plug circuit. The air gap between the rotor tip and the distributor cap is shown in Fig. 5–5.

The *spark plugs* are plugs that allow high voltage from the coil to flow into the combustion chamber in the form of a spark to ignite the air-and-fuel mixture. This electricity flows into the chamber through a hole and it is very important that the spark plug seal this hole or the compressed air-and-fuel mixture could escape. Escaping of the mixture would result in decreased engine efficiency. To have proper combustion, there must be the proper mixture of air and fuel compressed in a sealed container and the proper amount of heat introduced to cause the immediate expansion of the mixture. Too often the spark of the spark plug is of concern and the plug part is ignored.

The spark at the spark plug occurs at an air gap between the center electrode tip and the ground electrode (Fig. 5–6). The *center electrode* is the conductor that passes through the center of the spark plug. The *ground electrode* is attached to the metal shell of the spark plug. When the spark plug is installed in the engine, the ground electrode becomes part of the common ground circuit of the engine. The two electrodes are typically adjusted to be 0.035 to 0.080 in. apart. The width of the gap will vary with different engines, and when the engine's spark plugs are replaced, the gap is adjusted to the manufacturer's specifications.

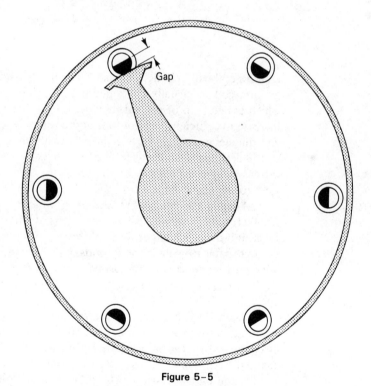

Figure 5–5

IGNITION COILS

The primary ignition circuit includes the battery, ignition switch, ballast resistor or resistance wire, primary ignition coil, breaker or contact points, and condenser. The secondary circuit includes the secondary coil, the coil wire, distributor cap, rotor, spark plug wires, and spark plugs. The *ignition coil* is part of both the primary and secondary circuit and is the item that causes the voltage to increase. The coil is the center or the "heart" of the ignition system. A look at the construction of the

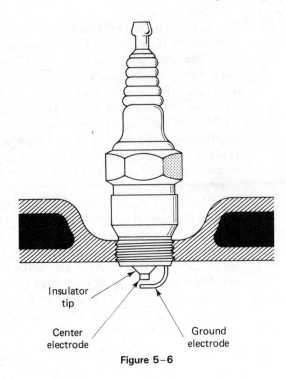

Insulator
tip

Center
electrode

Ground
electrode

Figure 5-6

ignition coil will help define what takes place inside it. The items that make up the ignition coil are usually contained in a can-like container. In the center is the *core,* which is made up of several thin strips of soft iron. The purpose of the core is to increase the efficiency and output of the coil by promoting faster and more complete coil magnetic saturation. Soft iron is easily magnetized and the operation of the coil is dependent on the effects of magnetism. The strength of a magnetic field is increased by the presence of the soft-iron core, which concentrates the magnetic field to the center of the coil.

Around the core is the *secondary winding,* which consists of approximately 20,000 turns of fine wire. The many layers of windings are insulated from each other by a high-dielectric paper. This paper opposes current flow and does not allow the current that is flowing in the secondary to leave the circuit. One end of the secondary winding is connected to the primary circuit at the coil, and the other is connected to the coil's high-tension tower. From this winding comes the high voltage for the spark plugs.

Around the outside of the secondary windings are the *primary windings.* These windings consist of approximately 250 turns of heavy wire. One end of the windings is connected to the same primary terminal as is the secondary. This terminal is marked "Bat" or "+" because it is connected to the battery through the ignition switch and ballast resistor. The other end is connected to the contact points and condenser in the distributor. The terminal of the coil that connects the coil to the distributor is usually marked "Dist" or "−" (Fig. 5-7).

The container that houses the windings and core is usually filled with oil to aid in insulating the windings. The oil also helps to dissipate the heat that is created when the voltage is stepped up. Heat tends to break down insulation and a breakdown of insulation can result in partial or total coil failure. An ignition coil that is leaking oil should be replaced with a new one because it can be the cause of a poor-running engine or eventually cause such a condition. If the ignition coil is not housed in a sealed container filled with oil, it is constructed in such a way as to allow air to pass through the windings and cool them.

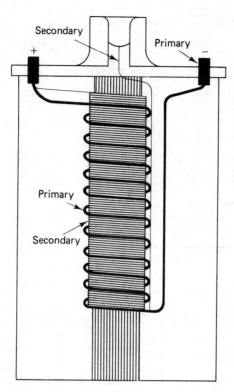

Figure 5-7

IGNITION COIL ACTION

When the ignition switch is closed or "on" and the points are closed, current flows through the primary circuit. This current creates a strong magnetic field around the primary windings. When the points open, the current flow is interrupted. This causes the magnetic field to move toward the core. As the field moves toward the core, it passes through the secondary windings, which induces a high voltage in the secondary windings. This high voltage forces current to flow through the secondary circuit to the spark plug, which ignites the mixture. The high voltage that arcs across the gap of the spark plug supplies the heat needed to start combustion. An explanation of how all this occurs in the ignition coil will further explain the purposes of the parts in the ignition system.

When current flows through a conductor, it will immediately reach its maximum value as allowed by the circuit resistance. If the wire or conductor is wound into a coil, maximum current will not immediately be present. As a magnetic field begins to form as current begins to flow, the magnetic lines of force of one part of the winding pass over another part of the winding, and this tends to cause an opposition to current flow. This occurrence is called *reactance* or *counter EMF*. Reactance causes a temporary resistance to current flow and delays the flow of current from reaching its maximum value for a period of time (Fig. 5-8). When maximum current flow is present in a winding, a maximum magnetic field is present and the coil or winding is considered to be saturated.

When the points open, the primary current flow is interrupted and the magnetic field around the windings begin to collapse. The magnetic field previously held around the windings by the flow of current now moves toward the soft-iron core. The magnetic lines of force cut through the windings as they move to the core and cause a higher voltage to be induced in primary windings. This higher voltage tends to keep the primary current flowing, which is undesirable at this time. If current continues to flow, there will be an arc across the open points. This arc can cause a

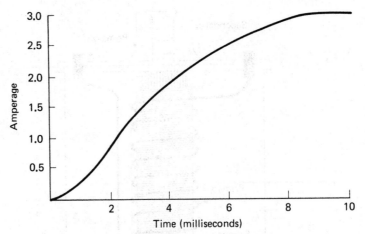

Figure 5-8   Current flow versus time in an ignition coil.

transfer of metal from one point contact to the other, which would quickly tend to destroy the contact points. Arcing would also prevent the flow of primary current from stopping quickly and allow the field to collapse gradually, thereby reducing the output of the secondary. To control the arc that takes place between the contacts of the points as they open and to quickly stop the flow of primary current to develop maximum coil voltage, a condenser is connected in parallel to the points.

Ignition *condensers* are used to prevent an arc at the points when they first open. This is accomplished by providing a place where current can flow until the points are fully opened (Fig. 5-9). This allows the primary current flow to be interrupted and the magnetic field to collapse quickly toward the secondary windings. When the field collapses through the primary windings, nearly 250 V are induced in the primary winding. This is more than enough voltage to cause current to flow across the gap of the points. The condenser absorbs the voltage to prevent this arcing.

Condensers are made of alternate sheets of metal foil and insulation. The sheets of foil are slightly smaller than the sheets of insulation. Every other sheet of foil is connected together at one terminal which connects to the points. The other sheets of foil are connected together and to the ground of the car. The ignition points also have two electrical connections, one connected to the coil and the condenser and the other to ground, through its mounting bracket. When the points are closed, this

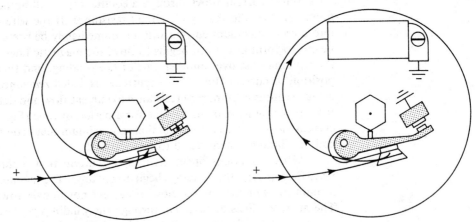

Figure 5-9   With the points closed, current flows to ground through the points; with the points open, current flows toward the ground potential of the condenser.

completes the circuit and allows current to flow. When the points are open, the path to ground is no longer available through the points, but the potential of ground from the negative sheets of the condenser attracts the voltage. Current does not flow through a condenser; the condenser merely absorbs the voltage. To discharge a condenser, a true path to ground must be present. This happens when the points close, to begin the firing of the next cylinder, preparing the condenser for the next time the points open.

Saturation of the primary windings can occur only if the points are allowed to be closed long enough to allow maximum current flow. For an ignition coil to have maximum output, it must have its primary windings saturated. The length of time that there is primary current flow or that the points are closed is called *dwell*. Dwell time is measured in terms of degrees of distributor rotation. One complete turn of the distributor shaft equals 360 degrees and represents one firing of all the cylinders of the engine. If the engine is a four-cylinder engine, every 90 degrees of rotation represents the amount of time available for the saturation of the coil and the firing of one cylinder. During this 90-degree period of time, the points should remain closed for enough time to saturate the coil and be open long enough to allow the spark plug to fire. The time needed for coil saturation is about 0.010 second. Dwell needs to be long enough to allow for this period. As engine speed increases, the distributor rotates faster and the time in seconds allowed for dwell decreases. The dwell time in degrees remains unchanged; however, the time in seconds that it takes the cam to rotate through these degrees decreases. Therefore, the time the points are closed is reduced as engine speed increases. Automobile manufacturers' recommended dwell specifications allow enough time for coil saturation at low speeds and also for the firing of the plug.

As engine speed increases, the time in seconds allowed for dwell decreases, causing the output of the coil to decrease. If the required firing voltage of the engine is great at idle speeds, then when the engine is at higher speeds, the coil may not be able to produce enough voltage to fire the cylinders. If this occurs, the engine will not run (Fig. 5-10). The amount of required firing voltage is determined by the resistance present in the secondary of the ignition. The higher the resistance, the higher the required firing voltage. Because the coil cannot be saturated at high engine speeds, the resistance in the secondary must be kept at a level that allows the decreased voltage output from the coil at high engine speeds to be able to complete the secondary and to allow current to flow across the gap of the plug.

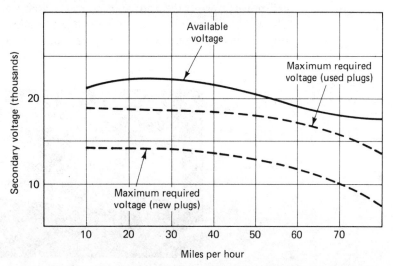

Figure 5-10   (Courtesy of AC Delco Remy, General Motors Corporation)

Usually, the recommended firing voltage for a cylinder is 40%, or two-fifths of the maximum coil output. This allows for enough voltage to cause current across the plug gap and enough coil voltage reserve to continue the firing of the cylinder at high speeds. Coil voltage reserve is the amount of voltage left in the coil after firing of a spark plug. There is usually enough time at idle and fast idle speeds to saturate the coil. During these engine speeds, the required firing voltages should be 40% of the total available output of the coil. If required firing voltages remain at this level through all engine speeds, this will allow the cylinders to fire effectively until the coil's output is reduced to less than 40% of its maximum output. If there is a problem in the primary ignition circuit, the amount of voltage available from the coil will be reduced. If the resistance is high in the secondary, the coil will not be able to supply the needs of the engine throughout all operating speeds. Ignition problems cause the engine to "miss" at certain speeds and could prevent the engine from starting.

BALLAST RESISTOR

To provide comparatively high current to the primary windings of the coil at high engine speeds, a ballast resistor (Fig. 5-11) or resistance wire is used to connect the battery to the positive terminal of the coil. At low speeds, current flows through the primary circuit for longer durations because the dwell time is longer. The current passes through the ballast resistor and heats it up. This causes the resistance of the ballast to increase and reduces the current and voltage going to the coil. This is not a problem at low speeds because there is more than enough time to saturate the coil, and the resistor prevents high current from burning away the contact-point surfaces. At high speeds, current flows for much shorter durations and the ballast resistor does not get as hot and therefore does not reduce the current flowing in the primary. In fact, the current flow is actually increased over the amount that flowed while the engine was at a low speed. This tends to increase the chances of coil saturation at high speeds.

The ballast resistor is usually bypassed during the starting of an engine, to allow as much voltage and current as possible to flow in the primary. Starter motor loads usually drop the available voltage of the battery while in use, and by bypassing the ballast, higher secondary voltages are available for easier starting. Figure 5-12 shows a typical bypass circuit.

When the points open, the magnetic field around the primary winding of the coil moves toward the core and induces a high voltage in the secondary winding. The points must remain open long enough to allow the voltage to establish current flow through the secondary. The points must also be opened wide enough to im-

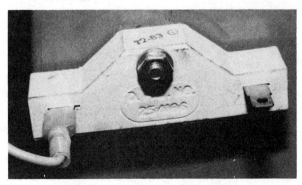

Figure 5-11 Typical ballast resistor.

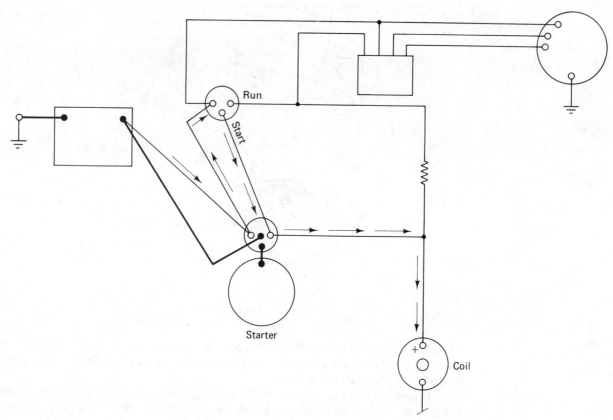

Figure 5-12 Notice that during starting, the ignition coil receives full battery voltage.

mediately stop current flow in the primary. If the gap or distance that the points are open is small, the voltage induced in the primary coil could jump the point gap when the points open. The condenser absorbs this high voltage and reduces the chances of arcing across the points. If the point gap is very small, the condenser could not affectively stop the arcing and current would continue to flow in the primary, which would prevent the secondary voltage from being produced. For the high secondary voltage to be induced, the points must interrupt the current flow.

A long dwell period is desirable, but the points would not open wide enough. The movement of the points is controlled by the cam on the distributor shaft. There is a lobe for each cylinder of the engine, and as the cam rotates, the lobe pushes the points apart. Ignition points are made up of a stationary bracket which holds a contact point in place and another contact point mounted on a movable arm. The movable arm has a projection on it called the *rubbing block* which rides against the distributor cam. When the high part of the lobe is against the rubbing block, the points are fully opened. As the cam rotates the points close, and this is the beginning of dwell. The wider the points are open, the longer it will take to close the points, thereby decreasing the dwell. The dwell of an engine is adjusted by setting the gap between the two point contacts when they are fully open (Fig. 5-13). A very narrow gap would give a long dwell period, but again, this would allow for some arcing across the points. Too wide a gap would decrease the dwell period. Manufacturers' recommendations for dwell allow sufficient time for dwell and a wide enough gap to control the arcing across the points.

When the points open, current flow is interrupted and this causes high voltage to leave the coil through the secondary to the spark plug. The occurrence of the opening of the points determines when the shock will be supplied to a cylinder.

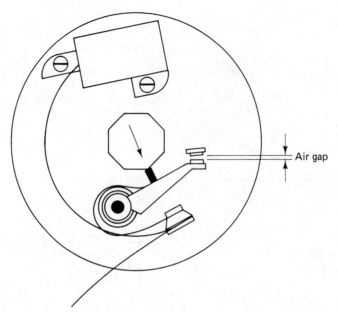

Figure 5-13

Therefore, the opening of the points controls ignition timing. If the dwell is too long, the points will open late and the cylinder will fire late. Similarly, if the dwell is short, the plug will fire early.

Figure 5-14 shows why the ignition timing should be set only after the dwell is adjusted. A short dwell means that the points close late and open early, causing the ignition timing to be early or advanced. A long dwell has the points closing early and opening late, causing a late or retarded firing of the spark plug. Dwell will affect timing, as the length of dwell dictates when the points will open. A change

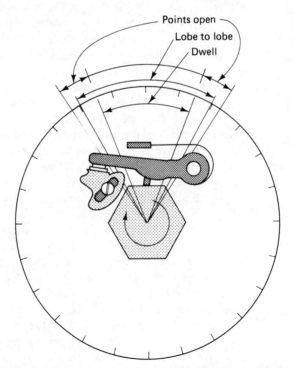

Figure 5-14 If the dwell is longer than the specifications call for, the points will close earlier and will open later than normal, thereby retarding the timing.

in dwell and in timing can occur because of wearing of the rubbing block. As the block wears from being in contact with the distributor lobe, the point gap decreases, causing an increase in dwell and retarded timing.

ELECTRONIC IGNITIONS

The biggest improvement in the ignition system has resulted from the use of advanced electronics. Electronic ignitions do exactly what breaker-point ignition systems did but do it better and tend to be more reliable. Periodic replacement of the points and condenser were necessary to maintain engine efficiency in the breaker-point system. With electronic ignitions there are no points or condensers to be replaced. In fact, electronic ignitions have no wearing parts. Things still go wrong with them, but not on a regular basis. Because there are no points to wear, pit, or burn, the initial ignition timing is not altered by driving. This allows the engine to operate efficiently for an extended period of time. The differences between the two types of ignition systems lie mainly in the primary circuit. Current flow still needs to be stopped and started in the primary circuit, but this is not accomplished by mechanical opening and closing of the points. All electronic ignition systems use two basic devices to control primary current flow. A switching device is used to control the current in the primary winding, doing what the points did. A triggering device signals the coil when it is the proper time to collapse the field and fire the spark plug, which was what was accomplished when the points opened.

Electronic ignitions do not use breaker points and are called *breakerless ignitions*. There are several types of breakerless triggering devices used. One breakerless system contains an optical device and a light beam (Fig. 5–15). When this light beam passes through a slot in a rotating disk attached to the distributor shaft, in place of the distributor cam, a trigger signal is sent to the switching device. This causes the current flow in the primary to be interrupted and the spark plug fires.

Another breakerless system uses a metal detector which senses the presence of a metal object moving by. The triggering device then sends an electrical signal to the switching device, which then controls the firing of the plug.

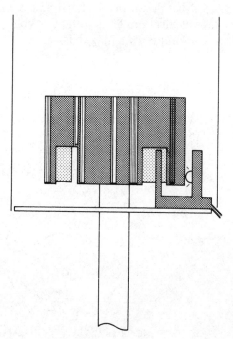

Figure 5–15 As the disk rotates past the light, the light beam shines on the optical device when the slot in the disk allows for it.

The most commonly used type of breakerless triggering device uses a magnetic *pickup* device (Fig. 5-16). The pickup senses changes in the magnetic field around it as it passes by another magnet. When this occurs, the triggering device sends a signal to the switching device, as did the other breakerless systems.

All the components of the magnetic triggering device are usually located in the distributor. The system consists of a magnet, a pickup coil, and a movable magnetic core. The magnetic core rotates with the distributor shaft in place of the cam. As the core rotates, it weakens and strengthens the magnetic field of the magnet. The presence of soft iron around a magnet increases its strength. The closer the soft iron is, the stronger the magnet. The core is made from soft iron and has projections from it that pass by the magnet in the pickup and change the strength of the magnet. The closer the iron core gets to the magnet, the stronger the magnetic field gets. The pickup senses this increase and signals to the switching device that it is time to fire the plug. These small projections on the core are called *teeth*. As the core rotates, the teeth alternately increase and decrease the distance between the core and the magnet.

The pickup coil is able to sense changes in the strength of the magnetic field because each time the field passes through the coil, a voltage is produced. The amount of voltage induced by a conductor passing through a magnetic field is determined by the strength of the field. Therefore, when the field strength is increased, an increased amount of voltage is induced in the pickup which is connected to the switching device. This higher voltage gives the switching unit the command to fire the plug. One spark is produced each time a tooth of the core passes by the magnet. There is one tooth per cylinder of the engine.

Another type of triggering device which is gaining popularity for use in automobiles, is the *Hall-effect pickup* (Fig. 5-17). This device uses the influence of magnetic fields on a semiconductor for sensing appropriate triggering times. In the distributor, a small magnet is mounted near a semiconductor. A rotor, with slots cut in it or shutters made of magnetic material, rotates between the semiconductor and the magnet. The shutters alternately pass and block the magnetic field of the magnet to the semiconductor. When semiconductor material has a current flowing through it and a magnetic field passes through the material, a voltage is induced in the semiconductor at right angles to the flow of current. This voltage is called Hall voltage after the scientist that discovered it. Each time the rotor passes between the semiconductor and the magnet, the Hall voltage is changed, and this change of voltage triggers a spark from the switching unit.

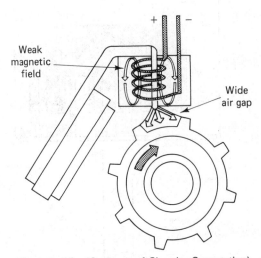

Figure 5-16   (Courtesy of Chrysler Corporation)

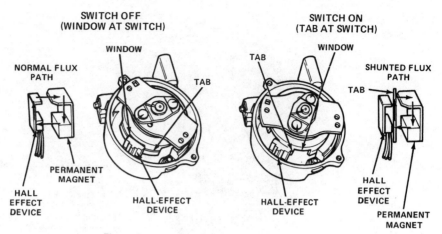

SWITCH OFF (WINDOW AT SWITCH)

SWITCH ON (TAB AT SWITCH)

NORMAL FLUX PATH

WINDOW

TAB

WINDOW

TAB

SHUNTED FLUX PATH

PERMANENT MAGNET

HALL EFFECT DEVICE

HALL-EFFECT DEVICE

HALL-EFFECT DEVICE

HALL EFFECT DEVICE

PERMANENT MAGNET

Figure 5-17 (Courtesy of Ford Motor Company)

The switching devices used are all quite similar, although they have different appearances. All switching devices use a transistor as the main switching component. The turning on and off of this transistor opens and closes the primary circuit. The switching transistor is contained within the electronic ignition module, which also contains the necessary electronic circuitry to allow the low-amperage triggering and switching devices to work on the engine. An amplifier that increases the signal sent by the triggering device is also in the module. The switching device or module not only appears different from manufacturer to manufacturer, but is also referred to by different names. General Motors uses an ECM or Electronic Control Module, Ford Motor Company uses an ignition or amplifier module, and Chrysler and American Motors use an ECU or Electronic Control Unit. In spite of the name differences, each is doing the same thing: receiving a signal from the triggering device, amplifying the signal, and then causing ignition to happen. Each automobile manufacturer also uses different terminology to identify the core and pickup units of its triggering devices. The core is called a trigger wheel by American Motors, a reluctor by Chrysler, an armature by Ford Motor Company, and a timer core by General Motors. General Motors calls the pickup a magnetic pickup, as does Ford Motor Company. Chrysler calls it a pickup coil and American Motors calls it a sensor. Not only do the names of the components differ, but also their basic appearance (Fig. 5-18).

Not only do the names and appearances of electronic ignition devices vary from manufacturer to manufacturer but also within car models and applications. The advances made by engineers in the electronic field have allowed manufacturers to customize ignition packages for particular applications. The basic operation of all electronic ignitions is the same regardless of what they look like or are called. All electronic ignition systems do the same thing that the breaker-point ignition system did, but they do it better and longer.

IGNITION TIMING

To have complete combustion, an engine must have the correct compressed mixture of air and fuel in a sealed container and be shocked by the correct amount of heat at the correct time. The timing of this delivery of heat is called *ignition timing*. The ignition is timed to allow the heat shock to occur at the right time to allow the combustion process in the combustion chamber to continue long enough to produce maximum power to push down the piston.

When the piston is at exactly TDC and BDC, it stays there for a short period

Figure 5-18   Ignition modules from Chrysler, Ford, and General Motors.

of time. This is caused by the connecting rod moving from an upward motion to a downward one. At TDC, the piston is not moving, while the connecting rod merely moves from one side to another. Maximum pressure from combusion should occur when the connecting rod and the piston are moving freely downwardly (Fig. 5-19). Ignition timing not only controls when the spark first occurs but also when it will end. Normally, the ignition system allows for the spark to continue across the gap of the spark plug for 0.003 second. The exact length of time is determined by the

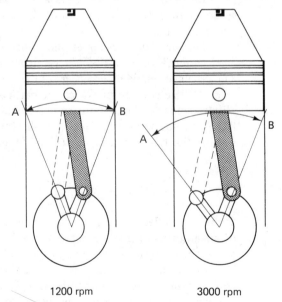

1200 rpm   3000 rpm

Figure 5-19   Advanced timing at increased speeds allows for combustion to be complete at the same desired time as it should during low speeds.

amount of energy stored in the ignition coil and the amount of resistance in the secondary circuit and spark plug gap.

Ignition must occur early enough to allow combustion to be complete when the piston is ready to move downward. The initial ignition timing for an engine is usually set to allow the spark to begin when the piston is slightly before TDC. This allows enough time for complete burning of the fuel. But as engine speed increases, the piston is moving through its strokes faster and the needed burning time of the fuel remains the same. To compensate for this increased piston speed, ignition must occur earlier. The interruption of the primary current controls the ignition timing.

Initial timing is the ignition timing specification for an engine when it is at its base or idle speed. By rotating the distributor body, the position of the triggering device on the distributor shaft is changed (Fig. 5-20). Because the body of the distributor moves independently of its shaft, the time at which the primary current flow begins and ends is changed without changing the length of time between each event. The initial timing specification has been determined by engineers to allow for fast engine starting and efficient operation at idle speeds.

Ignition timing specifications are given as a particular number of degrees *before top dead center* (BTDC) or *after top dead center* (ATDC). This refers to when spark should begin in relating to the position of the crankshaft. BTDC is anytime before the piston is at exactly TDC, and ATDC is when the piston is past TDC. ATDC is often referred to as *retarded timing* and BTDC as *advanced timing*. Actually, these terms are correct only when referring to TDC. If the timing specification called for 10 degrees BTDC and the timing was found to be a 6 degrees BTDC, the timing found is retarded from the desired specification. If 16 degrees were found, this would be advanced timing. Most automobiles have an initial timing specification at BTDC, and in this text any timing change that causes ignition to occur further from TDC than called for will be labeled as advanced and any occurring closer to TDC, as retarded.

Because the speed of the piston changes according to engine load and speed, the base timing of an engine does not allow for enough ignition advance to meet the needs of an engine at all speeds and loads. *Ignition advance mechanisms* are built into the distributors to compensate for the changes of load and speed. To change ignition timing in response to engine speed, a *mechanical* or *centrifugal advance system* is used. A *vacuum advance unit* is used to alter ignition timing according to engine load. These two systems work independently of each other and cause ignition timing to change as engine operating conditions change.

The mechanical advance system uses centrifugal force, which is developed by rotating small weights inside the distributor to move the distributor cam or reluctor (Fig. 5-21), which changes the ignition timing by triggering the interruption of primary current flow earlier. The distributor shaft consists of two sections: one connected to the engine's camshaft and the other connected to the distributor rotor.

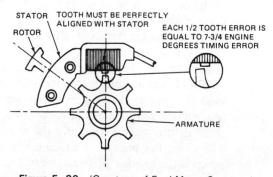

Figure 5-20   (Courtesy of Ford Motor Company)

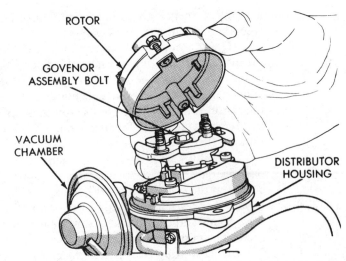

ROTOR

GOVENOR
ASSEMBLY BOLT

VACUUM
CHAMBER

DISTRIBUTOR
HOUSING

Figure 5-21 In this application, the mechanical advance assembly is referred to as the governor assembly. (Courtesy of Chrysler Corporation)

These two sections are connected to each other by small coil springs. At idle speeds, the lower section, driven by the camshaft, turns the upper section from the connection made by the springs at the same speed. The distributor cam or reluctor is located on the upper section and turns with the rotation of the shaft. In the assembly that houses these connecting springs are small weights mounted on a pivot. As engine speed increases, the weights are thrown outwardly by centrifugal force. The force caused by these weights overrides the strength of the springs and allows the rotation of the upper section of the shaft to go ahead of the rotation of the lower shaft. This causes the distributor reluctor to turn slightly ahead of distributor rotation, allowing the primary current flow to stop earlier. As speed increases, the weights will move out more, causing the reluctor to move more and allowing for more ignition advance. To limit the amount of total mechanical advance there is a limiter built in which stops any further movement of the weights when it is reached. As speeds are decreased, the springs draw in the weights, causing the timing to retard until the engine is at an idle speed and base ignition timing is present. It should be noted that the position of the rotor changes with the change in distributor cam location.

Vacuum advance units use engine vacuum to move the breaker plate assembly, which causes the points to open earlier. By tapping into the intake manifold or carburetor, the vacuum present can be used to operate accessories without having any ill effects on the engine. The vacuum advance unit uses this vacuum to move the distributor's breaker plate assembly. The breaker plate assembly has the triggering unit mounted to it. If the plate is rotated, the position of the unit in respect to the distributor shaft is changed. The breaker plate is mounted to the distributor body in such a way as to allow only a slight movement (Fig. 5-22). This movement can affect the ignition timing by advancing it as much as 40 degrees.

The amount of vacuum advance is determined by the amount of vacuum present at the unit. The amount of vacuum advance required and the amount of engine manifold vacuum present at the vacuum advance unit are determined by engine load. The higher the load on the engine, the lower the manifold vacuum is and the less ignition timing advance is required. Similarly, manifold vacuum is high when there is little engine load and the engine can use more ignition timing advance. Therefore, the amount of vacuum advance is governed by engine load. The movement of the breaker plate assembly does not change the initial dwell setting or the mechanical advances. The vacuum advance unit is composed of a can which con-

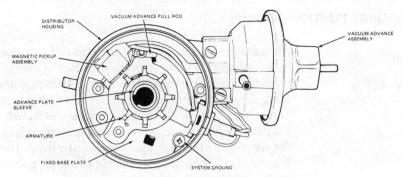

Figure 5-22 (Courtesy of Ford Motor Company)

tains a rubber diaphragm that separates the inside of the can into two sections. Attached to the diaphragm is a rod which is connected to the breaker plate. One side of the diaphragm is exposed only to engine vacuum and the other only to atmospheric pressure. When vacuum is present at one side, outside air pushes the flexible diaphragm into the vacuum. This movement pulls the rod, which moves the breaker plate and advances the timing (Fig. 5-23). A higher pressure will always move to a lower pressure or vacuum. The presence of a vacuum on one side of the diaphragm and air pressure on the other causes it to stretch into the vacuum as the higher pressure attempts to equalize with the lower pressure. The amount of movement of the rod is determined by the vacuum present and by a calibrated spring that will resist movement of the diaphragm until a certain amount of vacuum is present. This spring also allows the timing to return to base as the vacuum decreases. Specifications are given as to how much advance should be caused by a certain amount of vacuum. If the vacuum advance unit is not within manufacturer's specifications, the correct amount of ignition advance is not being supplied for the engine. This can result in poor driveability and/or fuel economy.

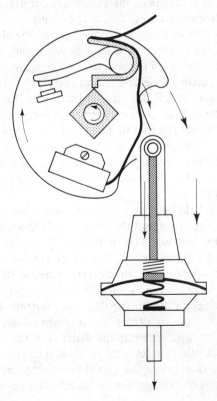

Figure 5-23 The movement of the vacuum advance diaphragm causes the high point of the cam lobe to contact the rubbing block of the points earlier.

SECONDARY IGNITION CIRCUITS

All ignition systems, with the exception of one that is discussed in later chapters, have basically the same secondary ignition system. Electronic ignitions have not yet provided a better means of getting secondary voltage to the spark plugs than by using the rotor, distributor cap, and spark plug wires. When discussing the secondary ignition system, all ignition types are included. Because an engine is equipped with electronic ignition or even a computerized ignition system does not mean that the secondary voltage is distributed differently. The primary gets the coil ready to fire the plug. The coil sends the high voltage to the distributor cap, where the rotor sends it to the individual spark plugs one at a time. From the coil, the secondary voltage enters into the center terminal of the distributor cap through a conductor. This conductor is made of the same material as that used for the spark plug wires; these will be discussed shortly. Exceptions to this are General Motors cars with a V-6 or V-8 engine equipped with electronic ignition. These engines have the coil mounted directly above the rotor and have no need for a wire to connect it to the rotor (Fig. 5–24).

The rotor, which rotates with the distributor shaft, has a blade on the top of it. This blade is in constant contact with the coil wire through a carbon brush fitted in the distributor cap. The other end of the rotor blade points to the inside connections or terminals of the distributor cap for each spark plug circuit. Contact is not made by the blade and the terminals; rather, there is a small air gap that separates the two. When secondary voltage is present on the rotor blade, voltage will jump across the gap to the terminal. To jump this gap, the coil provides for an increase of voltage. Remember that the resistance in the secondary determines the amount of voltage that is required to cause current flow across the plug gap. The resistance from the rotor to the distributor cap and the resistance present in the coil wire are the first resistances that need to be overcome by the coil in its effort to establish a current flow in the circuit (Fig. 5–25).

The individual cylinders of the engine are connected to the distributor cap by spark plug or high-tension wires. These wires must be able to keep the high voltage from escaping, as well as direct the movement of the electrons to the spark plugs. These wires are large conductors with a heavy shield of insulation around them. The conductive material used in these wires is typically carbon (Fig. 5–26). Carbon is not a good conductor of electricity and therefore offers increased resistance for the coil to overcome. This is not all that bad; once the circuit is completed by the voltage, current flows. With the increased resistance comes a lower current. Current is not needed to bridge the gap of the spark plug; voltage is. Current is used to maintain the spark. The use of carbon or resistance materials in spark plug wires became widely used because they limit the amount of magnetism that forms around the spark plug wire while current is flowing. The lower the current flow, the less magnetic strength there is.

This magnetic field became a concern when automobiles used copper as a conductor for spark plug wires and the magnetic field around them was found to interfere with radio waves. The magnetic field caused static on the radio communications of airlines, police, and other emergency stations. Also, these resistive suppression wires allow for clearer reception of FM radio waves in automobiles. Resistor wires are not as durable as the copper types were, but the benefits of longer spark plug life due to the decreased current flow and the lack of radio interference seem to outweigh the slight problem of durability.

The spark plug wire connects the distributor cap to the spark plug. A spark plug is a plug that allows electricity to flow into the combustion chamber. The amount of voltage that jumps the gap is determined by the total resistance in the spark plug circuit. The voltage is what shocks the air-and-fuel mixture; however,

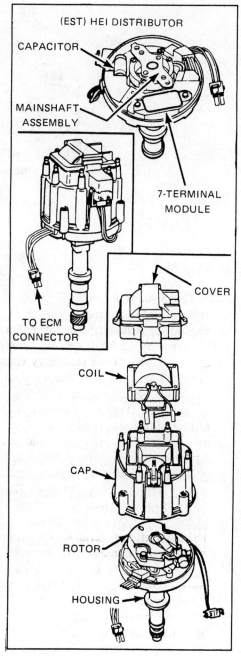

Figure 5-24   (Courtesy of Chevrolet Motor Division)

there are many features of a spark plug that determine how well it will perform in continuous delivery of the spark. To examine these features, let's look at the construction of the typical spark plug.

SPARK PLUGS

Figure 5-27 shows the construction of a typical spark plug. Notice that the terminal is the electrical connector that connects the spark plug wire to the spark plug. Connected to the terminal is the *center electrode,* which is the conductor that the voltage

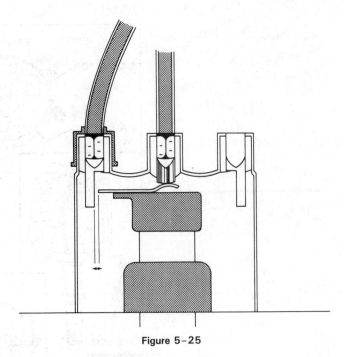

Figure 5–25

and current flows through to get inside the combustion chamber. The other end of the center electrode sticks out of the spark plug into the combustion chamber. Often, a glass binding material is used to seal the terminal to the center electrode to prevent the compressed gases in the chamber from escaping.

Around the center electrode is an *insulator.* This insulator prevents the high voltage in the spark plug from escaping. The insulator usually has several ribs formed on it to prevent water and dirt from entering between the terminal and the end of the spark plug wire. Spark plug wires are fitted with a boot at their ends to form a good seal; the ribs aid in this sealing. The ribs also increase the distance from the terminal to the shell to avoid *flashover* (a leakage of voltage to the engine block). The insulator is made of a very hard ceramic material that is highly resistant to heat and will easily dissipate the heat it does have.

The steel body or shell of the spark plug is threaded to hold the plug in the engine. The shell is formed to be tightly against the insulator; this allows for the transfer of heat from the insulator to the cylinder head through the threads

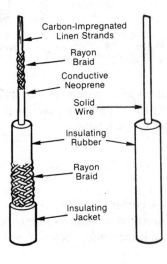

Carbon-Impregnated
Linen Strands

Rayon
Braid

Conductive
Neoprene

Solid
Wire

Insulating
Rubber

Rayon
Braid

Insulating
Jacket

Figure 5–26 Solid-wire secondary cable compared to resistor-type spark plug cable. (Courtesy of AC Delco Remy, General Motors Corporation)

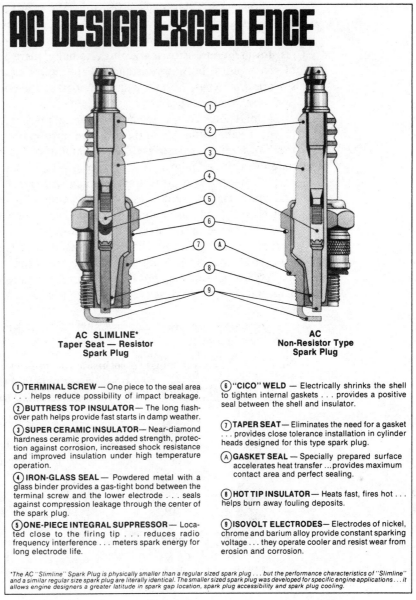

AC DESIGN EXCELLENCE

AC SLIMLINE*
Taper Seat — Resistor
Spark Plug

AC
Non-Resistor Type
Spark Plug

(1) **TERMINAL SCREW** — One piece to the seal area . . . helps reduce possibility of impact breakage.

(2) **BUTTRESS TOP INSULATOR** — The long flash-over path helps provide fast starts in damp weather.

(3) **SUPER CERAMIC INSULATOR** — Near-diamond hardness ceramic provides added strength, protection against corrosion, increased shock resistance and improved insulation under high temperature operation.

(4) **IRON-GLASS SEAL** — Powdered metal with a glass binder provides a gas-tight bond between the terminal screw and the lower electrode . . . seals against compression leakage through the center of the spark plug.

(5) **ONE-PIECE INTEGRAL SUPPRESSOR** — Located close to the firing tip . . . reduces radio frequency interference . . . meters spark energy for long electrode life.

(6) **"CICO" WELD** — Electrically shrinks the shell to tighten internal gaskets . . . provides a positive seal between the shell and insulator.

(7) **TAPER SEAT** — Eliminates the need for a gasket . . . provides close tolerance installation in cylinder heads designed for this type spark plug.

(A) **GASKET SEAL** — Specially prepared surface accelerates heat transfer . . . provides maximum contact area and perfect sealing.

(8) **HOT TIP INSULATOR** — Heats fast, fires hot . . . helps burn away fouling deposits.

(9) **ISOVOLT ELECTRODES** — Electrodes of nickel, chrome and barium alloy provide constant sparking voltage . . . they operate cooler and resist wear from erosion and corrosion.

*The AC "Slimline" Spark Plug is physically smaller than a regular sized spark plug . . . but the performance characteristics of "Slimline" and a similar regular size spark plug are literally identical. The smaller sized spark plug was developed for specific engine applications . . . it allows engine designers a greater latitude in spark gap location, spark plug accessibility and spark plug cooling.

Figure 5-27 (Courtesy of AC Delco Remy, General Motors Corporation)

of the plug. The insulator tip surrounds the center electrodes and is exposed to the heat of combustion. The retention of all this heat is not desirable and the contact between the insulator and the shell helps move this heat to the engine's cooling system through the cylinder head. The shell also holds the ground electrode and electrically connects it to ground. The ground electrode is also located in the combustion chamber, and it is the space between this electrode and the center electrode that provides the spark plug gap.

Seals and gaskets are important in the construction of a spark plug because these keep the spark plug a plug. The spark plug seals the hole in the combustion chamber through the threads of the shell and the gasket or tapered seat at the top of the threads. The insulator must also be sealed to the shell. Spark plugs use O-ring seals to prevent the high-pressure gases of the combustion chamber from leaking out. Similarly, the center electrode is sealed to the insulator, totally plugging the hole in the chamber.

Some spark plugs have a resistor built into the center electrode. The purpose of this resistor is the same as that of the resistor spark plug wires. This type of spark plug is called a *resistor plug*. There are two types of spark plugs used today: the resistor and nonresistor spark plug. Actually, there are thousands of different types of spark plugs, but the resistance of the center electrode allows for the first classification of types. Certain characteristics of the construction of a spark plug determine the other types of plugs.

Spark plugs are offered in different reaches. The *reach* of a spark plug is determined by the length of the threaded portion of the shell (Fig. 5–28). The desired reach of a plug is determined by the amount of metal in the cylinder head that the plug must thread into to reach the combustion chamber. If the reach is too long, part of the threaded portion of the plug will extend into the combustion chamber and can cause preignition, as the body of the plug cannot dissipate the amount of heat present, and this residual (leftover) heat causes the air-and-fuel mixture to ignite prior to the firing of the plug. Too long a reach will also make it very difficult to remove the plug from the cylinder head, as deposits from combustion fill the exposed threads of the plug. Too short a reach can cause poor combustion because the center electrode may not be in the proper location or depth within the chamber. This can delay the propagation of heat needed to ignite the air-and-fuel mixture entirely.

Insulator tips are made in many different shapes and lengths. They extend from the shell to the end of the center electrode and are exposed to the heat and pressures inside the chamber. The size of the insulator tip or nose determines the spark plug's heat range. The *heat range* of a spark plug is no more than a statement of how well a spark plug dissipates its heat from the insulator to the cylinder head.

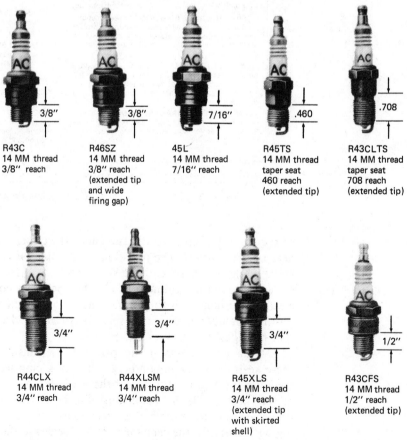

Figure 5–28 (Courtesy of AC Delco Remy, General Motors Corporation)

The heat-range system is a thermal classification of the spark plug based on the operating-temperature ranges of an engine. It is desirable for a spark plug to get rid of most of the heat it collects during combustion. However, some heat should be retained to burn off any oil and carbon that may collect on the center electrode between firings. An excessive amount of oil or carbon could insulate the center electrode from the ground electrode, preventing the spark. Heat needs to be removed so that there is ample shock provided by the sudden temperature increase of the high voltage. The amount of heat that should be retained and the amount that needs to be dissipated is determined by the design of the engine. Spark plugs are available in a number of heat ranges from cold to hot. A *cold plug* is one that easily dissipates heat and retains little. A *hot plug* is one that dissipates heat slowly and tends to retain more. A hotter plug does not provide for a hotter spark; in fact, the opposite is somewhat true. By retaining a lot of heat, the heat from the arc of voltage has less of a shock to the mixture. The heat range is determined by the shape and size of the plug's insulator nose (Fig. 5–29). Generally, the longer or larger the insulator nose, the hotter the plug.

A quick look through a spark plug manufacturer's catalog will give a good insight as to how many versions of spark plugs there are. Spark plugs also come in different sizes. The two sizes that are commonly used in today's engines reflect the diameter of the shell and the thread size. On the shell is the nut-like hex for the use of a wrench to tighten or loosen the plug from the cylinder head. Typically, automobile engines require plugs that have either a ⅝ or ¹³⁄₁₆-in. wrench size. Also, the diameter of the threads are either 14 or 18 mm. All the variations of size and design of a spark plug are given in its plug number as assigned by its manufacturer. These numbers contain information about the type, size, reach, thread size, heat range, and special features of the plug. To identify all the characteristics of a plug, manufacturers include tables in their catalogs for deciphering what each part of the plug number means (Fig. 5–30). Always use the plug recommended by the manufacturer until you know more about spark plugs than the manufacturers do.

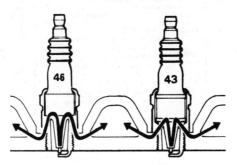

Figure 5–29 The longer the insulator nose, the longer the heat path, which raises the heat range. The shorter the insulator nose, the less distance the heat must travel to get to the cooling jackets in the cylinder head, and the faster the heat is transferred, lowering the heat range. (Courtesy of AC Delco Remy, General Motors Corporation)

SUMMARY

This chapter has covered a lot of basics dealing with the ignition system. The primary current has the responsibility for building and collapsing the magnetic field in the ignition coil. The secondary circuit produces and distributes high voltages to the cylinders to cause combustion. When operating normally, the ignition system supplies the heat necessary for combustion. The time the heat is delivered and the amount of heat delivered varies according to the engine's requirements. The output of the ignition system meets most of the needs of the engine, but like the air/fuel system, it cannot support total combustion for all engine operating conditions. In the next chapter, we will discuss those systems that help hide or correct the inefficiencies that have been mentioned so far.

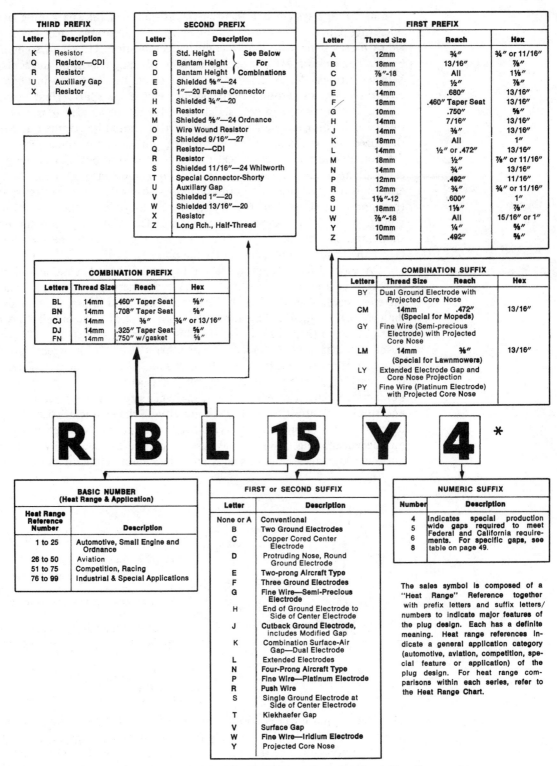

THIRD PREFIX	
Letter	**Description**
K	Resistor
Q	Resistor—CDI
R	Resistor
U	Auxiliary Gap
X	Resistor

SECOND PREFIX		
Letter	**Description**	
B	Std. Height	See Below For Combinations
C	Bantam Height	
D	Bantam Height	
E	Shielded ⅝″—24	
G	1″—20 Female Connector	
H	Shielded ¾″—20	
K	Resistor	
M	Shielded ⅝″—24 Ordnance	
O	Wire Wound Resistor	
P	Shielded 9/16″—27	
Q	Resistor—CDI	
R	Resistor	
S	Shielded 11/16″—24 Whitworth	
T	Special Connector-Shorty	
U	Auxiliary Gap	
V	Shielded 1″—20	
W	Shielded 13/16″—20	
X	Resistor	
Z	Long Rch., Half-Thread	

FIRST PREFIX			
Letter	**Thread Size**	**Reach**	**Hex**
A	12mm	¾″	¾″ or 11/16″
B	18mm	13/16″	⅞″
C	⅞″-18	All	1⅛″
D	18mm	½″	⅞″
E	14mm	.680″	13/16″
F	18mm	.460″ Taper Seat	13/16″
G	10mm	.750″	⅝″
H	14mm	7/16″	13/16″
J	14mm	⅜″	13/16″
K	18mm	All	1″
L	14mm	½″ or .472″	13/16″
M	18mm	½″	⅞″ or 11/16″
N	14mm	¾″	13/16″
P	12mm	.492″	11/16″
R	12mm	¾″	¾″ or 11/16″
S	1⅛″-12	.600″	1″
U	18mm	1⅛″	⅞″
W	⅞″-18	All	15/16″ or 1″
Y	10mm	¼″	⅝″
Z	10mm	.492″	⅝″

COMBINATION PREFIX			
Letters	**Thread Size**	**Reach**	**Hex**
BL	14mm	.460″ Taper Seat	⅝″
BN	14mm	.708″ Taper Seat	⅝″
CJ	14mm	⅜″	¾″ or 13/16″
DJ	14mm	.325″ Taper Seat	⅝″
FN	14mm	.750″ w/gasket	⅝″

COMBINATION SUFFIX			
Letters	**Thread Size**	**Reach**	**Hex**
BY	Dual Ground Electrode with Projected Core Nose		
CM	14mm (Special for Mopeds)	.472″	13/16″
GY	Fine Wire (Semi-precious Electrode) with Projected Core Nose		
LM	14mm (Special for Lawnmowers)	⅜″	13/16″
LY	Extended Electrode Gap and Core Nose Projection		
PY	Fine Wire (Platinum Electrode) with Projected Core Nose		

R B L 15 Y 4 *

BASIC NUMBER (Heat Range & Application)	
Heat Range Reference Number	**Description**
1 to 25	Automotive, Small Engine and Ordnance
26 to 50	Aviation
51 to 75	Competition, Racing
76 to 99	Industrial & Special Applications

FIRST or SECOND SUFFIX	
Letter	**Description**
None or A	Conventional
B	Two Ground Electrodes
C	Copper Cored Center Electrode
D	Protruding Nose, Round Ground Electrode
E	Two-prong Aircraft Type
F	Three Ground Electrodes
G	Fine Wire—Semi-Precious Electrode
H	End of Ground Electrode to Side of Center Electrode
J	Cutback Ground Electrode, includes Modified Gap
K	Combination Surface-Air Gap—Dual Electrode
L	Extended Electrodes
N	Four-Prong Aircraft Type
P	Fine Wire—Platinum Electrode
R	Push Wire
S	Single Ground Electrode at Side of Center Electrode
T	Kiekhaefer Gap
V	Surface Gap
W	Fine Wire—Iridium Electrode
Y	Projected Core Nose

NUMERIC SUFFIX	
Number	**Description**
4 5 6 8	Indicates special production wide gaps required to meet Federal and California requirements. For specific gaps, see table on page 49.

The sales symbol is composed of a "Heat Range" Reference together with prefix letters and suffix letters/numbers to indicate major features of the plug design. Each has a definite meaning. Heat range references indicate a general application category (automotive, aviation, competition, special feature or application) of the plug design. For heat range comparisons within each series, refer to the Heat Range Chart.

Figure 5–30 (Courtesy of Champion Spark Plug Company)

TERMS TO KNOW

Primary circuit   BTDC
Secondary circuit   ATDC
Ballast resistor   Retarded timing
Breaker points   Advanced timing
Condenser   Centrifugal advance
Firing order   Vacuum advance
Reactance   Flashover
Counter-EMF   Spark plug reach
Dwell   Insulator tip
Pick-up coil   Spark plug heat range
Hall-effect voltage

REVIEW QUESTIONS

1. What is the purpose of the ignition circuit?
2. What is the difference between the primary and secondary ignition circuits?
3. What is the purpose of the ballast resistor?
4. What part of the ignition system is part of both the primary and secondary ignition circuits?
5. What is the purpose of the ignition breaker points?
6. What is the purpose of the ignition condenser?
7. How are the spark plug wires arranged on the distributor cap?
8. What is the purpose of the rotor?
9. How does the ignition coil increase voltage?
10. Why must there be adequate dwell time?
11. Why must secondary resistances be kept at a particular level?
12. How does point gap affect dwell?
13. How does dwell affect ignition timing?
14. What controls ignition timing?
15. What is taking place in the secondary circuit during dwell?
16. What governs the amount of ignition timing advance allowed by the mechanical advance system?
17. During what conditions does the vacuum advance unit allow for full advance?
18. What are the advantages of electronic ignitions compared to conventional ignition systems?
19. Give examples of the typical triggering devices used in the ignition systems.
20. What takes the place of the breaker points in an electronic ignition system?
21. What determines the amount of secondary voltage required to fire the spark plug?
22. What is a "hot" spark plug?
23. What determines the heat range of a spark plug?

6

EMISSION CONTROLS

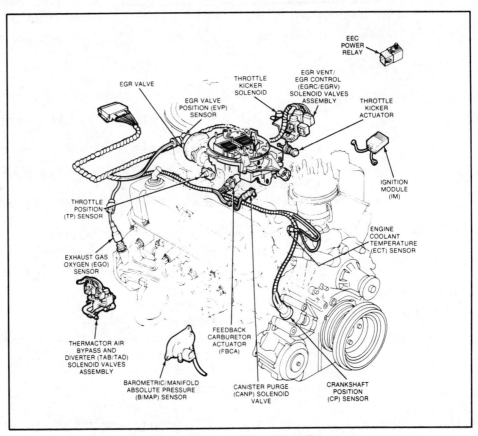

Figure 6-1 (Courtesy of Ford Motor Company)

Automobile manufacturers have constantly redesigned the engine and its related systems in an effort to allow it to support total combustion. Seeking this goal, engineers have developed new ignition and fuel systems. Major parts of the engine, such as the pistons, have even been redesigned in an attempt to achieve total combustion. This goal has not yet been achieved. The growing concern for efficiency was put into high gear when the government began to mandate that the exhaust from automobiles be of a certain quality and that automobiles achieve a certain gas mileage figure. To have exceptionally clean exhaust and to achieve very high gas mileage, automobile engines must improve their efficiency. All the fuel that enters the combustion chamber must be burned completely. There must also be the correct amount of air available to the engine to combine each fuel particle with enough air to form totally harmless gases and water after combustion.

As highly inefficient engines powered the cars of yesterday, attention was brought to the fact that the earth's atmosphere was being contaminated by pollutants. These pollutants were the result of exhaust from factories, home and office heating systems, trains, cars, and trucks. With the large number of cars on the road polluting the air, the exhaust from engines became a focal point of concern. The government reacted to this concern and passed a law requiring all automobiles produced after a particular date to release fewer polluting gases. This law set a standard for the maximum amount of poisonous gases that could exit the tailpipe of a car. With this law, the major pollutants of the automobile engine were defined. These pollutants are produced by the engine because of the basic inefficiencies of an internal combustion engine and because the engine does not operate under the most ideal conditions. The fuel or gasoline that is burned by the engine is not pure or perfectly cleaned, although refining processes have improved over the years. The air that the engine uses in the combustion process has little oxygen in it. Oxygen is that part of air that is used in combustion. The air drawn into the engine has some oxygen in it but is composed primarily of nitrogen (Fig. 6–2). These are some of the facts that prevent automobile engines from supporting total combustion.

The Environmental Protection Agency (EPA) has identified three major byproducts or pollutants of the internal combustion engine: *hydrocarbons* (HC), *oxides of nitrogen* (NO_x), and *carbon monoxide* (CO). HC emissions are not produced by the engine; rather, this pollutant is the result of the engine not being able to burn all the fuel it receives. HC particles are fuel particles. This pollutant is no more than unburned gases leaving the engine in the exhaust. CO is produced in the combustion process because of the lack of oxygen in the air supply. The high temperature of combustion combines the nitrogen in the air with the oxygen to form NO_x.

Photochemical smog results from the presence of HC and NO_x in the air and having bright sun shining on it (Fig. 6–3). The term *smog* was formed by combining two words: "smoke" and "fog." Smog was the first sign that there was an air pollution problem. Smog creates a brown haze in the sky. The other major pollutant of an engine is CO, which is an invisible and odorless harmful gas. Legislation has been passed that limits the maximum levels of these pollutants that can be present in the exhaust of newer cars and light trucks.

HC is emitted by an internal combustion engine that does not burn all of the fuel that enters it. NO_x is produced by the heat of combustion and consists of a variety of nitrogen and oxygen compounds. CO is a odorless, colorless gas that results from a lack of oxygen during combustion. This extremely poisonous gas accounts for nearly half of the air pollution problem. When the govenment set up

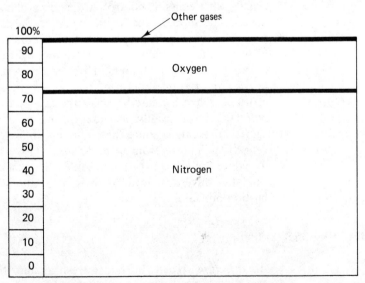

Figure 6–2　Composition of air in our atmosphere.

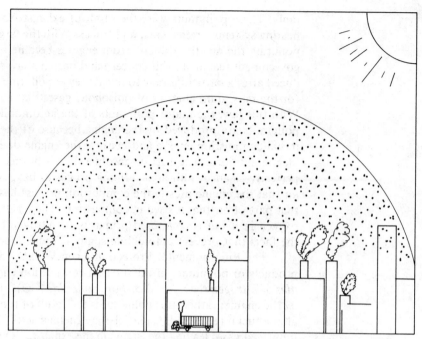

Figure 6-3 The heat of a city traps in the pollutants in the air and causes a blanket of smog to cover the city.

the maximum allowable limits for the existence of these pollutants in the exhaust of cars, it did not mandate how to meet these limits. Automobile manufacturers have developed many ways to allow engines to run more efficiently and many systems that reduce the amount of pollutants in the exhaust. These systems, emission control systems, are the topic of this chapter.

A study of this topic becomes rather complex because many factors were used to determine what emission control systems would be used. To meet the government standards on exhaust emissions, manufacturers have used, and still use, many different devices. These devices vary from manufacturer to manufacturer, model of car to model of car, year to year, and even state to state ("state" here refers to the state in which the car was built or in which it was expected to be operated). In this chapter we discuss the major groupings of emission control devices. To identify exactly what emission controls are on a particular car, it is necessary to refer to the appropriate shop manual or the emissions label (Fig. 6-4) on the car itself.

On newer cars there is a Vehicle Emission Control information label which identifies the devices and specifications used to meet the constantly decreasing allowable limits for pollutants. Through the development of more emission control devices and the integration of the systems of the engine, overall driveability and efficiency have increased despite the fact that the maximum allowable limits on pollutants have steadily decreased. Emission controls can be divided into two categories: those that reduce the pollutants emitted by the exhaust and those that prevent the pollutants from being formed. Two of the most commonly used systems to reduce the amount of pollutants in the exhaust are the air injection system and catalytic converters. Both of these alter the quality of the exhaust, not the combustion process.

AIR INJECTION SYSTEMS

Air injection systems introduce fresh air into the exhaust as it leaves the engine. This fresh supply of oxygen introduced into the hot stream of exhaust causes minor combustion to take place, by combining with the hot unburned fuel particles in the

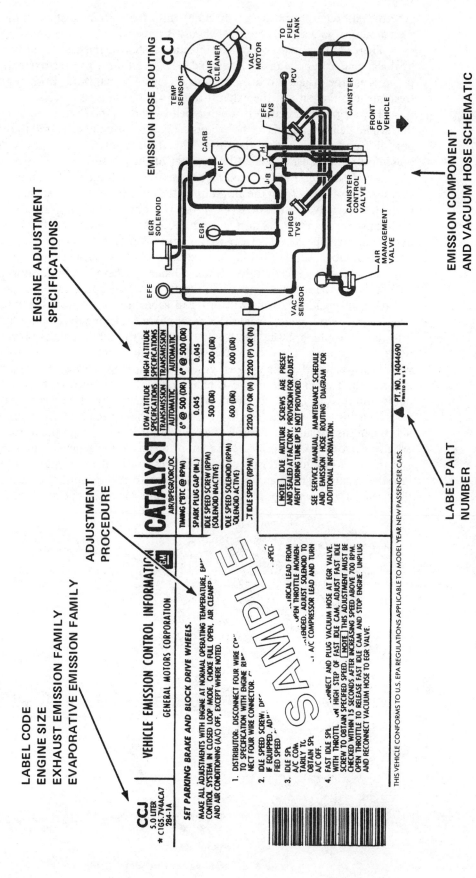

Figure 6–4 (Courtesy of Chevrolet Motor Division, General Motors Corporation)

exhaust. Although the power produced by this after-burning exits with the exhaust, the amount of CO and HC in the exhaust is reduced.

There are two basic types of air injection systems. One uses an air pump driven by the engine to supply air to the engine. The second, the *pulse air system,* uses the pulses from the exhaust to draw air into the exhaust manifold. Both systems draw fresh air into the path of the exhaust to control CO and HC levels.

The air pump is driven through a belt by the engine (Fig. 6–5). It pressurizes the air and delivers this air to the exhaust. The air is usually delivered to the exhaust port of each cylinder behind the valve. This fresh supply of air mixed with the fuel present in the exhaust is ignited by the heat of the exhaust.

An air injection rail is used to channel the fresh air to the cylinders. In the common passageway from the air pump to the air injection rail there are two valves. One of these valves is the *diverter valve.* The purpose of this valve is to prevent backfiring through the exhaust during deceleration. When the throttle is closed quickly at high speeds, an overly rich mixture is momentarily present in the combustion chamber. All the fuel in this mixture cannot be burned in the normal combustion process, so it exits through the exhaust. If there is a large supply of fresh air present at the exhaust port when these many fuel particles leave, combustion will happen quickly and this will result in a backfiring through the exhaust. The diverter valve stops the air from the air pump from reaching the exhaust by diverting it to the atmosphere. Stopping the flow of air to the exhaust ports prevents objectionable backfiring. The other valve used in the air injection rail is a *one-way check valve.* This valve prevents exhaust gases from moving toward the air pump. This protects the system itself, for if exhaust gases were allowed to back into the air pump and then be ignited by fresh air, the air pump would be destroyed.

The pulse air system (Fig. 6–6) uses the pulses of the exhaust to draw in fresh air. When the exhaust first leaves the cylinder, it is under high pressure. When the exhaust valve closes, the pressure present by means of the exhaust is no longer there, and this creates a low pressure. This low pressure is like a vacuum and draws air into the exhaust. The use of the pulses of high to low exhaust pressure eliminates the need for an air pump. These systems usually have a single air inlet for all cyl-

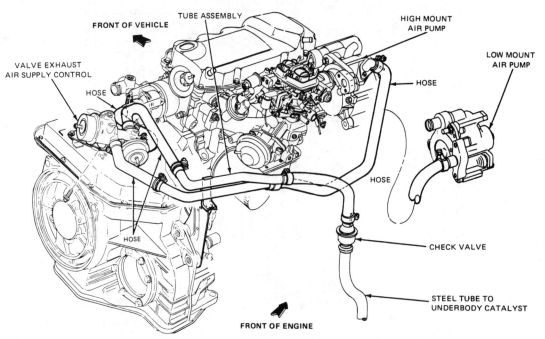

Figure 6–5 (Courtesy of Ford Motor Company)

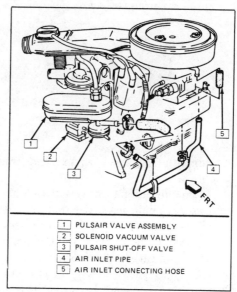

1	PULSAIR VALVE ASSEMBLY
2	SOLENOID VACUUM VALVE
3	PULSAIR SHUT-OFF VALVE
4	AIR INLET PIPE
5	AIR INLET CONNECTING HOSE

Figure 6-6 (Courtesy of Cadillac Motor Car Division, General Motors Corporation)

inders. The air is delivered to the cylinders through an *air injection rail,* which, like the air pump system, has a check valve to protect it from exhaust gases. The one-way check valve allows air to enter the exhaust only when the pressure is low.

CATALYTIC CONVERTERS

The other device used to reduce the levels of the pollutants in the exhaust is the *catalytic converter.* This device is placed in the exhaust system, so that exhaust gases must pass through it. The catalytic converter contains a catalyst to cause a chemical reaction which converts the pollutants in the exhaust to harmless gases. A catalyst is a substance that by its presence will cause a chemical reaction to occur between other substances present. In a catalytic converter, platinum and palladium are used as the catalyst for converting HC and CO into water and carbon dioxide.

The converter contains either a ceramic honeycomb "monolith" or aluminum oxide pellets coated with the catalyst. Converters that reduce the NO_x outputs, in addition to HC and CO, use rhodium or rhodium and platinum as a catalyst. This type of converter is called a *three-way catalyst* because it reduces the amount of all three pollutants in the exhaust. There are four basic types of converters used today: the conventional, the three-way, the dual-stage, and a warm-up converter. The purpose of all four is simply to reduce the exhaust emissions of an engine.

The *conventional converter,* called this because it is the most commonly used, may use pellets or a monolith coated with platinum and palladium or just platinum. This type of converter is designed to reduce HC and CO levels in the exhaust. Converters that use a monolith contain the catalyst-coated ceramic structure, a steel mesh blanket, and diffuser plates (Fig. 6–7). The *monolith* is a honeycomb-like structure with many small passages through it. As the exhaust gases enter the converter, diffuser plates tend to spread out and break up the flow of the exhaust. This prevents the exhaust from going through only a section of the monolith. Exhaust gases are spread out so that all the monolith has exhaust gases passing through it. As the gases flow through the monolith, they are exposed to the catalyst and the desired chemical reaction takes place. The steel mesh blanket is used to protect the

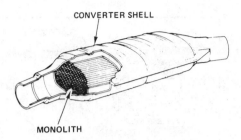

CONVERTER SHELL

MONOLITH

Figure 6-7 (Courtesy of Chevrolet Motor Division, General Motors Corporation)

monolith from excessive heat and from road vibration which can damage the monolith structure.

Conventional converters that use aluminum oxide pellets coated with the catalyst have built-in baffles that direct the flow of exhaust through the entire bed of pellets (Fig. 6–8). As the exhaust gases encounter the catalyst on the pellets, a chemical reaction takes place and the amount of HC and CO is reduced or changed into water and carbon dioxide.

Unleaded fuel must be used in all engines that are equipped with a catalytic converter. The lead that exits the engine is not only a poison, but tends to collect on the catalyst, causing it to become ineffective. The pellets or monolith would become covered with lead and the exhaust gases would no longer be exposed to the catalyst. The pellets, if damaged or subjected to much lead, can be replaced. However, if a monolith type of converter is damaged, the whole converter must be replaced.

Converter systems that reduce NO_x, HC, and CO are called *three-way converters*. Adding a rhodium catalyst to the system, either in an additional converter or enclosed in the conventional converter, will chemically reduce NO_x. If this additional catalyst is housed in a separate housing, it is mounted in the exhaust ahead of the conventional converter. If it is added to the conventional converter, this assembly is called a *dual-stage converter*.

As shown in Fig. 6–9, the two converters are housed in one assembly and are separated by an air space. Exhaust gases pass through the rhodium converter first, where NO_x is chemically reduced. Oxygen is separated from the NO_x and is used by the conventional converter to decrease the amounts of HC and CO in the exhaust. The chemical reaction that takes place in the conventional converter is the combining of oxygen to the HC and CO particles. This is a process called *oxidation*. The amount of oxygen supplied to the conventional converter by the front converter

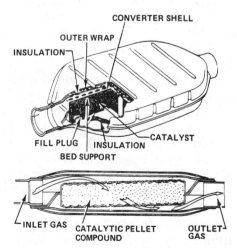

CONVERTER SHELL

OUTER WRAP

INSULATION

FILL PLUG   INSULATION
BED SUPPORT

CATALYST

INLET GAS   CATALYTIC PELLET   OUTLET
COMPOUND   GAS

Figure 6-8 (Courtesy of Chevrolet Motor Division, General Motors Corporation)

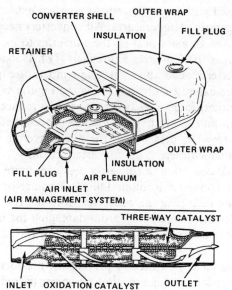

Figure 6-9 (Courtesy of Chevrolet Motor Division, General Motors Corporation)

is not enough to cause good oxidation of the HC and CO present. Therefore, the air space between the two converters is filled with a fresh supply of air from the air injection pump. This air also tends to cool the converter.

The fourth type of converter is the *warm-up converter* or *mini converter* (Fig. 6-10). This converter was developed to control exhaust emissions during those times when a regular converter is not effective, such as during engine warm-up. The mini-converter is placed right after the exhaust manifold of the engine, well before the

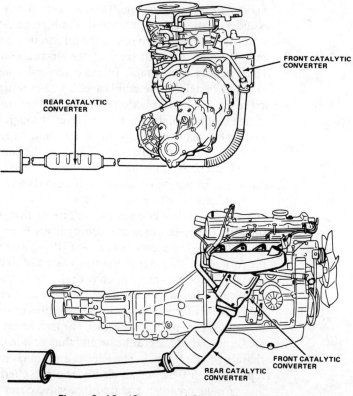

Figure 6-10 (Courtesy of Chrysler Corporation)

other converters used with it. Its placement near the exhaust manifold allows it to come quickly to operating temperatures. All converters need to heat up before they can be effective. Exhaust gases heat up the converter to its required operating temperature of more than 1000 degrees Fahrenheit (°F). Due to its location and small size, the miniconverter reaches this operating temperature quickly and provides for some control of the exhaust emissions while the regular converter is warming up. The miniconverter therefore serves two purposes: to reduce emissions while the engine is warming up and to start the oxidation process before the exhaust reaches the regular converter. The presence of the miniconverter in a system reduces the amount of time in which there is no converter action.

During normal operation, a converter produces quite a lot of heat. The inside temperature of the converter is about 1400°F. Because of this heat, metal shields are used to separate or isolate the converter from the rest of the car. These metal heat shields prevent the extreme heat from damaging the underbody or bottom of the car and stop the heat from warming the floor of the car to uncomfortable temperatures.

Many systems have been used to prevent the production of pollutants by the engine. The air injection system and catalytic converter reduce emission levels after combustion. Positive crankcase ventilation, exhaust gas recirculation, and spark control systems are commonly used to prevent the production of pollutants. Changes have also been made to the fuel and ignition system as well as to valve timing and engine compression ratios. It has been found that by controlling combustions, exhaust emissions can also be controlled.

POSITIVE CRANKCASE VENTILATION SYSTEMS

The positive crankcase ventilation system not only prevents the by-products of combustion from entering the atmosphere but also prevents mechanical damage to the engine. Piston rings are not able to completely seal the piston to the bore of the cylinder, and during combustion, the high pressure present pushes some of the exhaust past the piston rings into the oil pan or crankcase. Not only does some exhaust enter the crankcase but some of the air-and-fuel mixture enters from compression. These vapors and exhaust fumes enter under high pressure. If high pressure is allowed to remain in the crankcase, it will eventually find a way to move toward the lower outside atmospheric pressure. This exit of high pressure can be through any engine sealing point. Usually, it leaves through the valve covers or through the oil pan gasket. Not only does this escape allow the vapors and fumes to enter the atmosphere but also causes an oil leak at the point of escape. While the fumes and vapors are present in the crankcase, they tend to break down the engine oil and cause the formation of sludge in the crankcase. To avoid these potential problems, the crankcase must be vented. But these vapors cannot be vented into the atmosphere because they contain the pollutants that are of concern and carry with them small oil particles which are also hydrocarbons.

Positive crankcase ventilation (PCV) *systems* were first developed to prevent the formation of sludge in the crankcase and to extend engine life. But these systems are considered an emission control device because as a consequence of venting the crankcase, they also reduce the levels of engine emissions. As shown in Fig. 6-11, a positive crankcase ventilation system uses engine vacuum to draw in fresh air from the atmosphere to the crankcase. The movement of this air into the crankcase pushes the fumes out of the crankcase and the vacuum draws the fumes into the combustion chamber to be burned. The vapors and fumes in the crankcase are constantly being replaced with fresh, clean air under atmospheric pressure.

An *air breather* is placed in a valve cover or in the air cleaner assembly (Fig.

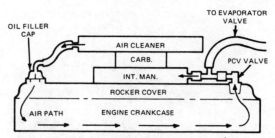

Figure 6-11 (Courtesy of Ford Motor Company)

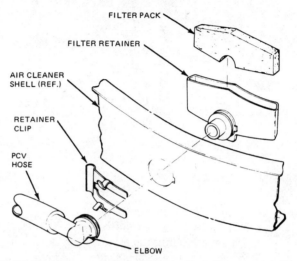

Figure 6-12 Crankcase ventilation system filter. (Courtesy of Ford Motor Company)

6-12) to allow the movement of outside air into the crankcase. At another point common to the crankcase, there is a one-way valve (PCV valve) which has one end connected to a source of engine vacuum. As engine vacuum builds, the valve opens and the vacuum draws the fumes and vapor out of the crankcase. As the vapors are drawn, outside air enters the crankcase to replace them. The crankcase vapors enter the engine's cylinders on the intake stroke and are burned with the intake of air and fuel.

The amount of vapors that pass through the PCV valve is determined by the amount of vacuum present at the valve, the strength of the spring inside the valve, and the design of the valve itself (Fig. 6-13). There is a particular design and spring strength of the valve for every engine application. The purpose of a PCV valve is to control the flow of crankcase fumes into the intake manifold and to prevent flames or gases, which may occur in the intake manifold, from entering the crankcase.

During periods of high intake manifold vacuum, at idle or cruising speeds, the PCV valve allows for a moderate flow of vapors from the crankcase to the intake manifold. As the engine speed and load increase, the engine vacuum decreases and the PCV valve opens fully to allow as much vapor as it possibly can to enter the intake manifold. When vacuum drops to zero, such as at full throttle or when the engine is off, the valve closes off all flow of vapors from the crankcase. At high engine speeds, the production of fumes and vapor in the crankcase is high. When the PCV valve is closed, these vapors are vented through the air breather.

There are two basic types of PCV systems: an open and a closed system. The difference between the two types is the breather. If the breather is vented to the

To intake
manifold

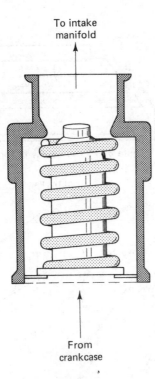

From
crankcase

Figure 6-13 Vacuum from the intake manifold
pulls the valve up against the spring; the weight
of the valve and the strength of the spring determine the amount the valve will open with a given
amount of vacuum.

atmosphere, the system is *open*. At full throttle the crankcase fumes vent through
the breather into the atmosphere. A *closed* system is one that seals the breather and
connects it to the air cleaner. There is a filter, called a *crankcase filter*, located in
the air cleaner to provide for clean air when air is drawn by vacuum through the
breather. When the vacuum is very low and the PCV valve closes, the crankcase
vents through the breather and these vapors are drawn into the engine through the
carburetor.

EXHAUST GAS RECIRCULATION SYSTEMS

Another system commonly used to reduce the production of harmful pollutants by
the engine is the *exhaust gas recirculation* (EGR) *system*. EGR systems control NO_x
emissions by lowering combustion temperatures. NO_x is produced only through high
heat and the EGR adds a small sample of exhaust gas to the incoming air-and-fuel
mixture to control the temperature of combustion. EGR systems typically consist
of an EGR valve, vacuum lines, a thermostatic switch, and an exhaust back-pressure
sensor.

EGR valves are mounted on the intake manifold. The valve consists of a top
housing, which is sealed and contains a spring-loaded diaphragm, and a valve which
is connected to the diaphragm (Fig. 6-14). The sealed housing and its diaphragm
react to the presence of vacuum. As the vacuum in the housing increases, the diaphragm moves, pulling the valve up and off its seat and allows exhaust gas to enter
the intake manifold. The vacuum signal for the EGR valve is from the carburetor
venturi or from a port on the carburetor.

Ported vacuum comes from an outlet on the carburetor, just above the throttle
plate. When the throttle is closed, there is little or no vacuum present at this port
and the EGR valve does not open. As the throttle opens and speed increases, the
vacuum at the port increases and the vacuum signal of the EGR valve is strong,
causing the valve to open. The vacuum decreases at full throttle and the EGR valve
closes.

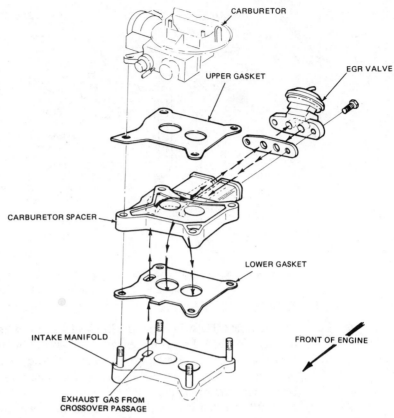

CARBURETOR

EGR VALVE

UPPER GASKET

CARBURETOR SPACER

LOWER GASKET

INTAKE MANIFOLD

FRONT OF ENGINE

EXHAUST GAS FROM
CROSSOVER PASSAGE

Figure 6-14 (Courtesy of Ford Motor Company)

The vacuum formed by the air passing through the venturi of a carburetor increases as the volume and speed of the air increases. Any change in airflow will effect a change in the vacuum produced by the venturi. Venturi vacuum is very sensitive to changes in engine load and speed and provides an excellent source for the control of the EGR valve. However, the vacuum signal at the venturi is very weak and not strong enough to open the valve of the EGR. To employ venturi vacuum as a source of control for an EGR, a vacuum amplifier must be used (Fig. 6-15). Through the use of an amplifier, the EGR valve is sent manifold vacuum, which is very strong. The vacuum amplifier allows the weak venturi vacuum signals to control the amount of manifold vacuum going to the EGR. The vacuum amplifier is much like an electrical relay, where a low current controls a high current. The vacuum amplifier is controlled by venturi vacuum and the amplifier controls the amount of manifold vacuum going to the EGR valve.

Regardless of the source of vacuum, all EGR valves have their vacuum lines interrupted by a temperature switch. The purpose of an EGR valve is to reduce combustion temperatures, thereby reducing the formation of NO_x. When the engine is cold and during its warm-up, the high temperatures that produce NO_x will not be present; therefore, the EGR system is not needed. Most EGR systems have a *coolant temperature override* (CTO) *switch,* which blocks the vacuum signal to the EGR valve until a particular engine temperature is reached. Temperature switches that monitor intake air temperature and ambient air temperature have also been used with EGR systems. The CTO switch is mounted where it can read engine coolant temperatures. It can be located in the intake manifold, cylinder head, top radiator tank, or the thermostat housing.

Exhaust back-pressure sensors (Fig. 6-16) have been added to some EGR systems to provide better EGR control. The sensors are located in the vacuum line

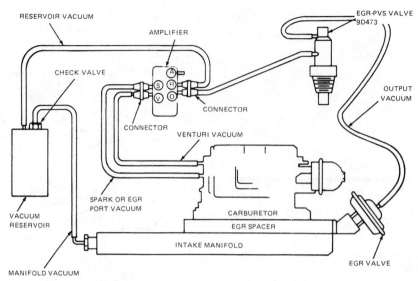

Figure 6-15 (Courtesy of Ford Motor Company)

between the CTO switch and the EGR valve and contain a spring-loaded diaphragm and an air bleed. From the sensor is a vacuum line that connects to a spacer plate between the EGR valve and the intake manifold. A tube in this plate is exposed to exhaust gases. The vacuum line from the spacer sends the exhaust back-pressure sensor an exhaust signal.

When exhaust pressure is low, the air bleed is open. This allows air to enter the EGR vacuum line, which reduces the vacuum signal to the EGR valve and closes it. When exhaust pressure is high, the air bleed is closed and the full vacuum signal is sent to the EGR valve, allowing it to open.

Many newer engines are equipped with EGR valves that have been modified to include the control of exhaust back pressure without having two separate units. These EGR valves include two diaphragms: one for the movement of the EGR valve and one to control the vacuum signal to the valve. These combination EGR and back-pressure valves operate much the same as an EGR valve with a separate sensor. There are two types of EGR valves used: positive back pressure and negative back pressure. The positive-back-pressure EGR opens only when exhaust pressure is great enough to close the air bleed and there is a strong vacuum signal to it.

As shown in Fig. 6-17, a negative back-pressure EGR valve modulates, or

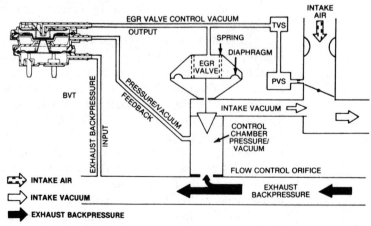

Figure 6-16 (Courtesy of Ford Motor Company)

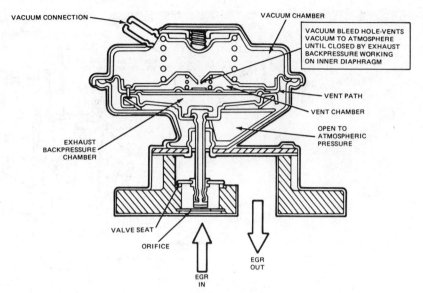

VACUUM CONNECTION

VACUUM CHAMBER

VACUUM BLEED HOLE-VENTS
VACUUM TO ATMOSPHERE
UNTIL CLOSED BY EXHAUST
BACKPRESSURE WORKING
ON INNER DIAPHRAGM

VENT PATH

VENT CHAMBER

OPEN TO
ATMOSPHERIC
PRESSURE

EXHAUST
BACKPRESSURE
CHAMBER

VALVE SEAT

ORIFICE

EGR
IN

EGR
OUT

Figure 6-17 (Courtesy of Ford Motor Company)

opens and closes, in response to the differences between ported and manifold vacuum. Low exhaust pressure opens the air bleed, and this weakens the vacuum signal from the port of the carburetor. When exhaust pressure is high, the air bleed is closed and full ported vacuum is received by the EGR valve; however, as the valve opens, the lower part of the valve's diaphragm moves, closing the valve. The closing of the valve shuts off the supply of manifold vacuum and the ported vacuum begins to open the valve. This causes the EGR valve to modulate when exhaust pressure is great enough to allow the valve to open.

SPARK CONTROLS

The purpose of the EGR system is to prevent the formation of NO_x by reducing combustion temperatures. Another method used to reduce NO_x by controlling combustion temperatures is to control the spark timing. By retarding the timing, the temperature of combustion can be reduced. Spark controls attempt to provide better control of ignition timing, which will also result in the production of the less HC and CO. There are two basic types of spark controls: one blocks the vacuum signal to the vacuum advance unit and the other delays the signal. Ignition timing or spark controls have evolved from systems that are quite basic to totally computerized systems. Computerized systems control ignition timing by monitoring many systems of the engine. The integration of these systems is a topic of its own, and these systems will be covered in a later chapter. The precomputer spark controls will be discussed here.

Early spark controls provided full vacuum advance during deceleration. This eliminates backfiring in the exhaust pipe. Upon deceleration, manifold vacuum is high. Vacuum control valves send the high manifold vacuum to the vacuum advance unit to cause full advance during deceleration. Normally, the vacuum source for the vacuum advance is ported. The ported vacuum is bypassed during deceleration to allow manifold vacuum to the distributor. Some manufacturers used both ported and manifold vacuum to control the vacuum advance, as shown in the dual-diaphragm setup in Fig. 6-18.

The initial purpose of spark control valves was to provide a timing change during deceleration. Later, spark control valves were designed to control vacuum

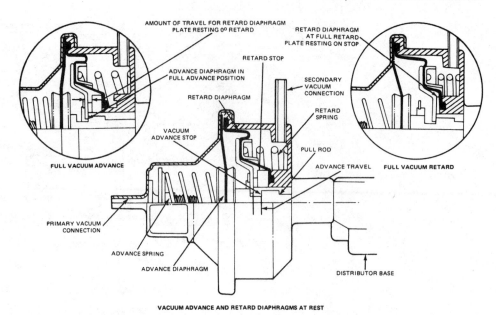

AMOUNT OF TRAVEL FOR RETARD DIAPHRAGM
PLATE RESTING 0° RETARD

RETARD DIAPHRAGM
AT FULL RETARD
PLATE RESTING ON STOP

ADVANCE DIAPHRAGM IN
FULL ADVANCE POSITION

RETARD STOP

RETARD DIAPHRAGM

SECONDARY
VACUUM
CONNECTION

RETARD
SPRING

VACUUM
ADVANCE STOP

PULL ROD

ADVANCE TRAVEL

FULL VACUUM ADVANCE

FULL VACUUM RETARD

PRIMARY VACUUM
CONNECTION

ADVANCE SPRING

ADVANCE DIAPHRAGM

DISTRIBUTOR BASE

VACUUM ADVANCE AND RETARD DIAPHRAGMS AT REST

Figure 6–18   (Courtesy of Ford Motor Company)

advance during low-speed operation. Two systems are used to block the vacuum signals to the advance unit during low speeds. The difference between the two is in the way they sense vehicle speed. One system uses a transmission switch that blocks the vacuum until the transmission is in high gear. The other system uses a vehicle speed sensor that is installed in the speedometer cable. The vacuum signal is blocked until a particular speed is reached.

CTO switches are also used to block off manifold vacuum from the vacuum advance unit during particular engine temperatures. These switches can be designed to allow full manifold vacuum to the advance unit when the engine is cold or hot. CTO switches are called many names by the various manufacturers and are used heavily with emission control devices.

Spark delay valves slow vacuum advance by delaying the vacuum signal to the distributor. The valve consists of a small plastic housing with two vacuum connections on it; a sintered metal delay orifice and a one-way check valve (Fig. 6–19). The delay orifice prevents the vacuum signal received at one connector from being delivered immediately to the advance unit. The one-way valve allows for the delay to occur in only one direction; therefore, vacuum signals in the other direction are not delayed. The vacuum signals going to the distributor are delayed and when the vacuum source is weak, the vacuum present at the advance unit is free to leave through the delay valve. This allows the advance unit to respond to decreases and increases of vacuum. All increases in vacuum are delayed, whereas all decreases are immediate. The length of the delay varies with the application but is usually between 15 and 30 seconds.

A common version of a delay valve that delays the vacuum signal only when the engine is warm is the *orifice spark advance control* (OSAC) *valve* (Fig. 6–20). During normal operating temperatures, the OSAC valve functions as a delay valve. But when the coolant temperature is low, the delay is bypassed, allowing the vacuum advance unit to respond directly to the vacuum signal. All of these spark controls provide a control for ignition timing. These controls limit the temperature that can be present with combustion. Spark control valves therefore control the formation of NO_x and they improve driveability.

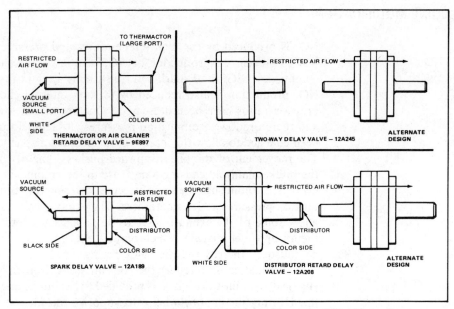

Figure 6–19 Four types of spark delay valves. (Courtesy of Ford Motor Company)

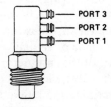

SECONDARY VACUUM BREAK T.V.S. (LOCATED IN AIR CLEANER)

2.8L V-6 VIN CODE I
BELOW 10°C (50°F) ALL PORTS CLOSED
BETWEEN 10°C 19°C (50°-66°F) PORTS 1 & 2 CONNECTED
ABOVE 15-19°C (59°-66°F) ALL PORTS CONNECTED

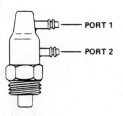

EGR T.V.S. (LOCATED AT FRONT OF INTAKE MANIFOLD)

2.8L V-6 VIN CODE i
ABOVE 54°C (130°F) PORTS 1 & 2 CONNECTED
BELOW 49°C (120°F) PORTS 1 & 2 CLOSED

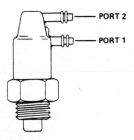

EFE T.V.S. (LOCATED AT REAR OF INTAKE MANIFOLD)
5.0L V-8 VIN CODE H
ABOVE 32°C (90°F) BOTH PORTS CLOSED
BELOW 26°C (80°F) BOTH PORTS OPEN

Figure 6–20 Types of thermal vacuum valves and switches. (Courtesy of Chevrolet Motor Division, General Motors Corporation)

ENGINE MODIFICATIONS

NO$_x$ is produced by the high temperatures and pressure of combustion. The best way to prevent the production of NO$_x$ is to lower the temperature and pressure of combustion. EGR and spark control systems attempt this through external means. NO$_x$ production is also reduced by making internal changes in the engine. Compression ratios have been decreased, which results in lower compression pressures and therefore lower combustion pressures and temperatures. Valve timing has also been changed to allow the intake to be open during part of the compression stroke. The movement of the piston upward pushes some of the intake mixture out into the intake manifold causing a decrease in the volume of air and fuel mixture. This decreased volume has the same effect as lowering the compression ratio: lower combustion pressures and temperatures. The exhaust valve is also kept open longer, to allow some of the exhaust gases to be pulled back into the chamber. This dilutes the air-and-fuel mixture, which reduces the maximum temperature reached during combustion, thereby reducing NO$_x$ formation.

In an attempt to reduce the HC emissions from an engine and to improve cold engine driveability, extra heat is provided to the intake manifold during engine warm-up. Heating the manifold helps to vaporize the fuel particles, which increases the chances of it all being burned. This heat is typically provided by two methods. One method routes exhaust gases into passages in the intake manifold below the individual runners for the cylinders. The heat from the exhaust warms the floor of the runners and promotes fuel vaporization. Either a bimetallic spring or vacuum-operated *exhaust heat control valve* is used to send exhaust from the exhaust manifold to the passageway in the intake (Fig. 6–21). When the engine is warm, the control valve allows the exhaust to flow freely through the exhaust pipes.

Another method of heating the intake is through the use of an electrical grid (Fig. 6–22). The grid is a heater made of ceramic materials and is positioned under the carburetor as an integral part of the carburetor gasket. When the engine's coolant temperature is below a given value, electrical current is supplied to the heater.

Yet another method of warming the intake is by increasing the overall temperature of the engine. This not only increases the temperature of the intake manifold but also increases the temperature of the combustion chamber, which promotes the vaporization of any fuel particles that may be trapped in it.

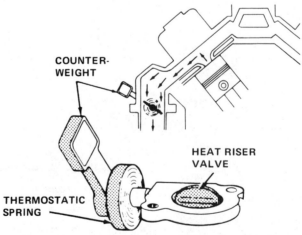

COUNTER-WEIGHT

HEAT RISER VALVE

THERMOSTATIC SPRING

Figure 6–21 (Courtesy of General Motors Corporation)

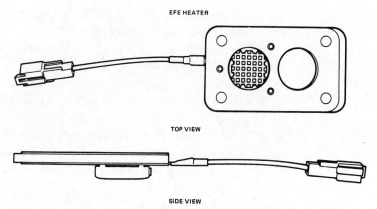

Figure 6-22 (Courtesy of Ford Motor Company)

SUMMARY

Changes have been made in the design of intake and exhaust manifolds, combustion chambers, and of course ignition systems. These changes provide for increased engine efficiency, which reduces the levels of exhaust emissions. Many of the early emission control devices were designed to reduce the levels of pollutants in the exhaust. These systems hid the basic inefficiencies of the internal combustion engine. Engineers today are concerned more with the prevention of these pollutants through increased engine efficiency. Redesigned engines with computer monitored fuel, ignition, and emission control systems are the present thrust of the automobile industry. It is important to understand what each emission control device is attempting to accomplish and how it works, because a faulty emission control system can drastically affect the way the engine runs.

TERMS TO KNOW

HC	Catalytic converter
NO_x	Oxidation
CO	PCV
Smog	EGR
Air injection	CTO
Pulse air	OSAC
Diverter valve	Spark delay valve

REVIEW QUESTIONS

1. What are the major pollutants present in an engine's exhaust?
2. What are the two categories for emission control?
3. What is the purpose of the air injection system?
4. What is the purpose of a catalytic converter?
5. What are the four major types of catalytic converters used today?
6. Why is unleaded fuel required for engines equipped with a catalytic converter?
7. What is a three-way catalyst?

8. Why is the air pump connected to a dual-stage converter?
9. What is the purpose of a miniconverter?
10. What are the purposes of the positive crankcase ventilation system?
11. What is the purpose of the EGR valve?
12. How does the EGR system control emission levels?
13. Why is a vacuum amplifier used with most EGR valves?
14. How does a positive-back-pressure EGR valve work?
15. How does a negative-back-pressure EGR valve work?
16. What effect does ignition timing have on exhaust emissions?
17. What modifications to the basic engine have been made to limit the amount of pollutants in an engine's exhaust?

7

SAFETY PROCEDURES

Figure 7-1

Like effective troubleshooting, safe work habits result from an understanding of the items with which you are working. Many systems and parts are often overlooked while troubleshooting an engine. Troubleshooting becomes a game of luck. Many play this game of luck with their behavior and attitude while troubleshooting and working on a car. Eventually, their luck runs out and they get hurt or hurt someone else. Safety is everyone's job. Too often someone says "I've done it this way all my life" or "I've seen others do it this way a million times and nothing ever happened to them." They have been lucky. As with anything you do on a car, you must think before you act. Not only can unsafe practices hurt yourself or someone else, it can damage something on a car.

A professional attitude is displayed by safe and neat work habits. Actually, neat habits are also safe habits. Cleaning up spills and keeping equipment and tools out of the path of others and yourself prevents accidents. This is only common sense, but too often, time is not taken to remove these safety hazards. It is the rush to complete a job that usually results in unsafe conditions. The truly professional technician takes the time to clean his or her tools and work environment. Typically, this technician will do a better job, make more money, and have fewer accidents than others who are not quite as professional. A professional also does not clown around in the repair shop, does not throw items in the shop, and does not create an unsafe condition as a result of saving time. Instead of ignoring basic safety practices to save time, a professional saves time by efficiently diagnosing and servicing the car. Most of the safety procedures that are included in this chapter are based on common sense. These should be followed by both the student and the professional. With neat and safe work habits, the road that leads to professionalism will be shorter.

The dress and appearance of technicians not only says something about their personalities but also about their attitudes, including their attitudes on safety (Fig. 7–2). Clothing that hangs out freely, such as shirttails, can create a safety hazard. These can get caught in something and cause injury. Jewelry can also get caught and cause serious injury. A ring can rip a finger off, a watch can cut a wrist, or a necklace can choke a person. Anything that a person wears should not be allowed to dangle freely in the engine compartment. It is best to make sure that shirts are tucked in and buttoned. Long sleeves should be buttoned or carefully rolled. Rings, necklaces, bracelets, and watches should not be worn while working on an automobile. It is also wise to protect your feet. Tennis and jogging shoes provide little protection if something falls on your foot. Shoes or boots made of leather, or a synthetic that approaches the strength of leather, offers good protection from falling objects. There are many brands of comfortable safety shoes and boots available which not only protect your feet from falling objects but also have soles that are designed to resist slipping on wet surfaces. Foot injuries are not only quite painful but can also prevent you from working for some time!

Some sort of eye protection should be worn whenever there is a possibility of dirt or metal particles flowing in the air. While working under a car, dirt, grease, or rust can fall into your eyes and cause serious damage. Safety glasses or goggles should be worn (Fig. 7–3). When using grinding wheels, chemicals, compressed air, or fuels, eye protection should be used. Chemicals, such as brake fluid, can cause

Figure 7–2

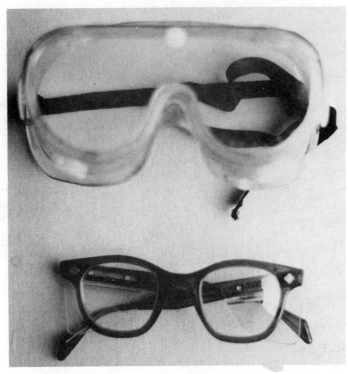

Figure 7–3

serious eye irritations which could lead to blindness. Let common sense dictate when the use of safety glasses or goggles is appropriate.

PROPER USE OF EQUIPMENT

When working with any equipment, make sure that it is used properly and is hooked up according to the manufacturer's instructions. All equipment should be properly maintained and periodically inspected for unsafe conditions. Frayed electrical cords or loose mountings can cause serious injuries. All equipment with rotating parts should be equipped with safety guards to reduce the possibility of the parts coming loose and injuring the operator (Fig. 7-4). Tools and equipment should never be used for purposes other than those they are designed for. Using the proper tool in the correct manner will not only be safer, but will also produce a better job. Do not depend on someone else to inspect and maintain equipment. Check it out before you use it! If the equipment is found to be unsafe, put a sign on it warning others not to use it and notify the person in charge. Safety is everyone's job. Take it upon yourself to ensure the safety of yourself and of others.

Out of fear, many people are reluctant to work with electricity. A certain amount of fear is advantageous if it leads to respect for electricty. A human body is made up mostly of water, and water is an excellent conductor of electricity. Electrical shocks can cause burns and other damage to the body. Remember that your body conducts electricity well, so to avoid injury, use insulated tools when working on an electrical circuit. Injury is often the result of the person jumping from the electrical shock. Quick movements of the body may send a head into the hood latch, which can leave a big hole in a head! Elbows may be damaged by forcefully hitting the engine or other parts under the hood. Quick movements caused by shocks are usually uncontrolled movements, and injury results.

Figure 7–4

AUTOMOTIVE BATTERIES

When possible, it is recommended that the battery be disconnected before wires or components are disconnected, cut, or handled. This prevents the possibility of electrical shock. Disconnecting the battery before making electrical repairs also eliminates the possibility of an accidental short which can ruin an electrical system. There is a proper procedure for disconnecting the battery. Disconnect the negative or ground cable first, then the positive cable. Since electrical circuits need a ground to be completed, removal of the ground cable eliminates the possibility of a circuit being completed accidently. When reconnecting the battery, connect the positive cable first and then the negative.

The active chemical in a battery, the electrolyte, is sulfuric acid. Sulfuric acid can cause serious burns if it contacts the skin. It may also cause blindness if it gets into your eyes. In fact, sulfuric acid can destroy nearly anything it touches (Fig. 7–5). Therefore, keep the electrolyte of the battery off yourself and the car. If the acid gets on your skin, wash it off immediately and flush your skin with water for at least 5 minutes. If the electrolyte gets into your eyes, flush out your eyes with water and go to a doctor or hospital immediately. Working with batteries is a time when common sense should tell you to wear safety glasses or goggles and rubber gloves. The gases that form in the top of a battery when it is being charged are very explosive. Never smoke or introduce any form of heat around a charging battery. The explosion not only will send the battery top into orbit but will spray sulfuric acid all over the shop and on you. If this were to happen, immediately rinse yourself off with water and continue to flush with water. If some of the acid went into your eyes—do not rub your eyes—just flush them and get to a doctor.

To avoid hurting yourself or the car's electrical system when using a jumper or booster battery to start a car, certain steps should be followed (Fig. 7–6). Remove the vent caps of both batteries and cover the holes with a wet cloth or rag. Most batteries manufactured today do not have vent caps; these batteries are sealed. Do not attempt to pry the tops off this type of battery. Make sure that the two cars are not touching each other. Turn off all the electrical accessories on both cars. Connect the positive cable of the jumper cables to the booster car battery positive terminal first, then connect the other end of the cable to the dead battery. Now connect the

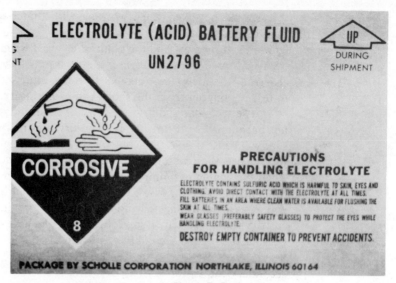

Figure 7–5

negative cable to the negative post of the booster battery. Connect the other end of the negative cable to the engine block of the car with the dead battery. Do not connect this to the negative terminal because this could cause electrical problems. Next, start the car with the booster battery. Finally, start the car with the dead battery. Once the engine has started, disconnect the cable going to the engine block first and then the other end of the negative cable. The positive cable can now be disconnected. Following this procedure and keeping in mind the other safety procedures for working with a battery, such as no smoking and the use of safety glasses, allows you to jump-start a car without hurting anyone or anything.

When working with the ignition system, remember that this is a very electrical system. The primary circuit should be dealt with as any other electrical circuit. Extra

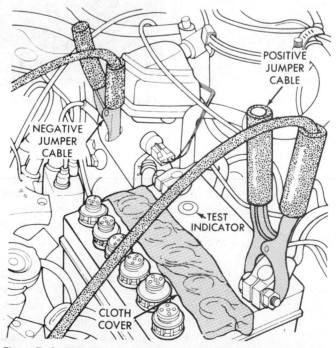

Figure 7–6   Use of jumper cables. (Courtesy of Chrysler Corporation)

care should be taken when working with the condenser, because it can become charged and shock you as you touch it. It is wise initially to make sure that the ignition is turned off and the points are closed when you touch the condenser. Electricity is stored in the condenser and you may serve as the ground it needs to discharge. The shock is not enough to cause injury to most people, but the reaction to or the jumping from the shock can cause injury.

The secondary ignition system has thousands of volts available. Voltage does not burn nor does it directly cause injury. But the jolt of voltage or electrical pressure will cause a quick, uncontrolled reaction which will almost certainly cause injury. When the ignition is off, there is little chance of electrical shock; however, when the engine is running, voltage is present and insulated pliers should be used to handle spark plug wires. Care should also be taken when spark plug or ignition coil wires are disconnected and the engine is cranking or running. It is quite possible for the voltage to ignite some fuel that might be present near the wire. This can cause a fire. When secondary wires are disconnected, they should be grounded in a safe place away from any fuel. A small jumper wire should be used to ground a secondary cable properly (Fig. 7-7).

Figure 7-7 Proper grounding of the ignition system.

GASOLINE

Gasoline is used so much in shops that the dangers from its presence are often forgotten. A slight spark or an increase in heat can cause an explosion or fire. Never smoke around gasoline; even the droppings of hot ashes could ignite the gasoline. If the engine has a gasoline leak, wipe it up immediately and fix the leak. The rags used to clean up the gasoline should be placed outside to dry. While correcting the leak, be sure not to cause any sparks.

Gasoline should always be stored in approved containers and never in a glass bottle or jar. If the glass jar were knocked over or dropped, a terrible explosion could take place. An explosion could also take place from the presence of gasoline vapors. If the shop has gasoline vapors in it from a gasoline leak or spillage, the doors and windows should be opened to vent the vapors outside. Remember, it takes only a little bit of fuel mixed with air to cause combustion. Oily rags can also be a source of fire. Such rags should be stored in an approved container and never thrown out with normal trash. Like gasoline, oil is a hydrocarbon and can ignite with or without a spark or flame.

FIRES

In the case of a fire you should know the location of the fire extinguishers in the shop and should also know how to use them. You should also know the different types of fire extinguishers and the types of fire for which they are used. Basically, there are four types of fire extinguishers (Fig. 7–8): the foam type is used for class A and B fires, the carbon dioxide type for class B and C fires, the dry chemical type for class A, B, and C fires, and soda-acid types for class A fires. The classification of fires is based on the fuels in of the fire. Class A fires are those in which wood, paper, and other ordinary materials are burning; class B fires are those involving gasoline, greases, oils, and other flammable liquids; and class C fires are electrical fires. The use of the wrong fire extinguisher will not put out the fire and under some circumstances may cause the fire to increase.

If no fire extinguisher is available, a blanket or fender cover can be used to smother the flames. Care must be taken when doing so, because the heat of the fire may still burn you and the blanket you are attempting to use. If the fire is too great to smother, move everyone away from the fire and call the local fire department. A simple under-the-hood fire can result in total destruction of the car and of the building. You must be able to respond quickly and precisely to avoid a disaster.

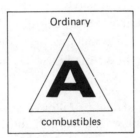  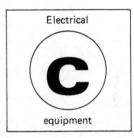

Figure 7–8

COMMONSENSE SAFETY RULES

There are many things that you should do and many others that you should not do to prevent accidents in the shop. Most of them are just common sense. What follows is a list of dos and don'ts which you should read and think about. Try to put these into your daily activities as an automobile technician. A professional technician performs work in a neat, orderly, safe, and systematic way. To be a professional and to prevent accidents, follow these simple rules:

1. Think about what you are doing at all times.
2. Double-check all your work to ensure that nothing is undone.
3. Keep your tools and equipment clean and in their proper place when they are not in use.
4. Check all equipment for possible safety hazards before using it.
5. Wipe off all excessive oils and grease from your hands so that tools will not slip out of them.
6. Do not play around in the shop.

7. Do not wear jewelry while working on a car.

8. Wear the right clothing for the job and make sure that no part of it is dangling free.

9. Wear shoes or boots that will protect your feet. Wear safety boots or shoes if possible.

10. Use eye protection whenever you are using fuels, chemicals, or compressed air or anytime dirt or metal particles are in the air.

11. Never use compressed air to blow dirt off your clothing or from your hair.

12. Never point a compressed-air gun at another person.

13. Do not put sharp objects into your pocket; this can cause injury to yourself or to the seats of the car.

14. Clean up any spills and keep all tools and equipment out of the walk paths of others.

15. Always use the right tool for the job. Always use it in the correct manner.

16. Keep hands and clothing away from the rotating fan blades and belts of an engine. Be especially careful of electric cooling fans, which can come on when the engine is off.

17. Never run an engine in a closed unventilated shop.

18. Never stand directly in line with a cooling fan; the blades may come loose and can cause serious injury.

19. Keep jack handles out of the way and always use jack stands to support a car when you are working under it.

20. Take extra care when moving a car in and out of a shop.

21. Never pour gasoline into a carburetor in an attempt to get an engine started.

22. Never continue to work in gasoline-soaked clothing.

23. Always vent the exhaust of a running engine out of the shop area.

24. Always set the parking brake of the vehicle you are working on when the engine is running.

These 24 rules of safety should be followed at all times. Some would say that these take too much time to follow and they will not follow all of them. Most people learn the rules of safety after they get hurt and sometimes that is much too late. Losing an eye because some brake fluid got in it would teach you a lesson, but that is an awfully hard way to learn safety practices! Don't be fools; follow the rules.

ACCIDENTS

If an accident does happen, the quicker you respond to it, the less the damage there will be. As with a fire, the quicker it is controlled, the less damage it will do. With an eye injury, the sooner the eye is flushed, the less chance of blindness there is. Never pass off an eye injury as something unimportant. Often, when a chemical enters the eye, it takes a few hours to cause great discomfort and by that time the damage may already be done. The supervisor should be informed immediately of any accident that may occur in a shop. This includes any accident involving you, another worker, or to a car. Safety hazards should also be reported immediately.

It is not a bad idea for you to become well versed in first aid. Knowledge of how to treat certain injuries can often save someone's life. The American Red Cross offers many low-cost but thorough courses on first aid. The first time you have to use the techniques of first aid will be the time you realize the worth of taking such a course.

SUMMARY

The purpose of discussing potential safety hazards and precautions in an automotive shop is not to scare you. The purpose is to convince you to think. Thinking about the conditions under and mannerisms with which you work can prevent injury to yourself and your fellow workers. Working on cars can be hazardous, but only if you let it be. Take the time to think and be careful.

REVIEW QUESTIONS

1. What are some general rules to follow while working on a car?
2. When should safety glasses or goggles be worn?
3. How should a battery be disconnected from a car?
4. What should you do if battery electrode gets into your eyes?
5. What is the correct procedure for jump-starting a car?
6. What should be done if there is a fuel leak in the engine compartment?
7. What are the types of fire extinguishers, and when are they used?
8. If there is no fire extinguisher available and there is an engine fire, what should you do?
9. If an accident happens, what is the first thing you should do?
10. What are some general dress rules to follow while working on a car?

8

MAINTENANCE TUNE-UP

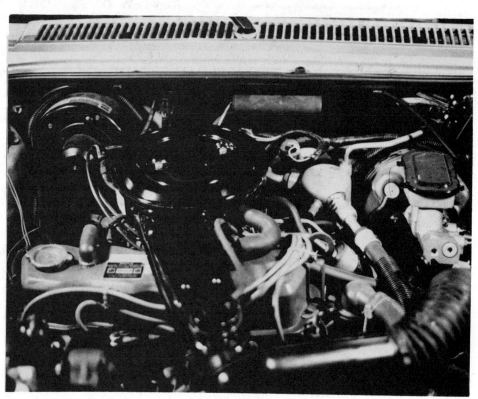

Figure 8–1

You will recall that the four-cycle engine goes through a series of events that prepare the engine for combustion. There are systems that have the purpose of supplying the needs of the engine and supplying them at the correct time and in the correct amounts. If one of these systems fails to deliver the correct amounts or is not delivering at the correct time, the engine will not perform the way it was designed to perform. Let's review the operation of the engine and detail the events that are taking place in each system. To avoid getting too complicated, assume that the engine is running and that the battery and its charging system are working well. The firing of one cylinder has just completed, and the piston is being pushed downward and turning the crankshaft, which is moving the car (Fig. 8–2).

When the piston reaches BDC, the exhaust valve begins to open. The ignition points have been closed for a short time since the firing of the spark plug was completed. Current is flowing through the primary wires of the ignition coil.

As the piston moves toward TDC, it pushes the exhaust out of the cylinder past the exhaust valve (Fig. 8–3). Just before TDC, the intake valve opens slightly, allowing some atmospheric pressure into the cylinder. The movement of air into the

Power stroke

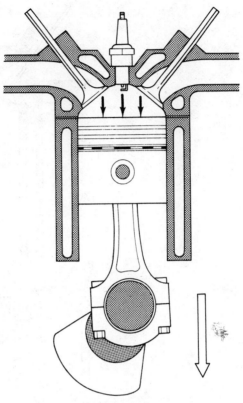

Figure 8–2

cylinder helps push the remaining amount of exhaust out of the cylinder, past the exhaust valve, which is also only slightly opened now. This period of time when the two valves are open is called *valve overlap* (Fig. 8–4).

When the piston reaches TDC, the exhaust valve is nearly closed. The piston changes its direction and moves down the cylinder (Fig. 8–5). This downward movement changes the volume of the cylinder, which creates a low pressure or vacuum in the cylinder. Atmospheric pressure, being higher than the pressure in the cylinder, attempts to equalize with the lower pressure. Air rushes into the intake manifold and past the intake valve into the cylinder. The rate of the airflow going into the cylinder is determined by the opening of the throttle plate of the carburetor. The wider the throttle plates are open, the more air that is flowing in. The rate of airflow determines the amount of fuel that will be delivered into the cylinder with the air. The movement of air through the carburetor draws out fuel from the carburetor at a rate proportional to the airflow. This air-and-fuel mixture enters the cylinder until the piston reaches BDC.

At BDC, the piston changes direction again and heads up to TDC. With this change of direction, the intake valve closes, sealing the cylinder (Fig. 8–6). The upward movement of the piston compresses the air-and-fuel mixture into a smaller volume, causing an increase in the pressure of the mixture. During all this time, the points have been closed and current has flowed in the primary of the coil.

Just before the piston reaches TDC, the points open. The current that was flowing across the closed points is absorbed by the condenser. Since it is not possible to have current flow through a condenser, the current in the primary stops immediately. The magnetic field that surrounded the primary windings of the coil while

Exhaust stroke

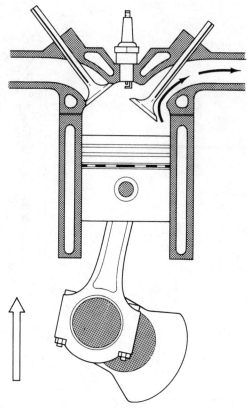

Figure 8-3

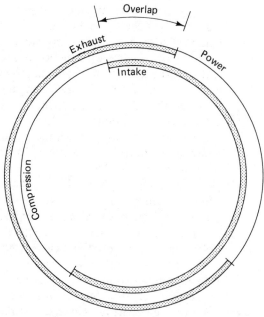

Figure 8-4

Intake stroke

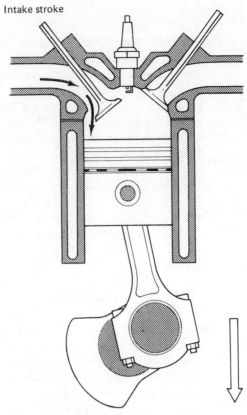

Figure 8-5

Compression stroke

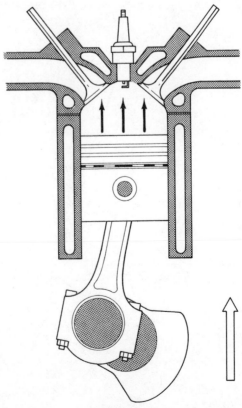

Figure 8-6

current was flowing in it no longer stays around the winding and moves to a soft-iron core in the center of the coil. As the magnetic field moves to the center, it passes through the secondary windings of the coil and induces a high voltage in it. This high voltage, seeking a ground, leaves the coil through the coil wire. The coil wire is attached to the distributor cap. Under the cap is the rotor, which has one end making contact to the coil wire terminal at the distributor cap and the other end rotating toward a terminal inside the distributor cap, which is connected to a cylinder of the engine.

The secondary voltage pushes across an air gap between the rotor end and the spark plug wire terminal in the distributor cap. This voltage then moves through the spark plug wire to the spark plug and through it. At the end of the spark plug there is another air gap. This air gap is inside the combustion chamber and the gap is filled with a compressed air-and-fuel mixture. Once the voltage pushes itself across the gap, the circuit is complete and current flows in the secondary. The current jumping across the gap supplies an immense amount of heat, which shocks the mixture and causes the pressure in the cylinder to rise quickly (Fig. 8-7).

By this time the piston has reached TDC and is ready to move downward. The current flowing in the secondary continues the spark across the plug gap and continues to burn the mixture. When combustion is complete, the pressure inside the combustion chamber is great and this pressure pushes the piston down with great force. At the same time, the rotor of the distributor moves away from the spark plug wire terminal, increasing the rotor gap to a point where the coil can no longer jump the gap. This ends the current flow in the secondary and allows for the beginning of a new dwell cycle. These events are followed by the exhaust stroke and the cylinder is being prepared for another firing.

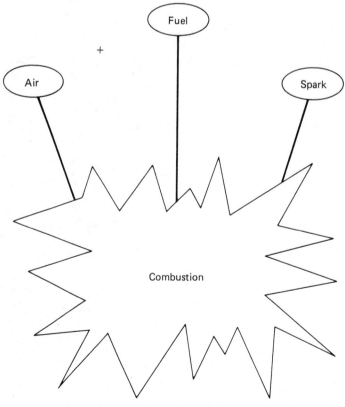

Figure 8-7

MAINTENANCE TUNE-UP

During the operation of the engine, there are many parts that move and because of the movement will wear. The wearing of some parts is more rapid than others. Engine oil is used to supply a lubricant on the engine parts, and this tends to reduce the amount of wear that takes place. However, there are parts that are not protected by engine oil and they tend to wear out. The wearing of these parts will affect the way they perform, and this affects the way the engine will perform. The replacement of these parts is the reason for performing a maintenance tune-up.

Besides the replacement of these wearing parts, the maintenance tune-up also includes the replacement of parts that do no more than protect the engine. Engine oil limits wear of the moving parts of an engine. If the oil wears or gets dirty, its lubricating ability is decreased. To help keep the oil clean and useful, an oil filter is added to the engine oiling system. With time, the oil and its filter gets dirty and should be replaced. Changing the oil and filter is a good start for a maintenance tune-up. Other filters should also be changed at this time; filters for the supply of fuel and air are examples of this. A maintenance tune-up should include the replacement of those parts that wear and collect dirt.

A maintenance tune-up should be done on a regular basis to ensure that the engine will perform well and also to extend the life of the engine. This tune-up consists of the replacement of parts of the ignition system, fuel and air filter replacement, maintenance of the emission controls, and making the necessary adjustments to ensure that events take place exactly when they should. Figure 8–8 shows a typical preventive maintenance (PM) schedule.

PRIMARY IGNITION SYSTEM

In the primary ignition system of a breaker-point ignition, the points wear and should be replaced as part of a maintenance tune-up. Although a lubricant is used, the points' rubbing block wears and decreases the point gap, which in turn increases the dwell time. As the points' gaps decrease, the chance of arcing when they open increases. This arcing causes the points to burn and become poor conductors of electricity when they are closed. For these reasons, points should be replaced on a regular basis.

Although the condenser is not a wearing item, it should also be replaced. A condenser can suddenly become defective without causing driveability problems, prior to its going out. Therefore, when the points are replaced, it is wise to replace the condenser with them. Electronic ignitions have no wearing parts and therefore have no items in the primary circuit that should be replaced routinely.

SECONDARY IGNITION SYSTEM

The secondary ignition system experiences wear from a different source from the parts of the primary circuit. Spark plug wires and spark plugs wear from the heat of the current that travels within them. Spark plugs also wear from the nearby continuous arcing of current across the gap. This arcing moves metal from one electrode to the other, causing the shape of the electrodes to change as well as the gap distance. The tip of a spark plug also is subjected to immense heat and pressure within the combustion chamber. This condition weakens the exposed electrodes of the plug and can eventually cause the plug to fail. Spark plugs are a key replacement item in a tune-up. Spark plug wires do not need to be replaced each time a main-

The series shown in the schedule up to 43,000 miles (72,000 km) are to be performed after 45,000 miles at the same intervals

Item no.	To be serviced	When to perform — Months or miles (kilometers), whichever occurs first	7500 mi 12,000 km	15,000 mi 24,000 km	22,500 mi 36,000 km	30,000 mi 48,000 km	37,500 mi 60,000 km	45,000 mi 72,000 km
1	Chassis lubrication	Every 12 months or 7500 miles (12,000 km)	●	●	●	●	●	●
2	Engine oil and oil filter change*	See explanation for service interval in section B						
3	Carburetor choke and hose inspection*	At 7500 miles (12,000 km) and every 24 months or 30,000 miles (48,000 km)	●†			●†		
4	Carburetor mounting bolt, torque*	At 7500 miles (12,000 km)	●†					
5	Engine idle speed adjustment (some models)*	At 7500 miles (12,000 km)	●†					
6	Vacuum or air pump drive belt inspection*	Every 24 months or 30,000 miles (48,000 km)				●†		
7	Cooling system refill*					●		
8	Wheel bearing repack (rear-wheel-drive cars only)	Every 30,000 miles (48,000 km)				●		
9	Transmission/transaxle service	See explanation for service interval in section B						
10	Vacuum advance system inspection (some models)*	Every 30,000 miles (48,000 km)				●		
11	Spark plug replacement*					●†		
12	PCV valve inspection*					●		
13	EGR system serivce*					●		
14	Air cleaner and PCV filter replacement*	Every 36 months or 30,000 miles (48,000 km)				●		
15	Engine timing check*					●		
16	Spark plug wires and distributor inspection*	Every 30,000 miles (48,000 km)				●		
1	Fuel tank cap and lines inspection*					●		

* An emission control service.

† In California, these are the minimum emission control maintenance services an owner must perform according to the California Air Resources Board. General Motors, however, urges that all emission control maintenance services shown above be performed. To maintain your other new car warranties, all services shown in this folder should be performed.

Figure 8-8 Scheduled maintenance services. (Courtesy of Chevrolet Motor Division, General Motors Corporation)

tenance tune-up is performed but should be replaced on a routine basis, as suggested by the manufacturer's PM schedule. Also, during each maintenance tune-up, the distributor cap and rotor should be carefully inspected for cracks, corrosion, and signs of arcing.

When a filter gets dirty, it can not only restrict the flow of whatever it is cleaning, but can also allow some of the dirt it has trapped in it to flow out. To avoid the plugging up of filters, they should be replaced on a regular basis and should be included in a maintenance tune-up. The filters of concern are the fuel, air, and crankcase or PCV filter. Because of the wear that takes place throughout the engine and its systems, certain adjustments must be made to ensure that things are delivered at the correct time and in the correct amounts. When the points are replaced, the dwell must be set. Because the end of the dwell is also the firing of the spark plug, ignition timing must be adjusted to allow the spark to occur at the right time. The slight but continuous wear of the engine and its pistons causes a slight unsealing of the cylinders. As a result, engine vacuum decreases, which may call for readjustment of the carburetor. The correct balance of fuel and air is also adjusted in a tune-up, together with an adjustment of dwell and ignition timing.

Emission control devices and how well they operate affect the driveability of a vehicle. Certain services to these devices are recommended by the manufacturers at particular intervals. Maintenance of these systems will not only have a positive effect on how the engine runs, but will help keep the exhaust emission levels within specifications. These services are part of a tune-up.

SPECIFICATIONS

The key to performing a good maintenance tune-up is to follow the procedures for doing so. One of the most important procedures is the meeting of manufacturers' specifications. In a tune-up, always use the specifications for that engine. Tune-up specifications are available from a number of reliable sources. Often, the desired specifications (specs) are given in the owner's manual of each car. More accurate information and specifications are found on the manufacturer's decal, which is under the hood (Fig. 8–9). This decal lists all the specifications that are needed to bring the exhaust of the engine within the desired exhaust quality standards. If the car is equipped with such a decal, use the specifications given.

Another source of specifications supplied directly by the auto manufacturers is a shop manual (Fig. 8–10). These books contain not only the desired specs for an engine but also give excellent procedures for testing and repairing individual components of the engine. These factory shop manuals are the best source for correct procedures. They are also good sources for finding the specs needed for a tune-up. Typically, service businesses that work on all makes and models of cars will not have factory shop manuals for each type. New car dealers, on the other hand, should have a complete series of manuals for the cars they sell.

To accommodate those service facilities that work on a variety of automobiles, publishing companies assemble lists of specs for domestic and imported cars. These manuals also often include many procedures taken from the various manufacturers' manuals and thus are good sources for specs. Some of the better known companies that publish such manuals are Mitchell, Motor, and Chilton. Part suppliers often assemble a list of specs for all cars to give or sell to their customers. These lists tend to be easier to read and use to identify particular specs, but they also contain much less information than do the other types of manuals. Whatever the source of specs, it is important that you find the correct specs for the engine. Prior to finding the specs, make sure that you know what engine is being used and how it is equipped. You should know the size of the engine, the type of fuel system, and the accessories

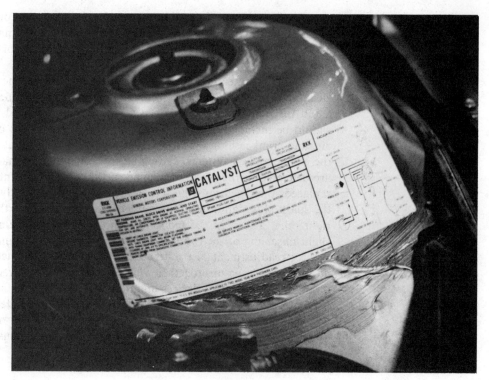

Figure 8-9

that are driven by the engine. Also, engine specs may vary with the type of transmission used.

Once you have gathered the appropriate specs for the engine on which you are working, you are ready to perform a maintenance tune-up. What follows is a description of the procedures involved in doing a good, thorough maintenance tune-up. The procedures will not be given for any particular engine unless noted and will not include the minor details of the procedure. If you experience difficulty in performing this type of tune-up, you should seek guidance from an experienced person such as your instructor, or you should follow the step-by-step procedures given in a manual.

Figure 8-10

INSTALLING POINTS AND CONDENSER

The first step is the proper installation of a new set of points and of a condenser. When installing a new set of points, you should wipe off the contact points with a rag to remove any oil that may be present. Often, manufacturers of points put oil on the contact surfaces to prevent rusting of the contacts after production. If this oil is not removed, it can cause unwanted resistance between the two contacts, which could limit the amount of current flow in the primary and therefore affect saturation. This oil will also be burned by the current flowing through the points and leave a coating behind. This coating offers resistance to primary current flow and is called *point frosting*. Frosting of points is caused not only by the oil protection but can also be caused by any oil or lubricant that happens to get on the contact surfaces. To prevent the frosting of points, wipe the contact surfaces well before installing them.

Points can be replaced with the distributor installed in the engine or with it removed. The preferred way is to remove the distributor; this allows you to do a better and quicker job and to properly inspect the rest of the distributor. To remove the distributor, you should first remove the distributor cap. The cap is fastened to the distributor body with screws, spring clips, or screwdriver lock clips. After the cap is loosened, it should be carefully placed to the side of the distributor. If possible, all the spark plug wires should remain attached to the cap. The position of the rotor and vacuum advance unit should be noted so that you can reinstall the distributor in the same position after you have serviced the distributor. You may find it advantageous to mark reference lines on the engine with a piece of chalk, to remind you of the positions of the rotor and advance unit. Next, remove the vacuum hose from the advance unit and the primary wire from the ignition coil. Then remove the distributor hold-down clamp so that the distributor can be lifted out. While removing some distributors, the distributor shaft may rotate slightly. The amount of this movement should be noted to make reinstallation easier. If the removal of the distributor is difficult, it may be the result of heavy deposits in the engine. Do not pry on the distributor to remove it; rather, carefully rock and turn it until it is free. The reinstallation of the distributor is the opposite of the removal procedure. With the rotor and vacuum advance unit positioned correctly, the ignition timing will be close to the correct timing to allow the engine to start.

To install a new set of breaker points, first remove the rotor. On distributors that have the centrifugal advance plates and springs above the points, the rotor is fastened by two screws. If the advance mechanisms are below the points, the rotor is simply pulled straight off the distributor shaft. The rotor should be inspected carefully before it is set aside.

With the rotor removed, the old set of points can now be removed. The primary coil wire and the condenser wire are attached to the points at the same terminal. This terminal may have a bolt and nut, a screw, or it may be held in place by spring tension. The wires should be disconnected and then the breaker-point hold-down screws removed, allowing for the removal of the points. If the condenser is not part of the point assembly, its hold-down screw should be loosened and the condenser removed. With the points and condenser removed, wipe the distributor cam clean. Installation of the new set of points and the condenser is the reverse of their removal.

Once the new set of points is installed, check to make sure that the contacts meet squarely when they are closed. If necessary, bend the stationary point bracket to provide for a full-face contact between the points. Use a point alignment tool or a pair of pliers to bend the bracket carefully. Never bend the movable contact arm or bracket. A bend in this arm can weaken it and cause the dwell to change as it flexes at the weak spot created by the bend. Full-faced alignment of the contacts is desirable because it provides for maximum continuity or conductivity from one point to another. This allows for a full current flow, which is necessary for coil saturation. Proper point alignment will also prevent the points from becoming pitted, as would be caused by current jumping the small gaps formed between the contacts due to poor alignment.

After the points have been aligned, the point gap needs to be set. The point gap should be set to the manufacturer's recommendations. To set the gap properly, the rubbing block should be resting on the high point of a lobe of the distributor cam (Fig. 8-11). The engine often needs to be cranked over slightly, to allow for this. With the rubbing block against the high point of the lobe, the points should be fully open. The point gap is the distance between the open contacts. A flat feeler gauge is used to measure the gap. Once the gap is set, make sure that the anchor screw for the contact point bracket is tight. Be careful not to overtighten this screw, but make sure that it is snug.

Lubricate the distributor cam with a film of light high-temperature grease. Usually, a new set of points is packaged with a small container of grease. Use this grease if supplied. The grease should be spread lightly all around the cam. The only place where the grease should be heavy is on the side of the rubbing block that the cam turns into. Extra grease at this point will allow the grease to be spread as the cam turns. Clean off excess grease from the cam and rubbing block. Excess grease could reach the points and cause them to frost or burn. Also, an excessive grease build up on the cam may affect the dwell.

With the point gap set and points aligned, the point spring tension should be adjusted. To do this properly, a point spring tension gauge should be used (Fig. 8-12). To adjust the tension on point assemblies with a slotted spring held in place with a nut, loosen the nut and move the spring toward the pivot to increase the tension or away from the pivot to decrease the tension. Some point assemblies have an adjustment at the pivot, which can change the point spring tension.

With the point replacement complete, it is time to install a new condenser. The replacement of a condenser is quite straightforward. Install the new in place of the old. With the old condenser out, it is wise to clean the surface of the breaker plate where the condenser sits. Dirt and grease at this mounting point can cause added resistance to the condenser circuit. When mounting the new condenser, make sure that the wires that run from the condenser to the points do not make direct

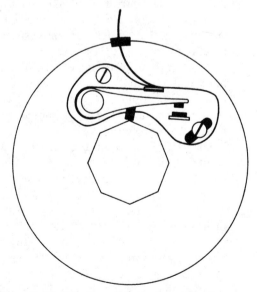

Figure 8-11 Breaker points ready to have the gap set.

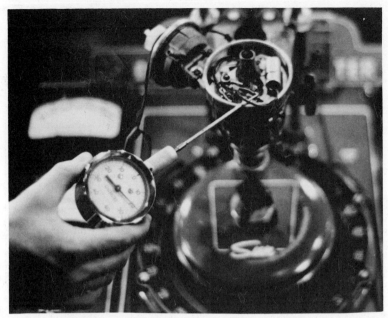

Figure 8-12 Spring tension gauge pulling the points open.

contact with the breaker plate (Fig. 8-13). Also make sure that the wire does not set in a place where it will be able to become pinched by the opening and closing of the points or by the distributor cap.

The distributor cap and rotor can now be put back on. Some manufacturers recommend that a small amount of silicone grease be put on the tip of the rotor. Check the shop manual. While doing so, the primary wires should be inspected for frays and cuts. Primary wires in poor shape can cause excessive resistance, which can reduce the maximum amount of current flow in the primary, which will affect the coil's output.

With the distributor assembled, hook up a dwell meter (Fig. 8-14) and run the engine at idle. Compare the dwell reading on the meter to the dwell specs and correct if necessary. To correct the dwell, the point gap must be adjusted. If the dwell is too long, the point gap is too narrow and needs to be widened. If the dwell is too short, the gap is too wide. Change the point gap in the same manner as you

Figure 8-13 It is important to make sure that the wires inside the distributor are neatly tucked inside the distributor so that the distributor cap does not pinch or cut them when the cap is installed.

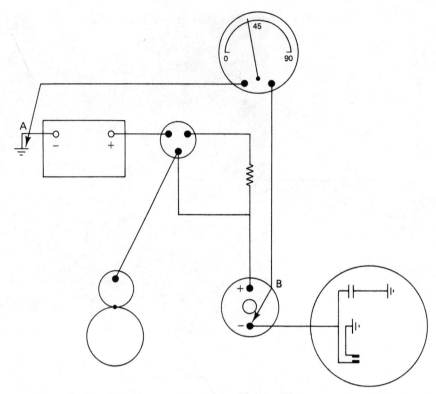

Figure 8-14 A dwell meter is hooked up between the negative terminal of the battery and the distributor (negative) side of the ignition coil.

originally set the gap. On General Motors V-8 and some six-cylinder engines, the point gap can be adjusted by inserting a hex wrench through a window or opening on the side of the distributor cap (Fig. 8-15). On Chryslers you should disconnect the vacuum advance hose when adjusting the dwell.

PRIMARY CIRCUITS OF ELECTRONIC IGNITIONS

On engines equipped with electronic ignitions, there are no periodic replacement items in the primary. During a maintenance tune-up it is good practice to inspect the primary circuit of the electronic ignitions. Certain manufacturers recommend that certain items be inspected, but for all types it is important to inspect all the primary wires and connectors carefully. Electronic ignition systems are very sensitive

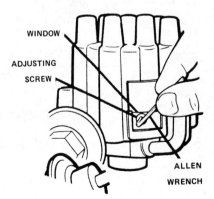

WINDOW

ADJUSTING
SCREW

ALLEN
WRENCH

Figure 8-15 Using an Allen wrench to adjust dwell on a distributor equipped with a window for adjustment. (Courtesy of Ford Motor Company)

to changes in current and voltage due to resistances. Any loose wires or corroded ends of the wires can cause the units to malfunction. Because these systems are so sensitive to resistance increases, American Motors, Chrysler, Ford, and many manufacturers of imports recommend that a coating of silicone dielectric compound be applied to the male and female connectors to prevent corrosion. A good maintenance tune-up should include applying this coating.

Together with inspecting the primary wiring, Chrysler recommends that the air gap between the magnetic pickup and the reluctor be checked. This adjustment is made through a means quite similar to adjusting the point gap. To check and adjust the gap, the pickup coil must be in line with a reluctor tooth (Fig. 8–16). Using a nonmagnetic flat feeler gauge, the gap should be set to 0.008 in. for all engines from 1972 to 1977, and set to 0.006 in. for 1978 and later engines. In 1980, Chrysler introduced a computer-controlled electronic ignition system that uses two pickups, one for starting and one for running. The gap of both of these pickups should be set to 0.006 in. It is extremely important that a nonmagnetic feeler gauge be used to check and adjust the air gap. A steel gauge will be attracted to the pickup coil because of its magnetic field, and this will result in an inaccurate reading.

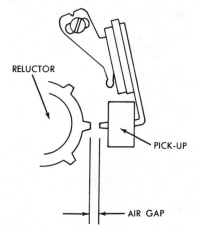

RELUCTOR

PICK-UP

AIR GAP

Figure 8–16 (Courtesy of Chrysler Corporation)

In 1980, Chrysler also introduced a Hall-effect type of ignition on some four-cylinder engines. This system does not have an adjustable air gap for the pickup. All that is required in a maintenance tune-up in the primary is an inspection of the primary wires at the distributor, the ballast resistor, and the electronic control unit.

Ford Motor Company has used basically the same system for electronic ignition since 1974. This system has been modified continuously, but the basic system remains intact. From 1974 through 1976, the system was called Solid State Ignition (SSI). In 1977, SSI was modified to allow for higher voltages and was called the Dura-Spark system. There have been three versions of the Dura-Spark system since 1977: Dura-Spark I, II, and III. The air gap of all Ford electronic ignitions is nonadjustable. Inspect all primary wires and connectors for corrosion, looseness, and damage. This inspection should include the tightness of the small ground tab inside the distributor from the wire harness. This tab is held to the breaker plate with a small screw that can easily loosen up from vibration (Fig. 8–17).

The High Energy Ignition system of General Motors has been installed on all GM engines since 1975. This system incorporates all of the primary circuit in the distributor housing on all engines with the exception of some four-cylinder engines. All HEI distributors contain the electronic control unit, the reluctor, and a pickup coil. Most distributors have a high-output coil mounted in a housing at the top of the distributor cap (Fig. 8–18). All Chevettes and some other four-cylinder models have the coil mounted externally to the engine. The air gap between the pickup coil

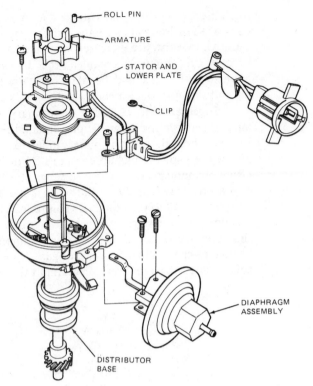

ROLL PIN

ARMATURE

STATOR AND
LOWER PLATE

CLIP

DIAPHRAGM
ASSEMBLY

DISTRIBUTOR
BASE

Figure 8-17 (Courtesy of Ford Motor Company)

and the reluctor is not adjustable. The only routine service to be done on HEI distributors is a careful inspection of the primary wires, advance plates, and connectors. The plates should be checked for rust, which can cause them to stick. If they are rusted, clean them and replace the rubber washer under the coil and coat it with silicone grease.

The electronic ignition systems used by imports are very similar in construction and design to those of the domestic manufacturers. In all cases the primary circuit should be inspected for corrosion, looseness, and damage. Most engines in Datsuns (Nissan), Hondas, and Toyotas are equipped with electronic ignition systems, which have adjustable air gaps for the pickup coil. Volkswagen uses a Hall-effect type of ignition system and the air gap of the pickup coil is nonadjustable. Refer to a good manual to identify whether or not the air gap is adjustable and to find out what the air gap should be if it is adjustable.

SECONDARY CIRCUITS

The secondary circuit of all ignitions is similar regardless of whether it is an electronic or a breaker-point system. There are two exceptions to this, and both are General Motors ignition systems. One exception is the HEI system, which has the ignition coil mounted in the distributor cap. The main difference between this system and other systems is the absence of a coil wire. All other parts of the secondary are like the others. The other exception was introduced in 1984 by GM on some engines; this system uses no distributor. In place of a distributor, it uses three computer-controlled ignition coils which are responsible for the firing of two cylinders each (Fig. 8-19). The distribution of the spark and the ignition timing is totally computer controlled. This system, like the HEI and all other secondary systems, uses spark plug wires and spark plugs. These two items are the main concern of a technician when performing a maintenance tune-up.

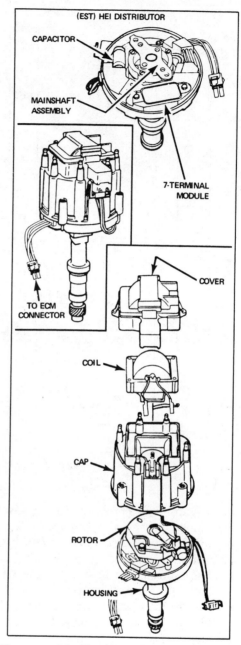

Figure 8-18 (Courtesy of Cadillac Motor Car Division, General Motors Corporation)

The secondary wires or spark plug wires have the responsibility of delivering the high voltage and current from the ignition coil to the spark plugs. The insulation of these wires must be capable of preventing the high voltage from leaking out. Remember that the voltage is pushing its way through the secondary, seeking a ground. The desired ground is across the gap of the spark plug. If the voltage is allowed to get through the insulation of the spark plug wire, the voltage may be able to find a ground at the engine. If this occurs, the spark plug will not fire and the engine would be performing without the power produced by that one cylinder. In a tune-up the condition of the spark plug wires should be checked. Visually inspect the wires for hardness, frays, cuts, and corrosion. If any abnormal condition is found, replace the wires. There are methods used to test the wires properly. These will be discussed later in this text. These methods are more appropriate during the

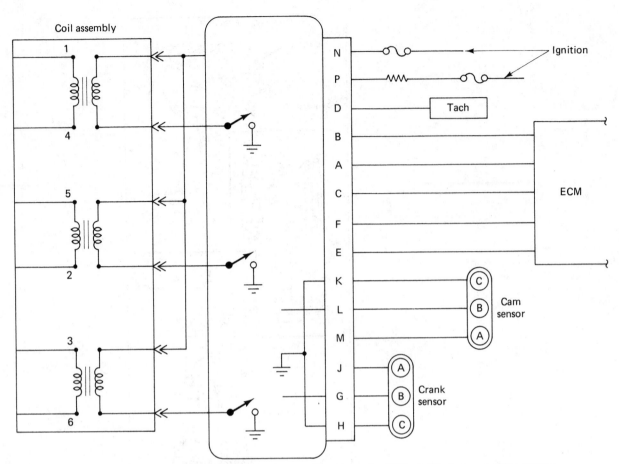

Figure 8-19 Basic system layout for GM's distributorless ignition system.

diagnostics of a problem. The purpose of a maintenance tune-up is to prevent engine running problems. A bad spark plug wire would cause one. The inspection of the wires may reveal a condition that may cause a performance problem later. When working with spark plug wires, be careful not to bend or pull on them. This can cause some internal damage to the wires. Spark plug wires should be removed from the spark plugs with a twist and a slight tug. The twist and tug should be on the boot of the wire, not on the wire itself. The twist is necessary to break the seal of the boot to the spark plug. If the seal is not broken, it is very difficult to remove the wires from the plugs.

If replacement of the spark plug wires is necessary, it is best to replace the entire set. To do so, you should locate the plug wire for the number 1 cylinder and find its location on the distributor cap. Check the manual you are using for specs to determine the firing order of the engine and the direction of distributor rotation. Often, the engine's firing order is printed on the intake manifold. The proper arrangement of the wires in the distributor cap follows the firing order of the engine (Fig. 8-20). Starting with cylinder 1 and moving in the direction of distributor rotation, place the appropriate spark plug wire in the cap. Keep in mind that the cylinder numbers used in the firing order refer to the location or cylinder arrangement of the engine. After the plug wires are installed in the cap, fasten the other end of each wire to the appropriate spark plug.

This procedure for replacing spark plug wires applies to most engines. An exception to this is some Ford engines equipped with the Electronic Engine Control (EEC) system. Because of the configuration of the rotor and the matching distrib-

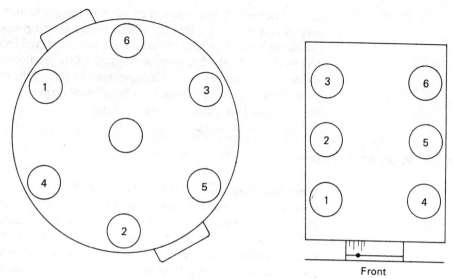

Figure 8-20

utor cap, the spark plug wires are not installed in the cap according to the engine's firing order. As shown in Fig. 8-21, the wiring order is a variation of the firing order. When working on an EEC-equipped engine, check the manual for the proper wiring sequence.

Most engines are equipped with spark plug wire hold-down brackets. These should be used to prevent the wire from contacting highly heated parts of the engine, which could melt the insulation. These brackets also are used to separate the wires. When current flows through a conductor, a magnetic field is formed around that conductor. Also, as a magnetic field passes through a conductor, a voltage is induced in that conductor. For these reasons it is important to route the spark plug wires so as to separate the wires of cylinders that are located next to each other and are consecutive in the firing order. Doing this prevents the firing of one cylinder being caused by the firing of another. This occurrence is called *induction cross-firing*. If the engine is not equipped with a separator bracket, the wires of those cylinders should be twisted around each other to prevent the wires from being parallel to each other for any distance. This is typically a problem on V-8's. The cylinders of concern on Ford engines are numbers 7 and 8; on some Chrysler, Pontiac, and Chevrolet engines are cylinders 5 and 7; and on some Oldsmobile engines are cylinders 2 and 4.

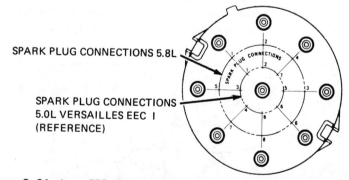

SPARK PLUG CONNECTIONS 5.8L

SPARK PLUG CONNECTIONS 5.0L VERSAILLES EEC I (REFERENCE)

Figure 8-21 In an EEC distributor, upper- and lower-level electrodes on the rotor fire alternately in a pattern that jumps from one side of the cap to the other. The correct firing order for 5.8-liter engines is 1-3-7-2-6-5-4-8. Compare this sequence to the wiring sequence on the distributor cap. (Courtesy of Ford Motor Company)

When reinstalling spark plug wires to the distributor cap or to the spark plugs, you should make sure that the terminals are seated properly. You should also give the boots a pinch or squeeze after they are connected to the terminals. This expels any moisture or air that may be trapped under the boot. Some manufacturers recommend that silicone dielectric compound be put on the ends and terminals of spark plug wires to prevent corrosion. Check your manual to see if this is recommended on the engine on which you are working.

SPARK PLUG REPLACEMENT

When installing spark plugs, make sure that the spark plugs to be installed are of the correct type and heat range for the engine make, model, and year on which you are working. Most spark plugs are pregapped at the factory but should be checked to make sure that they are at the specs recommended by the manufacturer. To measure the distance between the center and ground electrodes, it is recommended that you use a round wire gap gauge (Fig. 8-22). If the gap is too narrow or too wide, the ground electrode needs to be bent to achieve the desired gap. Care should be taken not to weaken or distort the ground electrode while adjusting the gap. The wire gauge should pass through the gap with a slight drag if the gap is correct. The ground electrode should be parallel to the base of the center electrode. Carefully twist the ground electrode to accomplish this alignment.

Newer engines require wider gaps than do older engines. To provide for this gap, manufacturers supply spark plugs with wider production gaps. Never attempt to adjust the gap of a spark plug more than 0.010 in. from the production gap. This causes an improper gap angle, which can increase center electrode wear and possible misfiring of the spark plug. Spark plugs should be replaced as a set.

To remove the spark plugs properly, carefully remove the spark plug wires from the spark plugs with a twist and a slight pull. With the correct spark plug socket wrench, apply steady pressure until the plug loosens. If bursts of pressure are applied, the plug may break upon removal. If the plug is very difficult to loosen, apply some penetrating fluid around the threads. Once the plugs are lossened, blow or brush the area around the plug, and clean to prevent any dirt from entering the cylinder when the plug is removed. After the plugs are removed, wipe the threads in the cylinder head with a soft rag. While doing this, be careful not to allow dirt to enter the cylinder. Anytime the plugs will be out of the engine for a long period, you should cover the holes with a rag until the plugs are reinstalled.

It is good practice to place the plugs that are removed in the order of cylinder location. The inspection of spark plugs is an excellent diagnostic technique, and knowing which cylinder a particular plug came out of will aid in the diagnosis of the engine. Spark plug diagnosis is discussed in Chapter 13.

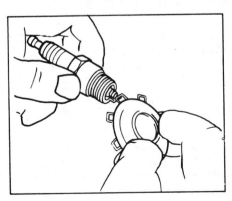

Figure 8-22   (Courtesy of Ford Motor Company)

When installing new spark plugs, you should make sure that the gaps are correct and should wipe off the threads of the new plugs to remove any dirt or metal filings that may have collected on them since they were packaged. Install the plugs by hand until they are tight. Be careful not to get the plug dirty and not to hit the ground electrode on anything. Once the plug is fully seated, use the appropriate socket wrench and tighten the plug to its proper torque value. Installing the plugs to their proper torque is an important step that is often ignored by persons doing a tune-up. There is a common tendency to overtighten spark plugs. Not only does this make it more difficult to remove the plugs, but it can damage the plugs or affect their operation.

Overtightening a plug can also cause the seal between the insulator and the shell of the plug to be damaged. Compressed gases in the cylinder can leak out of the cylinder through this broken seal (Fig. 8–23). This causes a compression leak, which reduces the power output of the cylinder. When this occurs, black streaks will appear up the side of the plug. A spark plug is a plug that allows electricity to flow into the combustion chamber. Improper torque can cause the spark plug not to be able to plug the hole in the cylinders.

To determine the proper torque spec for an engine, you must know the size (the diameter) of the plug, the type of metal the cylinder head is made of, and whether the plug is a tapered seat or gasket type. Most automobile engines use 14- or 18-mm plugs. Cylinder heads are made of iron or aluminum. If you do not know if a head is aluminum, an easy check can be made. If a magnet sticks to the head, it is an iron head. If it does not stick, the head is made of aluminum. If a spark plug has a washer fitted to it above the threaded portion, it is a gasket-type plug. This washer or gasket seals the plug into the hole. Plugs that do not use a gasket have a taper above the threads that seats into the cylinder head to form a seal. Aluminum is much softer than iron and therefore the plugs in an aluminum head are not tightened as much as the plugs in an iron head. Overtightening in an aluminum head can actually pull the threads out from the head. For this reason, plugs should never be removed or installed in a hot aluminum cylinder head.

Gasket-type plugs require more tightening than the tapered type, because the gasket must be crushed in order to form a good seal. Use a spark plug torque wrench whenever possible to tighten the plugs into the head (Fig. 8–24). If a spark plug

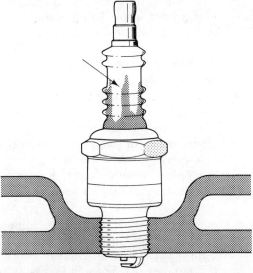

Figure 8–23 The arrow is pointing to the streaks formed on the insulator of the plug by the compressed mixture that escapes past a broken seal in the spark plug caused by overtightening.

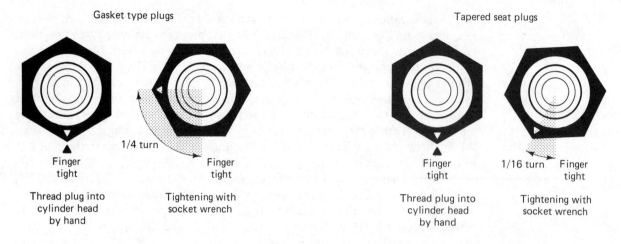

Figure 8-24 Torque recommendations. (Courtesy of Champion Spark Plug Company)

Spark plug Thread size	Cast-iron heads		Aluminum heads	
	w/torque wrench (lb-ft)	w/o torque wrench (turn)	w/torque wrench (lb-ft)	w/o torque wrench (turn)
Gasket type (mm)				
10	8–12	1/4	8–12	1/4
12	10–18	1/4	10–18	1/4
14	26–30	1/4 to 3/8	18–22*	1/4
18	32–38	1/4	28–34	1/4
Tapered seat (mm)				
14	7–15	1/16 (snug)	7–15	1/16 (snug)
18	15–20	1/16 (snug)	15–20	1/16 (snug)

*Mazda rotary engine requires only 8 to 13 lb-ft (1/4 turn).

torque wrench is not available, proper torque can be accomplished with a spark plug wrench. To do this, tighten gasket-type plugs with a quarter (¼) turn of the wrench after the plug is finger-tight. Tighten taper-type plugs ¹⁄₁₆ of a turn after the plug is finger-tight. This procedure may not bring the tightening exactly to the correct specs, but will allow you to install the plugs at a safe torque value.

During a tune-up, the rest of the secondary should be inspected. This includes the distributor cap and the rotor. These should be checked for cracks, burn spots, and corrosion. If any abnormal condition exists, these should be replaced. It is wise to replace both the rotor and cap if one needs to be replaced. When installing a new rotor, make sure that it lines up with the notch or notches on the distributor shaft before it is tightened completely. Some rotors are held in place by screws, and others merely fit snugly over the distributor shaft. GM engines usually have screw hold-downs for the rotor, because the rotor also serves as a shield or cover for the mechanical advance weights and springs which are mounted on top of the distributor shaft. These rotors have one round and one square alignment peg to ensure proper alignment of the rotor (Fig. 8-25). Before tightening the hold-down screws, make sure that the pegs are seated in the correct hole.

Some Ford engines are equipped with a version of their Electronic Engine Control (EEC) system. This version uses a rotor that must be aligned properly when it is installed. As shown in Fig. 8-26, the EEC rotor has two pickup arms and two electrode tips. The arms and tips are on two different levels, and so are the distributor cap electrodes, to provide a match with the rotor. The rotor does not fit

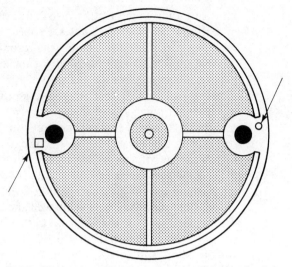

Figure 8–25   The square and round pegs of a GM-type distributor rotor.

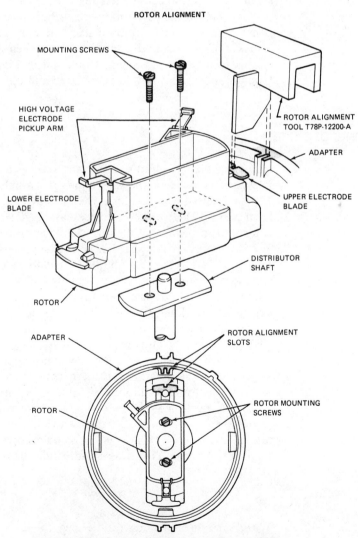

Figure 8–26   (Courtesy of Ford Motor Company)

into a locating notch on the distributor shaft. Instead, it is held in place by two mounting screws in slotted holes. Proper alignment requires the use of an alignment tool. The procedure for aligning the rotor is shown in Fig. 8–26 and should also be in your shop manuals. Distributor caps are also held in place by screws on some engines. Other engines use clamp-type springs to hold the cap snugly onto the body of the distributor. Before snugging the distributor cap down, make sure that the cap does not pinch any of the primary wires. Typically, there is an alignment notch on the cap to ensure that it is mounted in the correct position. Distributor caps and rotors should be replaced on a regular basis, but not as frequently as the spark plugs. Check the manufacturers' replacement intervals for their replacement.

A maintenance tune-up typically includes replacement of the PCV valve, the crankcase filter, the air filter, and the fuel filter. Periodic replacement of these items ensures that they will do their job well and not hinder the performance of the engine. The replacement of these items is rather simple. Care needs to be taken to install each of them fully into their housing. It is good practice to wipe down the housings for the filters and valves prior to installing new ones. Extra care should be taken when replacing the fuel filter. When the old filter is removed, there will be some spillage of fuel. Remember the safety rules and clean up the fuel as soon as possible. Make sure that the new filter is properly sealed and that it does not leak. Even a small fuel leak can cause a terrible fire.

So far, all the parts that need to be replaced and inspected during a maintenance tune-up have been discussed. The remainder of the tune-up includes some necessary checks and adjustments. Often when new parts are installed, the engine idle speed, idle air-and-fuel mixture, and ignition timing change. To complete the tune-up, these items should be checked and adjustments should be made if necessary.

ADJUSTING IDLE SPEED

Proper engine idle speed is important to the general operation of the engine. If the idle speed is too slow, the engine may run rough and tend to stall or quit running. If the idle speed is set too high, the car may tend to keep running when the ignition is turned off. This is often referred to as *dieseling*. The idle speed adjustments merely control the amount that the throttle plate is open at idle speed.

To adjust the idle speed of an engine properly, the engine must be at operating temperature. The spec for idle speed usually includes some conditions that must be met. For example, cars with automatic transmissions often have a recommended idle speed with the transmission in "drive." The transmission must be put into drive before any readings are taken or adjustments made. Often these conditions for proper idle speed include the plugging of some hoses or the disconnection of some electrical wires, which is usually the case with computerized ignitions. These conditions must be met to have the engine at the proper idle speed.

A tachometer is hooked to the engine to measure the engine's speed. This meter is usually hooked to the distributor lead on the coil and to a ground (Fig. 8–27). Some electronic ignition systems require a different hook-up, and this will be stated in the shop manual. Adjust the engine's idle speed to the specs, using the idle adjusting screw (Fig. 8–28) or the idle stop solenoid (Fig. 8–29). Refer to your manual to find the exact way of adjusting the idle speed of a particular engine.

IGNITION TIMING

With the engine at its desired idle spec, the ignition timing can now be checked. To check the timing of the ignition, a timing light must be used. A timing light is powered by the battery of the car and flashes a light each time cylinder 1 fires. The

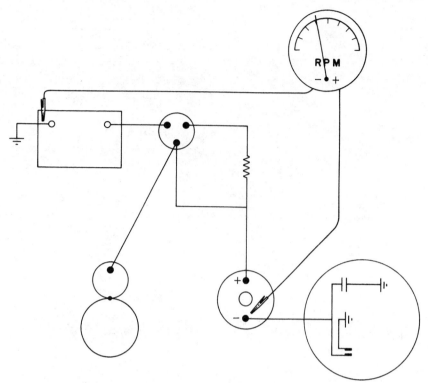

Figure 8-27 Tachometer hook-up is the same as the hook-up for a dwell meter.

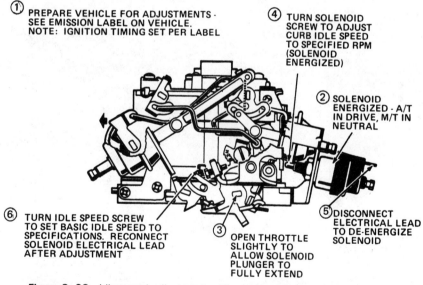

① PREPARE VEHICLE FOR ADJUSTMENTS - SEE EMISSION LABEL ON VEHICLE. NOTE: IGNITION TIMING SET PER LABEL

④ TURN SOLENOID SCREW TO ADJUST CURB IDLE SPEED TO SPECIFIED RPM (SOLENOID ENERGIZED)

② SOLENOID ENERGIZED - A/T IN DRIVE, M/T IN NEUTRAL

⑥ TURN IDLE SPEED SCREW TO SET BASIC IDLE SPEED TO SPECIFICATIONS. RECONNECT SOLENOID ELECTRICAL LEAD AFTER ADJUSTMENT

③ OPEN THROTTLE SLIGHTLY TO ALLOW SOLENOID PLUNGER TO FULLY EXTEND

⑤ DISCONNECT ELECTRICAL LEAD TO DE-ENERGIZE SOLENOID

Figure 8-28 Idle speed adjustments. (Courtesy of Cadillac Motor Car Division, General Motors Corporation)

typical hook-up for a timing light to an engine is shown in Fig. 8-30. This light is pointed to the crankshaft pulley, which will have timing marks inscribed in it. By rotating the body of the distributor, the ignition timing will advance or retard, depending on the direction of the rotation.

The procedure for checking and adjusting the ignition timing for most engines is as follows:

1. Check the under-the-hood decal or the shop manual for the timing specs of the engine. These specs will include the degrees of timing, the desired speed

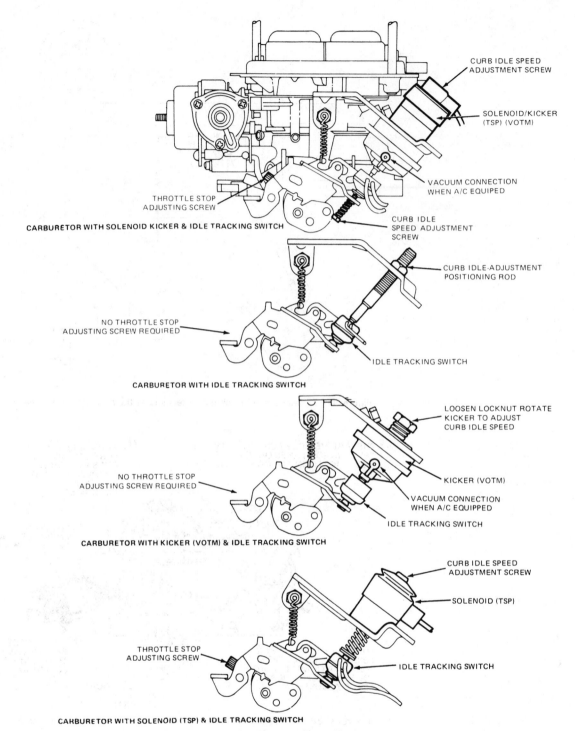

CURB IDLE SPEED
ADJUSTMENT SCREW

SOLENOID/KICKER
(TSP) (VOTM)

VACUUM CONNECTION
WHEN A/C EQUIPED

THROTTLE STOP
ADJUSTING SCREW

CURB IDLE
SPEED ADJUSTMENT
SCREW

CARBURETOR WITH SOLENOID KICKER & IDLE TRACKING SWITCH

CURB IDLE-ADJUSTMENT
POSITIONING ROD

NO THROTTLE STOP
ADJUSTING SCREW REQUIRED

IDLE TRACKING SWITCH

CARBURETOR WITH IDLE TRACKING SWITCH

LOOSEN LOCKNUT ROTATE
KICKER TO ADJUST
CURB IDLE SPEED

NO THROTTLE STOP
ADJUSTING SCREW REQUIRED

KICKER (VOTM)

VACUUM CONNECTION
WHEN A/C EQUIPPED

IDLE TRACKING SWITCH

CARBURETOR WITH KICKER (VOTM) & IDLE TRACKING SWITCH

CURB IDLE SPEED
ADJUSTMENT SCREW

SOLENOID (TSP)

THROTTLE STOP
ADJUSTING SCREW

IDLE TRACKING SWITCH

CARBURETOR WITH SOLENOID (TSP) & IDLE TRACKING SWITCH

Figure 8-29   (Courtesy of Ford Motor Company)

of the engine, and the conditions in which the timing should be checked. Often, the conditions include the disconnection of the vacuum advance and the plugging of the vacuum line; or disconnecting an electrical terminal (On GM cars with CCC, unplug the four-wire connector from the distributor.)

2. Using the manual or decal, identify the markings on the crankshaft pulley and the pointer installed on the engine for timing purposes. It is often helpful to put a chalk mark on the pulley to make it easier to see while checking the timing.

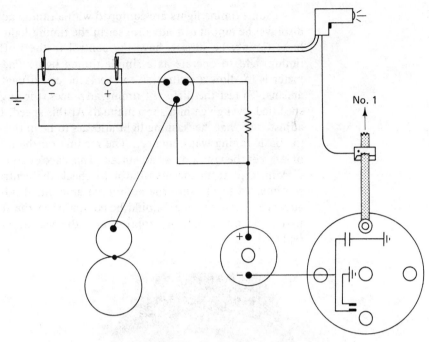

Figure 8–30 Timing light hook-up.

3. Connect the timing light to the battery and to the number 1 spark plug wire. Always follow the hook-up procedure given by the manufacturer of the timing light.

4. Set the engine to the required engine speed and meet the other conditions required for setting basic or initial timing.

5. Point the timing light so that it flashes at the timing marks. The light should flash and show an alignment between the timing marks and the pointer (Fig. 8–31). If it does, the timing is correct. If the marks do not line up, the timing must be corrected.

6. To correct the timing, loosen the hold-down bolt on the body of the distributor and rotate the distributor until the marks line up. This should not require much movement, so rotate the distributor a little at a time.

7. When the marks line up, tighten the hold-down bolt or clamp and recheck the timing. If it is not correct, the distributor moved when you were tightening it down. Loosen it again and readjust until the marks stay lined up.

8. When completed, reconnect the hoses and readjust the idle speed, if necessary.

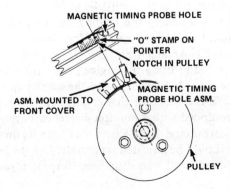

Figure 8–31 Magnetic timing probe hole assembly with timing marks and probe hole. (Courtesy of Cadillac Motor Car Division, General Motors Corporation)

Some timing lights are equipped with a timing advance control and meter that displays the amount of advance set in the timing light (Fig. 8–32). When adjusting basic timing, be sure to have the control in the "off" position. This allows the timing light to operate as a simple timing light. The purpose of the control and meter is to allow a technician to check the centrifugal and vacuum advance mechanisms. To test the different timing advances, the engine speed is increased to the specified level given in a shop manual. At this speed, the timing advance control is adjusted so that the flashing light appears to be in the same position as it was while the basic timing was being set. The reading on the meter will indicate the number of degrees the ignition has advanced. This check can be done with the vacuum hose disconnected from the distributor to check the centrifugal advance and the hose reconnected to measure the additional amount of advance caused by the vacuum advance. These readings should be compared to the specs given for the engine and if they do not agree, the appropriate advance mechanism should be repaired or replaced.

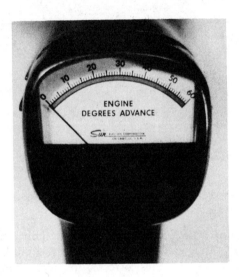

Figure 8–32

ADJUSTMENT OF IDLE MIXTURE

After the engine idle speed has been adjusted and the ignition timing is correct, the idle mixture can be adjusted. On newer engines this cannot be done because there may be no idle mixture screws or they may be sealed to prevent adjustment. Engines equipped with idle mixture screws should be adjusted to the proper mixture (Fig. 8–33). There are many techniques that can be used to adjust the idle mixture properly. The lean-drop and propane enrichment method will be discussed here. These are the more commonly used and recommended methods.

Using the *lean-drop method* requires only a tachometer and a screwdriver. To adjust the idle mixture, the engine must be at its operating temperature. With the engine idling and you watching the reading on the tachometer, turn the mixture screw clockwise until the engine begins to slow down. Now, slowly turn the screw counterclockwise until you reach the highest steady engine rpm. The idle speed screw should now be adjusted to bring the engine's idle speed to specs. The idle mixture screws will now be turned clockwise until the engine rpm drops to the lean-drop engine speed given in the specs for that engine.

On engines equipped with a multiple-barrel carburetor, each barrel should be treated as a single carburetor. Each barrel may have its own adjustment for the idle mixture, and each should be adjusted separately. Use the lean-drop method at each mixture screw. If the desired lean drop was 30 rpm, adjust each mixture screw to

Figure 8-33   Carburetor removed from engine to show placement of idle mixture screw.

cause a drop of 15 rpm. This should give a total lean drop of 30 rpm after both mixture screws are adjusted.

The *propane enrichment method* is recommended for carburetors whose normal idle mixture is too lean to use the lean-drop method. By using a special metering valve attached to a propane torch, propane is directed into the carburetor through the air cleaner or a vacuum port. The propane will enrich the idle mixture during adjustment. Shop manuals give the exact procedure for the adjustment of a carburetor using the propane enrichment method for a given engine. Following are some general guidelines for using this method.

1. Disconnect the PCV inlet hose to the air cleaner and other hoses as specified by the manufacturer.
2. Set the parking brake.
3. Run the engine at fast idle until normal operating temperature is reached.
4. Adjust the idle speed to the normal slow idle rpm or specified idle-mixture-adjusting rpm.
5. Place the transmission in neutral or drive, as specified.
6. Insert the adapter of the propane hose into the specified air cleaner connection (Fig. 8–34).
7. Make sure that the propane metering valve is closed and then fully open the main propane bottle valve.
8. Hold the propane bottle upright and slowly open the metering valve until the engine reaches its highest idle speed.
9. Compare this engine speed to the specs given for idle rpm gain.
   a. If the rpm gain noted is greater than the specified gain, the carburetor mixture is too lean.
   b. If the rpm gain is less than the specified amount, the mixture is set too rich.
10. Remove the propane bottle hook-up.
11. Turn the mixture screws slowly until the rpm changes by the exact same amount

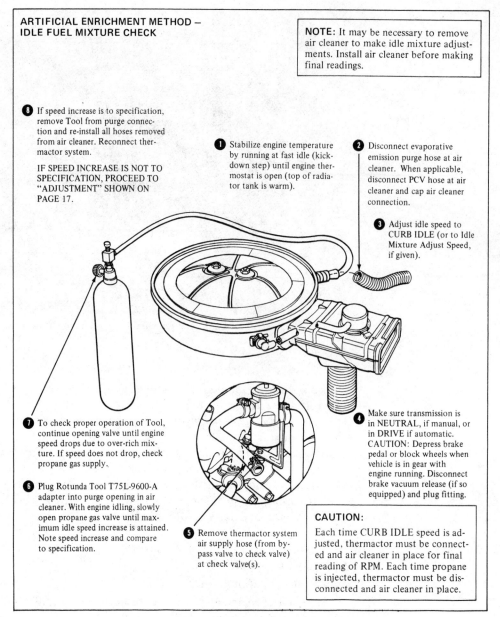

**ARTIFICIAL ENRICHMENT METHOD —
IDLE FUEL MIXTURE CHECK**

NOTE: It may be necessary to remove air cleaner to make idle mixture adjustments. Install air cleaner before making final readings.

8 If speed increase is to specification, remove Tool from purge connection and re-install all hoses removed from air cleaner. Reconnect thermactor system.

IF SPEED INCREASE IS NOT TO SPECIFICATION, PROCEED TO "ADJUSTMENT" SHOWN ON PAGE 17.

1 Stabilize engine temperature by running at fast idle (kickdown step) until engine thermostat is open (top of radiator tank is warm).

2 Disconnect evaporative emission purge hose at air cleaner. When applicable, disconnect PCV hose at air cleaner and cap air cleaner connection.

3 Adjust idle speed to CURB IDLE (or to Idle Mixture Adjust Speed, if given).

7 To check proper operation of Tool, continue opening valve until engine speed drops due to over-rich mixture. If speed does not drop, check propane gas supply.

6 Plug Rotunda Tool T75L-9600-A adapter into purge opening in air cleaner. With engine idling, slowly open propane gas valve until maximum idle speed increase is attained. Note speed increase and compare to specification.

4 Make sure transmission is in NEUTRAL, if manual, or in DRIVE if automatic. CAUTION: Depress brake pedal or block wheels when vehicle is in gear with engine running. Disconnect brake vacuum release (if so equipped) and plug fitting.

5 Remove thermactor system air supply hose (from bypass valve to check valve) at check valve(s).

CAUTION:

Each time CURB IDLE speed is adjusted, thermactor must be connected and air cleaner in place for final reading of RPM. Each time propane is injected, thermactor must be disconnected and air cleaner in place.

Figure 8–34   (Courtesy of Ford Motor Company)

that the measured idle speed gain was greater or less than the gain rpm specification.

12. Readjust the idle speed screw to obtain the proper idle speed.
13. Using a propane bottle hook-up, recheck the idle mixture.

EMISSION CONTROL SYSTEM MAINTENANCE

Very much part of a maintenance tune-up are the inspection and servicing of the engine's emission control devices. The frequency for services to these devices is given in a shop manual and may not be required with each tune-up. However, with each tune-up, a complete inspection of the PCV, AIS, EGR, and EEC systems should be included. The PCV hoses should be inspected for cracks, restrictions, and deterioration during each tune-up. Unless the PCV valve is defective, it needs to be replaced only at the intervals recommended by the manufacturer. If the engine is equipped with a PCV or crankcase filter, the recommendation for the replacement of the filter usually corresponds with the replacement of the PCV valve. Like other filters, if the PCV filter becomes excessively dirty, it should be replaced ahead of the recommended time interval. The hoses and connections for the AIS (Air Injection System) should also be routinely inspected. The belt that drives the air pump should be inspected and its tension checked.

EGR system maintenance consists of inspecting the hoses for deterioration, inspecting the orifices and passageways for carbon buildup, and checking the vacuum signal to the valve. Faulty components are typically replaced. Some manufacturers recommend that you clean the EGR passageways if they become filled with carbon, while others recommend that the valve be replaced. Inspection of the system should be included with each tune-up.

The EEC (Evaporative Emission Control) system should have all its components and hoses inspected at each tune-up. One of its components is the charcoal cannister, which should be inspected for cracks and fuel leakage. Inside the cannister, on most vehicles, is a replaceable filter. This filter is typically the most ignored filter on a car. Seldom do technicians replace them, even though automobile manufacturers recommend that they be replaced at a certain interval.

TUNE-UP TIPS AND SUMMARY

During the execution of the tune-up you have looked at a number of systems and parts of the engine. Studying and thinking about the condition of these parts will help you master the most difficult tune-up; the diagnostic tune-up. The condition of some of these parts may give you some insight into how well other systems of the car are performing.

One of the parts worthy of detailed inspection is the ignition breaker points. The color of the contacts of the points can tell you much about the primary ignition system and the engine's charging system. If the contacts are a dull blue and black color, it indicates that excessive current has been flowing through them. This is commonly caused by a high-voltage regulator setting; low resistance in the primary, which can be caused by a defective ballast resistor bypass circuit; using the wrong type of ignition coil; or simply by many short trips during cold weather. If the points have a solid black appearance, it indicates that low current has been flowing across them. The common causes of this are too short a dwell time or excessive resistance in the primary circuit, which can be due to grease and oil on the contacts, excessive resistance in the condenser circuit, or poor or loose connections at the contacts.

While working inside the distributor, the vacuum and mechanical advance systems can be quickly checked. With your hand you should be able to move the distributor cam against the advance springs. If there is movement you know that the springs and weights are free and are allowing for advance to take place. When you release the distributor shaft, it should snap back. This indicates that the spring tension is great enough to allow the timing to return to normal. This same technique can be applied to the vacuum advance. To check this, move the breaker plate assembly. There should be movement, and it should also snap back in place when released. These are quick ways to check the advance mechanisms in the distributor. These checks do not determine if the advances are within specs; they only determine if the systems are operational.

While conducting a tune-up, a quick under-the-hood inspection should also be included. Often, the belts and hoses of the engine are ignored. It takes only minutes to check them and completes a total care package for the engine.

REVIEW QUESTIONS

1. What is the purpose of a maintenance tune-up?
2. What happens to a set of points as its rubbing block begins to wear?
3. Why should the condenser be replaced at the same time as the points?
4. What should be known before beginning a tune-up?
5. When installing a new set of points, what five things should be done?
6. Why is a coating of silicone dielectric compound on electronic ignition connectors recommended by many automobile manufacturers?
7. What is the correct procedure for removing spark plug wires from the spark plug terminals?
8. Why are spark plug wire separation brackets used on engines?
9. Why should a round wire gauge be used to measure the gap of a spark plug?
10. Why should spark plugs be installed and tightened to the proper torque setting?
11. Why is it extremely important to set the engine to the conditions outlined in the manual, prior to setting the ignition timing?
12. What is the proper procedure for adjusting the idle mixture using the lean-drop method?

9

FUNDAMENTALS OF TROUBLESHOOTING

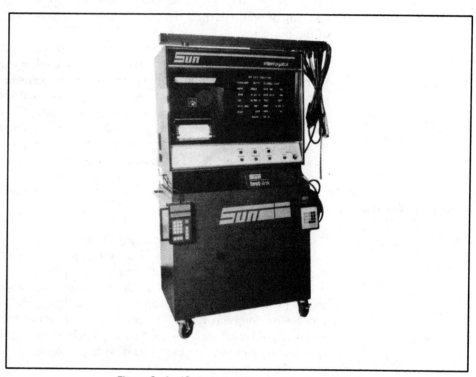

Figure 9-1 (Courtesy of Sun Electric Corporation)

In a maintenance tune-up, parts are replaced that typically wear out or get dirty. The failure of these parts can cause engine starting or running problems. The usual approach in a tune-up is to replace those parts and hope that if a problem exists, the replacement of these parts will correct it. Unfortunately, this is not always the case. A maintenance tune-up should be performed as a part of preventive maintenance, to prevent problems. When a problem exists, the cause of that problem should be found and corrected before continuing with the replacement of parts!

DIAGNOSTIC TUNE-UP

A diagnostic tune-up is the sort of tune-up that should be performed when an engine problem exists. The entire engine is analyzed and tested to identify the system that is failing. Then that system is tested to identify the part or parts that have failed. There are many systems and parts involved in the starting and running of the engine.

While conducting a diagnostic tune-up, it is ridiculous to think that all systems and parts are tested to identify the problem. Only the system that can cause the problem is tested and only the parts in that system that would cause the engine to behave the way that it does will be tested. To make these types of decisions, it should be obvious that an understanding of the systems is necessary. From this understanding, a systematic and logical approach to determining the cause will result. This logical approach is even more important today than it was a few years ago.

The technology of cars today is advancing at a rate never before witnessed in the automotive industry. The expanded use of electronics and computers in cars has increased the need for systematic diagnosis. Computers monitor the operation and results from a variety of systems, and use this information to control these systems as well as others (Fig. 9-2). The result of this integration is the less obvious separation of the systems of the engine. The best tool for diagnostics, especially with computer systems, is precise and logical problem-solving techniques. An understanding of how a particular system will affect another will aid in locating the cause of the problem.

By understanding what the problem is and what can cause the problem, the solution to the problem and the method to determine the solution become much simpler. An understanding of the problem always helps in solving the problem. To gain this understanding it is not necessary to know what the problem is! A problem exists when there is a difference in the way something is operating from the way it should be operating. When there is an engine problem, something about the way it runs, feels, or sounds makes one aware that there is a problem. Thinking about the way an engine runs, feels, and sounds leads to an understanding of the problem.

LOGICAL APPROACH

There is a logical approach to gaining this understanding. It starts with listening to the complaint of the driver (Fig. 9-3). Find out as much as possible about the problem. Find out what happens and when it happens! Does the engine start poorly when it is cold, hot, or all the time? Does the engine run rough when it is cold, hot, or all the time? Does the engine run rough when it is at idle, on the freeway, or all the time? How long has it had this problem? The answers to these questions can be very valuable information when you are troubleshooting. Knowing when the problem occurs, you will be able to start sorting through the various systems that can cause the problem. For example, if the engine runs rough only on freeways, it is very unlikely that the problem is a lean idle mixture. The engine is not running at idle at freeway speeds. However, the engine is on the main metering circuit of the carburetor, and that could be the problem. Emphasis here is on the word "could"! It is best never to approach a problem with one possible cause in mind. Much time could be spent testing and playing around with the carburetor only later to find out that the problem is in the ignition system! Ignition problems could also cause the engine to run rough only at freeway speeds. In fact, there are many things that can cause a problem only at high speeds. From the customer's information you should eliminate only the unlikely causes!

VERIFY THE COMPLAINT

The next step in this systematic approach at troubleshooting is to verify the complaint. Attempt to experience the problem as stated by the customer (Fig. 9-4). Take notice of all things that happen. Often you will be able to discover other things that go wrong with the engine that the driver does not realize. Often, the complaint is

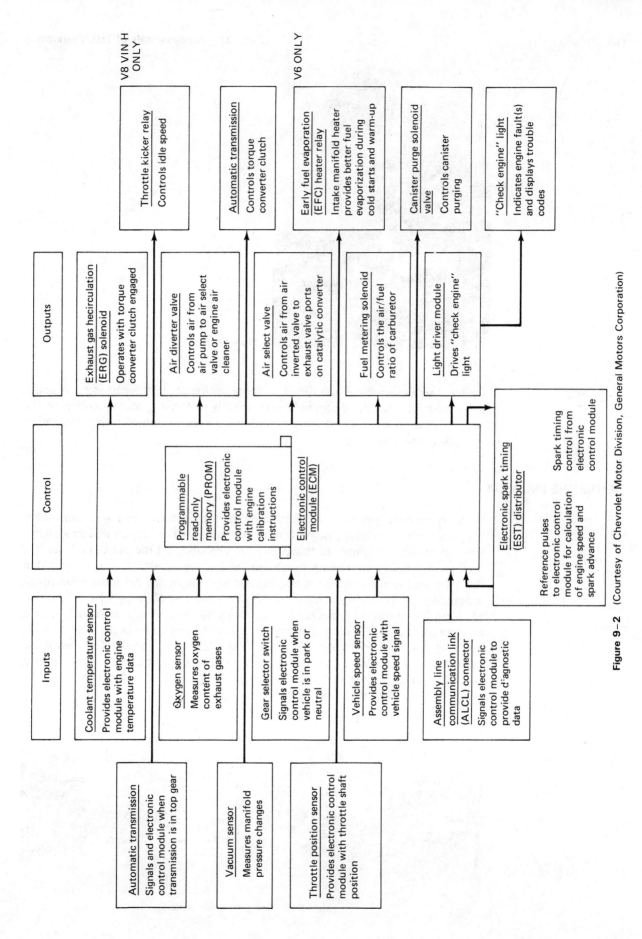

Figure 9–2 (Courtesy of Chevrolet Motor Division, General Motors Corporation)

Figure 9-3

the most serious problem and therefore more noticeable to the customer. Attempt to understand what the customer's complaint is and pay special attention to the engine's operation at other driving conditions. All the information that you gather will make your task simpler.

DEFINE THE PROBLEM

With this information you can start to define the problem. You will know what happens and when it happens. With an understanding of the engine and its systems, you can further determine the things that are probably not the cause. Let's use the example of the engine running rough only at freeway speeds. If you noticed that there was also a hesitation upon acceleration and the engine did indeed run rough at high engine all speeds, you have a better understanding of the problem. The cause of the problem is probably not the accelerator pump, because it operates only during an increase of speed. Similarly, the problem is not likely to be in the float or main metering system. These systems operate during all engine speeds. The conclusion can be that the problem is not likely to be related to the carburetor. Knowing what is not the cause can help in determining what is the cause. This happens through a simple process of elimination.

Figure 9-4 Going for a test drive.

ISOLATE THE PROBLEM

With an understanding of what the problem really is and an understanding of the engine and its systems, you can begin to isolate the cause of the problem. Isolating the problem involves knowing which systems can cause the problem. During this step of troubleshooting, the approach of not having a preconceived notion of what went wrong makes the isolation easier. With preconceived notions of what the problem is, it is only human nature to try and justify why the system we "feel" is the problem can cause the poor running condition. It is important to identify those systems and parts of the engine that could cause the problem (Fig. 9–5). Doing this allows you to have an idea of what you are looking for when you begin your troubleshooting.

Once again, take the example of an engine that runs poorly at high speeds. We have already discounted the likelihood of the carburetor being at fault. The problem could be in the ignition. But we need to determine which part of the ignition. If the engine runs well at idle and low speeds, we know several things. It is unlikely that the primary has many problems; the points could be floating or bouncing but the problem also occurs at low speeds when accelerating. Upon acceleration, the throttle plates are suddenly opened and a large mass of air enters the cylinders. The accelerator pump supplies the engine with a fuel charge to compensate for this lean condition. Perhaps the accelerator pump is not working? The problem also exists at constant speeds with no sudden opening of the throttle. More voltage is required to jump the spark plug gap with a lean mixture than with a rich mixture. Perhaps the secondary voltage is too low to meet these higher requirements. At high speeds it is more difficult to achieve coil saturation, and normally, the voltage available from the coil is less at high speeds than a low speeds. The secondary ignition system can have too great a requirement for voltage at all speeds and the coil simply cannot meet the needs. The problem may be excessive resistance in the secondary. The problem could also be caused by a faulty mechanical or vacuum advance unit, both of which would have a drastic affect on combustion at higher speeds. There could also be a vacuum leak above the throttle plates. This leak would be noticed only when the throttle plates are opened wide, as in acceleration and high speeds. These should be the basic areas tested during the initial troubleshooting on the car (Fig. 9–6).

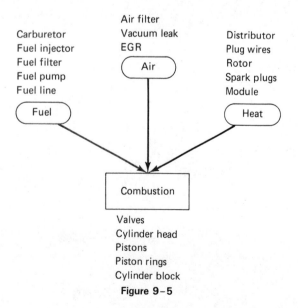

Figure 9–5

Figure 9-6

CUSTOMER COMPLAINT	PROBABLE CAUSES
1. Engine will not crank	A. Dead or run-down battery. B. Open in starting control circuit C. Locked-up starter motor D. Locked-up starter motor drive E. Siezed engine F. Transmission is not in neutral or neutral safety switch is defective
2. Engine cranks slowly	A. Run-down battery B. Defective starting motor C. Bad connections in starting circuit D. Engine is overheating
3. Solenoid clicks, but starter does not spin	A. Excessive drag on the starter motor by the engine. B. Defective starter motor. C. Undercharged or defective battery. D. Loose battery cables. E. Loose or defective wiring at starter. F. Defective solenoid.
4. Engine cranks normally, but will not start	A. Improper starting procedure B. Engine flooded C. Defect in engine D. Ignition bypass resistor burned out E. Plugged fuel filter F. Plugged or collapsed exhaust system. G. Binding linkage, choke valve or choke piston. H. Restricted choke vacuum and hot air passages. I. Improper float level. J. Dirty, worn or faulty needle valve and seat. K. Float sticking. L. Exhaust manifold heat valve stuck. M. Incorrect choke cover N. Inadequate unloader adjustment. O. Faulty ignition coil. P. Improper spark plug gap. Q. Incorrect initial timing. R. Incorrect valve timing S. No fuel in carburetor T. Carburetor jets clogged U. Air leaks into intake manifold or carburetor
5. Hard starting with warm engine	A. Choke valve closed B. Manifold heat control stuck C. Vapor lock D. Engine parts binding E. Carburetor leaks fuel F. Engine overheating
6. Engine idles rough or stalls	A. Incorrect curb or fast idle speed. B. Incorrect initial timing. C. Improper idle mixture D. Damaged tip on idle mixture screw(s). E. Improper fast idle cam adjustment. F. Faulty PCV valve or inadequate air flow. G. Exhaust manifold heat valve inoperative. H. Choke plate binding. I. Improper choke setting. J. Vacuum leak. K. Engine overheating L. High float level M. Faulty distributor rotor or cap. N. Leaking engine valves. O. Incorrect ignition wiring. P. Faulty coil. Q. Clogged air bleed or idle passages. R. Restricted air cleaner. S. Faulty EGR valve operation, if equipped.
7. Misfire at all speeds	A. Faulty spark plug(s). B. Faulty spark plug cable(s). C. Faulty distributor cap or rotor. D. Faulty coil. E. Primary circuit shorted or open intermittently. F. Leaking engine valve(s). G. Faulty hydraulic tappet(s) H. Faulty valve spring(s)

Figure 9-6 (*cont.*)

7. **Misfire at all speeds** (cont'd)

I. Worn lobes on camshaft.
J. Vacuum leak.
K. Improper carburetor settings.
L. Fuel pump volume or pressure low.
M. Blown cylinder head gasket.
N. Carburetor jets or lines clogged or worn
O. Fuel level not correct in bowl
P. Clogged exhaust
Q. Engine overheating

8. Poor low-speed operation

A. Clogged idle transfer slots.
B. Restricted idle air bleeds.
C. Restricted air cleaner.
D. Improper fuel level.
E. Faulty spark plugs.
F. Dirty, corroded, or loose secondary circuit connections.
G. Faulty ignition cable.
H. Faulty distributor cap.

9. Poor high-speed operation

A. Incorrect ignition timing.
B. Faulty distributor centrifugal advance.
C. Faulty distributor vacuum advance.
D. Low fuel pump volume.
E. Improper spark plug gap.
F. Faulty choke operation.
G. Partially restricted exhaust manifold, exhaust pipe, muffler, or tailpipe.
H. Clogged vacuum passages.
I. Improper size or obstructed main jets.
J. Restricted air cleaner.
K. Faulty distributor rotor or cap.
L. Worn distributor shaft.
M. Faulty coil.
N. Leaking engine valve(s).
O. Faulty valve spring(s).
P. Incorrect valve timing.

10. Poor Acceleration

A. Loose or broken vacuum hose
B. Air leaks around carburetor
C. Leaking engine valves.
D. Improper valve action
E. Excessive carbon in engine
F. Heavy oil
G. Wrong or bad fuel
H. Accelerator pump out of adjustment
I. Improper pump stroke.
J. Inoperative pump discharge check ball or needle.
K. Worn or damaged pump diaphragm or piston
L. Float level low
M. Power step-up on metering rod not clearing jet
N. Power piston or valve stuck
O. Choke stuck or not operating
P. Antipercolator valve stuck
Q. Fuel system defective; secondary throttle valves not opening
R. Throttle valve not opening fully.
S. Dirt in filters or in line, or clogged fuel tank vent
T.
U. Manifold heat-control valve stuck
V. Engine cold and choke too lean.
W. Restricted exhaust
X. Distributor spark advance faulty
Y. Incorrect ignition timing.
Z. Faulty spark plug(s).
AA. Faulty coil.
AB. Excessive rolling resistance from low tires, dragging brakes, wheel misalignment, etc.
AC. Transmission not downshifting, or torque converter defective

11. Engine lacks power when hot and cold

A. Incorrect ignition timing.
B. Faulty distributor rotor.
C. Worn distributor shaft.
D. Incorrect spark plug gap.
E. Faulty fuel pump.
F. Incorrect valve timing.
G. Faulty coil.
H. Faulty ignition cables.
I. Leaking engine valves.
J. Blown cylinder head gasket.
K. Leaking piston rings.

12. Engine lacks power when hot

A. Engine overheats
B. Choke stuck partly closed
C. Sticking manifold heat-control valve
D. Vapor lock

Figure 9-6 *(cont.)*

13. Engine lacks power when cold	A. Automatic choke improperly adjusted B. Manifold heat-control valve stuck open C. Cooling-system thermostat stuck open D. Engine valves not closing
14. Engine backfires through carburetor	A. Improper ignition timing B. Defective Accelerator Pump C. Lean air fuel mixture D. Improper choke operation
15. Engine backfires through exhaust	A. Vacuum leak B. Defective AI S diverter valve C. Improper choke operation D. Exhaust leaks E. Rich air fuel mixture
16. Engine pings or has a spark knock	A. Engine overheating B. Incorrect ignition timing. C. Distributor centrifugal or vacuum advance malfunction. D. Excessive combustion chamber deposits. E. Carburetor set too lean. F. Vacuum leak. G. Excessively high compression. H. Fuel octane rating excessively low. I. Heat riser stuck in heat on position.
17. Engine uses an excessive amount of fuel	A. Hard driving B. Short-run and start-and-stop operation C. Excessive fuel-pump pressure or leakage D. Choke not opened properly E. Clogged air filter. F. High carburetor float level or soaked float. G. Stuck or dirty float needle valve H. Worn carburetor jets I. Stuck metering rod or power piston J. Too-rich or too-fast idle K. Stuck accelerator pump check valve L. Vacuum leaks. M. Faulty ignition system N. Loss of engine compression O. Defective valve action P. Excessive rolling resistance from low tires, dragging brakes, wheel misalignment, etc. Q. Clutch slipping R. Transmission slipping or not upshifting
18. Engine runs on or diesels when turned off.	A. Idle-stop solenoid adjustment incorrect B. Engine overheating C. Hot spots in cylinders D. Loose or broken vacuum hose E. Incorrect timing
19. Engine surges at cruising speeds	A. Low fuel level. B. Low fuel pump pressure or volume. C. Improper PCV valve air flow. D. Lean mixture E. Vacuum leak. F. Dirt in carburetor. G. Undersize main jets. H. Clogged fuel filter screen. I. Restricted air cleaner.

20. Smoky Exhaust:
    A. White Smoke
    B. Blue Smoke
    C. Black Smoke
	A. Water in exhaust
	
	A. Oil in exhaust
	
	A. Rich air fuel mixture

PINPOINT THE PROBLEM

With the possible causes identified, tests can be conducted to pinpoint the problem. Before identifying the exact problem, each suspected system is tested. If a fault is discovered in the system, detailed tests should be conducted to determine the exact problem (Fig. 9-7). If no faults are found in the suspected systems, others should be tested. The more accurate your understanding of the engine and its systems, the easier it will be to find the fault in a suspected system. Never complete your trou-

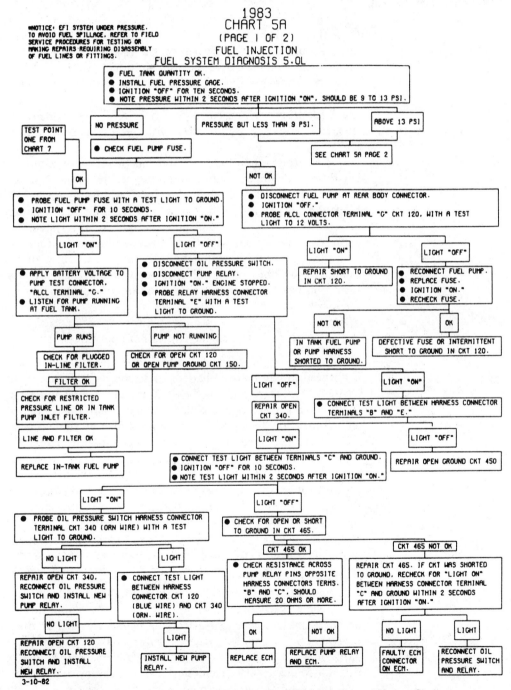

Figure 9-7 (Courtesy of Chevrolet Motor Division, General Motors Corporation)

bleshooting after a single fault is discovered! It is wise to continue your testing and keep in mind the effects of that one faulty part. Doing so will enable you to know exactly what should be done to correct the problem. Your detailed testing should prove or disprove your process of problem isolation. Thinking about what you find from testing not only enables you to correct the problem, but also continues your learning process.

IDENTIFY THE PROBLEM

Once the problems have all been identified, they can be corrected. Proper procedures should be followed when replacing any component and when making any adjustment. Carelessness can cause additional problems. With the repairs and adjustments made, run the engine and bring it to the same condition that the problem occurred in before. If the problem no longer exists, run the engine at all speeds to check for proper response and performance. If the engine runs fine, you have done fine. If a problem exists, proceed with the solving of the problem as you did originally.

This troubleshooting procedure has two main divisions: isolation of the problem through an understanding of the engine, and proper testing to identify the problem. Testing and other diagnostic methods will be discussed in the remainder of this book. The isolation process is based solely on an understanding of how things are affected by other things, and how the fault of the system or part can affect the entire engine. To help with this understanding, what follows in this chapter are examples of customer complaints and the likely causes of these complaints. Although the possible causes will be given, as well as the reasons why they could cause the problem, it is important for you to think about each case and try to understand the reasoning.

Complaint: The engine cranks but does not start.

Definition of the Problem: Because the engine turns with the starter motor, there is no need to consider the battery or starter as the cause of the problem. For the engine to run, there must be the right amount of fuel and air in a sealed container. If the container is properly sealed, the mixture in the container will be compressed enough to allow for ignition. There must also be enough heat supplied by the ignition system to shock the compressed fuel-and-air mixture. This heat must also be supplied at the correct time. The problem can therefore be caused by a fault in the engine, the fuel system, the ignition system, or the air delivery system. The latter possible cause is the least likely, because for this to stop an engine from running, there would have to be nearly no air available to the engine. However, if there is too much air, the engine will be difficult to start. Too little or no fuel will prevent the engine from starting. If there is a weak spark or no spark at the spark plugs, the engine will also not run. A lack of compression can also prevent the engine from starting.

Isolation of the Problem: The problem can be a fault in the ignition or fuel system. There can be vacuum leaks allowing unmetered air into the engine. Also, there can be an engine problem that affects all cylinders, such as the valve timing. If the problem is in the ignition system, it will probably be in the primary system or those parts of the secondary that affect all cylinders. Possibilities include the coil wire being disconnected or broken, the points not closing or opening, the condenser being bad, the pickup or ignition module being bad, or simply no electrical power available to the primary system because of broken or disconnected wires. If the problem is in the fuel system, the problem is likely to be a lack of fuel being delivered to the engine, due to a defective fuel pump, a plugged fuel filter, or no fuel in the tank. A plugged exhaust system or catalyst will prevent the cylinders from filling, and this will prevent the engine from starting. The most likely causes and first areas to test are the fuel and ignition systems. Testing for fuel delivery and for voltage available to the primary are the best places to start. If the engine was hard to start only when the engine was cold, the most likely cause would be in the only system that has the responsibility for cold starts: the choke system or the cold-start injector. These could also prevent the engine from starting when the engine was warm, by operating at a time when they were not supposed to, causing too much fuel or too little air. Again, when the problem occurs will dictate where to begin your testing. The detailed explanation of probable causes was based on a problem that occurred in all conditions.

Complaint: The engine runs rough at all speeds and seems to be missing on one cylinder.

Definition of the Problem: Because the engine runs rough and seems to be lacking the power produced by one cylinder, the problem must be a system or part that is not common to all cylinders. One of the basic needs of a cylinder to allow for combustion is missing or is present in the wrong quantity. To support combustion, there must be a compressed air-and-fuel mixture in a sealed container, which is shocked by the correct amount of heat at the correct time. Systems that provide for the individual delivery of these items or the individual sealed container should be suspected.

Isolation of the Problem: If the engine is equipped with ported-type fuel injection, the injector of that cylinder could be faulty and not providing the correct amount of fuel. Ported fuel injection would be the only type of fuel system that has accommodations for fuel delivery to individual cylinders, independent of the others. Therefore, other types of fuel systems cannot be suspected because they tend to be common to all cylinders. The fuel-and-air mixture does enter each cylinder through the cylinder's intake valve. If this valve does not open far enough or long enough, the cylinder cannot receive as much mixture as the others and will produce less power. This can be caused by a worn camshaft lobe, a bent valve push rod, or a worn rocker arm. On the other hand, if the intake or exhaust valve stays open too long or does not provide a good seal, the mixture will not be compressed well enough to support complete combustion. The prolonged opening of a valve can be caused by a broken valve spring or a burned valve and seat. Anything that would cause a lack of compression in the cylinder, such as worn piston rings, a bad head gasket, or a defective piston, would cause less power to be produced by that cylinder. It is unlikely that the primary ignition system is at fault, because this system is common to all cylinders. Exceptions to this are the individual lobes on the distributor cam and the teeth of the reluctor or trigger wheel. The secondary system is divided into two sections: common and noncommon. The common consists of the circuit from the coil to the distributor cap and should not be suspected for this problem. The rest of the secondary consists of individual circuits to each cylinder and may contain the problem. A spark plug may be fouled or worn out. The spark plug wire may be defective or disconnected or the distributor cap terminal may be corroded or defective. The most likely causes, and first areas to test, are the secondary ignition system and the cylinder's compression. If the engine is equipped with ported fuel injection, this system should also be suspected and the individual injectors tested. The fault may be discovered through a good visual inspection of any of these systems.

Complaint: The engine lacks power during acceleration and high speeds when hot or cold.

Definition of the Problem: Because there is a lack of power and not a rough condition, the problem must be a common system. The fuel, ignition, and related engine systems should be considered as part of the problem. If the engine seems to run well at idle, the rest of the drivetrain of the car should be suspected. There is a possibility that the brakes of the car are dragging or that there is a problem with the transmission. These should be checked before continued diagnosis on the engine.

Isolation of the Problem: Any part of the ignition system can be suspect, but it is unlikely that all the spark plugs or plug wires are defective! More likely is the possibility of the ignition timing being off or not set to specs. The vacuum or mechanical advance could also be functioning improperly. If there is insufficient electrical power to the primary ignition, the coil's output would be decreased and this would affect the engine's power output. Any problem in the primary circuit would affect coil saturation, which would also affect the output of the engine. If the fuel is being restricted as it is being delivered to the engine, there could be a lack of fuel, which can cause decreased power. There may be a lack of air caused by a dirty or plugged air filter. A possibility also exists that the throttle plates are not opening wide enough to allow the engine to produce maximum power. If the exhaust cannot leave the cylinder quickly enough, the cylinders cannot be filled with a fresh charge. A plugged catalyst or a restricted exhaust system will prevent the exhaust from leaving. It is also possible that the valve timing is incorrect, caused by a loose or worn timing chain or belt. The most likely causes, the first areas to test, are the ignition, fuel, and exhaust systems. Because a lack of air can cause the problem, before testing these systems, inspect the air filter. Start the testing with the inspections that are part of the maintenance tune-up and by checking the ignition adjustment for timing and dwell.

If the problem existed only when the engine was warm, the problem may be caused by a choke plate stuck in the closed position, the overheating of the engine, or the manifold heat-control valve being stuck closed. If the problem was present only when the engine was cold, suspect that the choke may not be closing or that the heat-control valve is stuck open. These two systems that could affect the performance of the engine when it is hot or cold are also the systems that are responsible for the transition of cold operation to hot operation. Any malfunction of these systems will affect the way the engine operates at various temperatures.

SUMMARY

To summarize this systematic approach to troubleshooting, here are the steps to follow, whether it is your first attempt to determine the problem or your last.

1. Listen to the customer's complaint. What is experienced, and when does it occur?
2. Verify the complaint. Check to see if other systems or conditions are affected.
3. Isolate the problem. Determine possible responsible systems.
4. Identify the problem. Use area and pinpoint tests.
5. Make the repair. Follow proper procedures.
6. Verify the repair. Check complete engine operation.

REVIEW QUESTIONS

1. What is the purpose of a maintenance tune-up?
2. What is included in a maintenance tune-up?
3. What is the purpose of a diagnostic tune-up?

4. While performing a diagnostic tune-up, what systems should be tested?
5. To find or identify the problem, what should your approach be?
6. What must be known before you can solve any problem?
7. What is the first step you must take in order to understand a problem?
8. Why is it important to test all the systems that could be the cause of the problem?
9. Before you can define the problem, what do you need to do after you have heard the customer's complaint?
10. How does defining the problem help reduce the possibilities of causes of a problem?
11. To isolate the problem, what two things do you need a good understanding of?
12. What is wrong with having a preconceived notion of what the cause of a problem is?
13. What should be identified before you begin to test an area?
14. When one problem is found, what should you do?
15. List the six steps of a good systematic approach to problem solving and troubleshooting.

10

ENGINE DIAGNOSTICS

Figure 10-1

In the preceding chapters, the fundamentals of how an engine works have been discussed. Chapter 9 was an introduction to the remainder of the book. Diagnostics requires an understanding of the fundamentals and a systematic approach to determining what the problem is. When conditions indicate that the problem may be in the engine itself, the engine must be tested. The cylinders of the engine house the combustion process. If the engine has a problem, the other systems cannot function properly or compensate for the fault in the engine. The other systems are responsible for the delivery of the fuel, air, and electricity to the engine. The engine is responsible for sealing the cylinders, compressing the mixture, and conversion of the power produced in the cylinders into the power needed to move the car. The engine also produces the signals for precise operation of the other systems.

There are four primary tests that are conducted to determine the condition of the engine and its parts. The compression test measures the amount of pressure produced by the piston moving up the cylinder on the compression stroke. The cylinder leakage test measures how well the cylinder is sealed when the valves are closed. The cylinder power balance test compares the amount of power produced by each cylinder when the engine is running. The fourth test does not really measure a performance characteristic of the engine, but rather the engine's ability to circulate the oil it needs to lubricate its parts. The engine oil pressure test measures the pressure at which the oil circulates in the engine. This test can be used in performance testing when the complaint indicates that insufficient or excessive oil pressure is causing the problem.

COMPRESSION TEST

The compression test measures the amount of pressure that is formed by the piston moving up on its compression stroke. The amount of pressure produced by the piston depends on the amount of air present in the cylinder, the compression ratio of the cylinder, and how well the cylinder is sealed. If the cylinder is not well sealed, the pressure produced in the cylinder will be low. As the piston moves upward, the volume of the cylinder is reduced, and this increases the pressure of the air contained in the cylinder. The compression ratio of an engine is a statement of how much the cylinder's volume will change. A compression ratio of 8:1 means that the piston will compress a volume of air into one-eighth of the space it previously occupied, as shown in Fig. 10–2. The pressure produced by this change of volume will be over eight times the amount of pressure that the air had before it was compressed. The air enters the cylinder under atmospheric pressure. Atmospheric pressure is 14.7 psi (pounds per square inch). Prior to the compression stroke, the pressure of the air in the cylinder is at atmospheric pressure. With a compression ratio of 8:1, the

Compression stroke

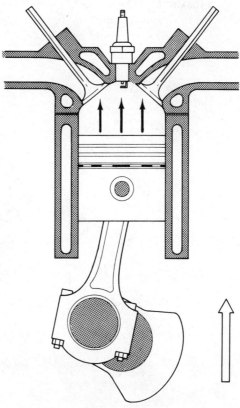

Figure 10–2

pressure will be increased to more than eight times atmospheric pressure or more than 117.6 psi. The reason the pressure will be greater than 117.6 psi is due to the increase of pressure that results from the increase of temperature as the air pressure increases. This temperature increase can cause the pressure to increase about another 25 psi, for a total of about 142 psi.

The expected cylinder pressures of an engine are usually given in the specs for the engine. If these specs are not available, use the following formula to determine the amount of pressure expected from the cylinder during the compression stroke:

atmospheric pressure times the compression ratio plus
atmospheric pressure plus 10 psi

or

14.7 psi × comp. ratio (8 for a ratio of 8:1) + 14.7 + 10 psi
= compression pressure

To measure the compression pressure in a cylinder, a compression tester is used (Fig. 10–3). The tester is inserted into the spark plug hole of the cylinder and the pressure of that cylinder is registered on the gauge of the tester. The pressure, moving a diaphragm inside the gauge, moves the needle of the gauge to indicate the pressure on the diaphragm. The compression tester is equipped with a one-way valve which allows the pressure to build up in the tester to determine the maximum pressure produced by the cylinder.

To use the tester, run the engine until it is warm. Do not let the engine get too hot because you will be getting your hands in and around the engine. Next remove all the spark plugs, being careful not to allow dirt to enter the cylinders. Now lock the throttle plates and the choke plate in a wide-open position. Doing this will allow a maximum flow of air to enter the engine. The ignition should now be disabled. To do this, disconnect the battery lead to the coil.

The tester should be placed in the spark plug hole (Fig. 10–4) of one cylinder.

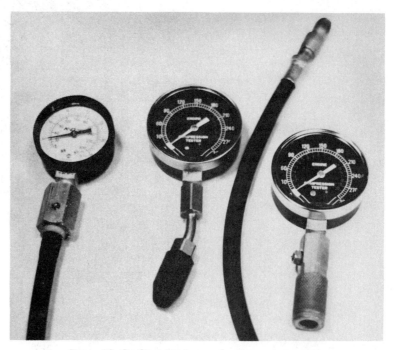

Figure 10–3 Common types of compression gauges.

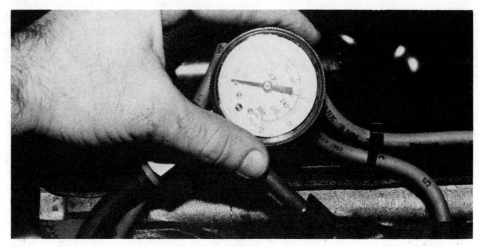

Figure 10-4 Compression gauge in place.

The preferred type of tester threads into the hole, while others are held there by hand. With the tester connected to the cylinder, crank the engine through four complete compression strokes. At each compression stroke you will notice that the needle of the gauge jumps to a higher reading. The reading of the fourth compression stroke is the reading that you will compare to the specs and to the other cylinders. Take a compression test of each of the cylinders in the engine and record the final reading of each. While the engine is rotating through the four compression strokes, the gauge reading should increase with each stroke. If the gauge does not increase with each compression stroke, this indicates that the valves of that cylinder are sticking. Sticking valves are caused by carbon or gum deposits built on the stem of the valves, which cause them to resist closing or opening.

The specs of the manufacturer tell you what the compression reading should be. Compare your findings to the specs. If the readings of one or more cylinders are below the spec, there is a problem with the valves, piston rings, or the head gasket (Fig. 10-5). Any reading lower than the specs indicates that the cylinder is not producing as much pressure as it should. However, a small difference between your readings and the specs is no cause for alarm. Any engine, no matter how well

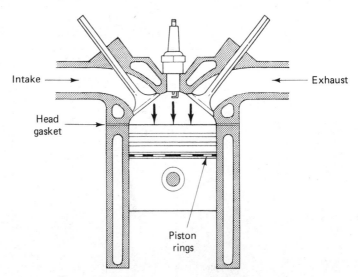

Figure 10-5 Areas of potential compression leaks.

it is maintained, will experience some wear. A reading that is less than 75% of the spec is too low. More important for diagnosis is the evenness of the readings among all the cylinders. If the cylinders have about similar readings that are low, the engine is simply worn evenly. If one or more cylinders are low and the others normal, this unevenness indicates that the compression losses in that engine will cause it to run roughly. Low compression readings indicate that the valves are not sealing because they are burned or the valve stems are bent, the piston rings are worn, or that the head gasket is leaking.

To identify the cause of low compression, two further steps can be taken. Look at your readings; if the readings of two cylinders that are located next to each other in the engine are low, it indicates that the head gasket is leaking. Usually, the area between cylinders is small and the head gasket is responsible for the separation of the cylinders. If the head gasket is defective, the compressed air from one cylinder can leak into the cylinder next to it through the leak in the gasket (Fig. 10–6). This is a quick method of determining the condition of the head gasket.

Low readings can also be caused by bad valves or worn piston rings. To determine which is the cause for the low readings, squirt a small amount of oil into a cylinder which has low compression. With the compression tester, retest that cylinder. If the reading increased over the original reading, bad rings are indicated. The oil that was put into the cylinder will temporarily seal the piston rings to the cylinder walls, causing an increase in the readings. Had the reading remained about the same, the cause of the low readings would be the valves. Valves lose their seal from a burning away of the metal they seat in or seal on, or of the valve itself. Oil will not provide a seal for the valves if they are burned.

A compression test conducted with oil in the cylinder is called a "wet" compression test. It is important to realize that the compression readings are not a true indication of the condition of the valves or piston rings of an engine. High readings can result from the buildup of carbon on the piston top and in the combustion chamber. As shown in Fig. 10–7, this buildup decreases the volume of the chamber, which in turn increases the compression ratio of the engine. With higher compression ratios come higher compression readings. Therefore, high readings may indicate that there is an excessive amount of carbon in the cylinder.

A high-mileage engine will have a lot of carbon in the cylinder, and the rings and valves will have worn some. Compression readings on these cylinders may show the engine to be in good shape by being within the specs for that engine. This may

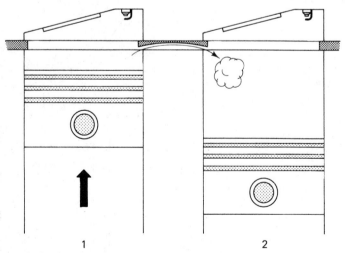

Figure 10–6 Compressed air leaking into the cylinder next to it through a defective head gasket (from cylinder 1 into cylinder 2).

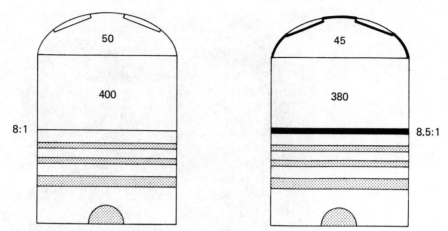

Figure 10-7 Two identical cylinders, one free of carbon buildup and the other with considerable carbon buildup. Notice the volume changes and the resulting compression ratio changes.

not be a true indication of the condition of the engine. With a carbon buildup, we would expect higher readings on the compression gauge. The wear of the valves and piston rings will cause a decrease in the pressure readings. The combination of the increase and the decrease may balance each other and give a normal reading.

The compression test is, however, very valuable in diagnostics. If the compression of some cylinders is lower than the rest, the engine will run rough. To correct the problem, more will be required than a tune-up. The engine must be disassembled and the problem taken care of. If all the cylinders have low compression, this may be the cause of poor performance of the engine. A general lack of power will result from the lack of compression in all cylinders. Excessive carbon in the cylinders may be the cause of preignition. The compression gauge should be used to determine whether this is the cause for the preignition.

A compression gauge is used to determine how well a cylinder is sealed, when the valves are closed, and how well the air is compressed in the cylinder. Whenever a problem indicates that either of these might be affected and causing the problem, this test should be conducted. Compression losses will normally affect the engine's performance at all speeds. If you are certain that the engine's problem exists at only certain speeds, there is no need to conduct a compression test. When conducting a compression test and when comparing your findings to the specs, think about your findings. Apply the findings of the test to the problem of the engine, not to the engine in general.

CYLINDER LEAKAGE TEST

Another test that measures how well the cylinders seal is the cylinder leakage test. This test will also identify the parts that are worn or defective, causing the leakage. Although the cylinder leakage tester (Fig. 10-8) does about the same thing as the compression tester, it does it in a different way. During the leakage test, the engine does not have to be cranked. Instead of measuring the air pressure formed on the compression stroke, the leakage tester applies air into the cylinder and measures the amount of air that leaks out the cylinder past the valves, rings, or head gasket. With the piston at TDC in the compression stroke, both valves are closed. This should allow very little air to escape when air is introduced by the leakage tester. It is normal for some air to leak past the rings, but the amount should be small. The cylinder leakage tester can precisely locate the points of leakage. These points of

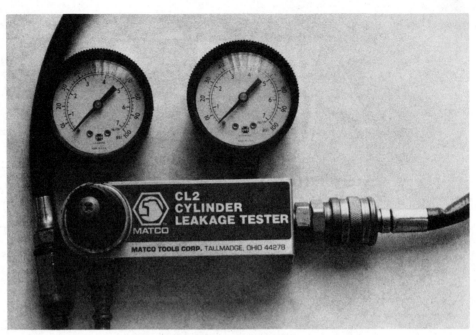

Figure 10-8 (Courtesy of Matco Tools Corp.)

leakage are also points of compression losses. The points of excessive leakage are found by feeling or listening to the air escaping. Where the air is felt or heard is the location of the leakage.

The cylinder leakage tester compares the amount of air pressure applied by the tester to the cylinder to the amount that the cylinder is able to hold or contain. An outside air supply is needed to use this tester. Usually, automotive shops are equipped with an air supply. If the air supply has 100 psi pressure and the cylinder is able to hold 80 psi, the cylinder allows 20 psi or 20% of the air to escape. The escaping air can be felt or heard from the carburetor or throttle body if it is escaping past the intake valve, from the tailpipe or exhaust pipe if it is escaping from the exhaust valve, from the oil filler tube if it is escaping by the piston rings, and from the radiator if it is escaping from the head gasket or a cracked cylinder head or block.

To use the tester, warm up the engine, then remove all the spark plugs. Now remove the air cleaner assembly, oil filler cap or dipstick, and radiator cap, and block the throttle to a full wide-open position. Check the fluid in the radiator and fill it to the proper level. When removing the radiator cap, make sure that the engine is warm and not hot. Use care to release any pressure present in the radiator before completely removing the cap.

The tester is equipped with a hose that threads into the spark plug hole and fastens to the tester. Install the appropriate size hose into the number 1 cylinder. Most testers are also equipped with a whistle to help locate TDC. If a whistle is available, connect it to the tester end of the hose. Crank the engine over until the whistle sounds, which means that the piston is moving toward TDC on the compression stroke. If no whistle is available, put your thumb over the end of the hose to feel the pressure building up. Continue to rotate the engine until the timing marks on the crankshaft align at TDC. At this point the engine is at TDC on the compression stroke and the valves are closed. Disconnect the whistle from the hose and connect the hose to the tester.

Now, connect the shop air supply to the tester. The gauge on the tester should show a reading. This reading indicates the amount of air that is leaking out of the cylinder. Listen at the radiator, oil filler, carburetor, and tailpipe for the escaping air. Check also to see if air is leaking from a cylinder next to the one being tested. If air is escaping from an adjoining cylinder, the head gasket leaks between cylinders.

Test the other cylinders in the same way. Make sure that each cylinder is on TDC of the compression stroke before applying air into the cylinder. You can use the distributor's rotor and its alignment with the terminals of the distributor cap to help locate TDC of the different cylinders. If the piston is not on TDC, the air applied to the cylinder may push the piston down, and this will cause the valves to open as the crankshaft turns. To measure the amount of leakage in a cylinder accurately, the valves must be closed. If high amounts of leakage are recorded, disconnect the air supply and attempt to find TDC again to retest.

Gauge readings of less than 20% are normal unless there is a noticeable amount of air leaking through the carburetor, exhaust, oil filler, radiator, or adjacent cylinder (Fig. 10–9). Any reading over 20% is excessive and the location of the escaping air identifies the problem. Obviously, the higher the leakage, the greater the problem.

The cylinder leakage tester has an advantage over the compression tester because it allows for precise identification of the problem area. Carbon buildup also does not influence the readings, which allows for a truer picture of the mechanical condition of the engine. The biggest disadvantage of this test is that it requires more time to conduct than does the compression test.

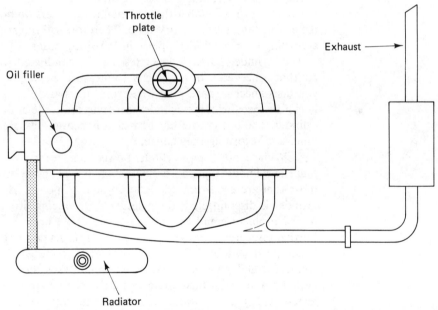

Figure 10–9 If during a cylinder leakage test, air is heard coming from the exhaust, this indicates that the air is leaking past the exhaust valve. If the air is coming from the throttle plates, this indicates that the air is leaking past the intake valve. If the air is coming from the oil filler, air is leaking past the piston rings. When the air is causing bubbles to form in the coolant in the radiator, air is leaking past the head gasket or through the cylinder block or head.

CYLINDER POWER BALANCE TEST

A cylinder power balance test measures the amount of power that a particular cylinder contributes to the total power output of the engine. The total output of an engine is the sum of all the power produced by the individual cylinders. The cylinder power balance test uses the cylinder's effect on the engine's speed to determine this power. While the engine is running, either a cylinder's spark plug or primary circuit is grounded. This prevents the spark plug from firing in the cylinder, which prevents combustion in that cylinder. Engine speed should decrease when it is minus the cylinder. The drop of engine speed is recorded and compared to the drops that occur when the other cylinders are eliminated. If all the cylinders of an engine produce the same amount of power, all engine rpm drops will be equal during a cylinder power balance test. Unless it is necessary to do so, more than one cylinder should not be grounded at one time during this test.

If one cylinder is not contributing much to the total power output of an engine, the drop in engine speed during the power balance test will be much less for that cylinder than for the other cylinders. When an engine is running rough, it is usually caused by one or more cylinders not producing as much power as the others. Performing the cylinder power balance test will identify the less productive cylinders.

Most engine analyzers are equipped with a power balance tester (Fig. 10-10). Some have buttons which, when depressed, short out the cylinders. Other testers have the feature of an automatic power balance test. The automatic test shorts out cylinders one at a time and displays the change in engine rpm as each cylinder is shorted. Power balance testers on engine analyzers short out the cylinders according to the firing order. The analyzers with button controls for the selection of the cylinders have those buttons numbered from one to eight. Number 1 is the first cylinder in the firing order, and 8 is the eighth cylinder of the firing order (Fig. 10-11). It is important to remember that the engine analyzers do not know the firing order of the engine that is being worked on. Therefore, all references to a cylinder are given according to the cylinder's place in the firing order.

A cylinder power balance test can be conducted without an engine analyzer. All that is needed is a tachometer to measure the engine rpm. To conduct the test, you disconnect a spark plug wire from the spark plug and ground it. The change in engine speed with the cylinder disconnected is the amount of drop caused by that cylinder. The test is continued by disconnecting the wire to each cylinder, one at a time, and recording the change.

While conducting a cylinder power balance test, it is important to keep in mind that the cylinders that cause the least change in engine speed when disconnected are those that are producing the least amount of power. If there is no change when a cylinder is disconnected, the cylinder is producing little, if any, power. The cause of the problem could be mechanical or could be in the fuel or ignition systems. Further tests are necessary to determine the exact cause. If the engine you are testing is equipped with an EGR valve, the vacuum line going to the valve must be disconnected and plugged. Changes in engine vacuum may cause the EGR valve to cycle on and off, which could cause the engine speed to vary, and this would interfere with your readings. The changes in engine vacuum are caused by the elimination of cylinders that are producing power.

If the car is equipped with a catalytic converter, each cylinder should not be shorted for longer than 15 seconds at a time. Without combustion taking place in the cylinders, raw fuel will be leaving the cylinders through the exhaust. Excessive amounts of raw fuel in the catalyst can cause damage to the converter. Between the shorting out of each cylinder, the engine should be run on all cylinders for at least

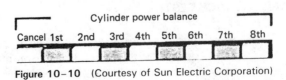

Figure 10-10 (Courtesy of Sun Electric Corporation)

Firing order   1   6   4   3   5   2

1	2	3	4	5	6	7	8

Actual cylinders   1   6   4   3   5   2

Figure 10-11

30 seconds to allow the raw fuel to leave the converter, preventing fuel buildup in the converter.

To conduct a cylinder power balance test with an engine analyzer, first connect the analyzer according to the manufacturer's procedure. Disconnect and plug the vacuum line to the EGR valve if the engine is equipped with one. Now start the engine and let it run until it is at normal operating temperature. When the engine is warm, set the engine speed to its fast idle speed using the fast idle cam. To do so, open the throttle just wide enough to move the fast idle cam freely, so that the adjustment screw rests on the next step of the cam (Fig. 10-12). This will allow the engine to run at about 1000 rpm.

With the engine running at its first idle speed, press the button on the analyzer marked "1." This will short out or "kill" the first cylinder of the firing order. Looking at the tachometer on the analyzer, note the drop in engine speed. Release the button and record the rpm drop. The engine is now running on all cylinders again. After a short period of time, press the 2 button. This will kill the second cylinder in the firing order. Record the rpm drop and continue the test, shorting out the other cylinders one at a time. While testing each cylinder, keep the cylinders shorted for only the amount of time that is needed to get an accurate reading of the drop.

If all cylinders receive the same amount of fuel, air, and heat, the rpm drops for each will be about equal. The drops of all cylinders should be within 30 rpm of each other. If the difference between cylinders is greater than 30 rpm, the engine's power is out of balance. The cylinders that decreased the engine speed the least are the cylinders that are not producing their share of power.

The cylinder power balance test is quick and easy to perform. It is valuable in identifying the exact cylinders that have problems. It can also be used to determine the system that is responsible for the loss of power and/or rough running. The results of this test alone will not tell you where the problem is. But when the test results are compared to the test results of a compression test and cylinder leak-

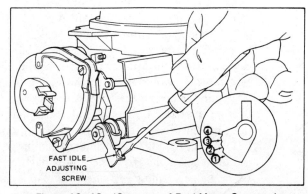

Figure 10-12 (Courtesy of Ford Motor Company)

age test, much can be found out about the problem. The compression test measures the amount of pressure that is formed in the cylinder on the compression stroke of the piston. The better sealed the cylinder is, the higher the pressures produced will be. The cylinder leakage tester measures how well the cylinder seals. Both of these tests are conducted on the compression stroke when the intake and exhaust valves are closed. The cylinder power balance test compares the contribution of each cylinder toward maintaining a particular speed. More is examined than the sealing of our container. By comparing the results of each one of these tests to each other, we can begin to identify what can be and what cannot be the problem.

If an engine has good compression, good cylinder leakage, and good power balance, we know that the cylinders are well sealed and that there are no problems with the systems that are not common to all cylinders. The engine would probably run smoothly, but if it had a problem, it would not be running efficiently.

If the results of the tests on an engine show that it has good compression, good leakage, but poor power balance, we know that the imbalance problem is not caused by leaky cylinders. We also know that the problem is not in the fuel system unless the car is equipped with port-type fuel injection. The fuel system of most engines is common to all cylinders and therefore would not cause a power balance problem. The problem in this engine must be caused by a noncommon system or part. The poor power balance could be caused by an ignition problem (typically in the secondary) or by a vacuum leak that does not affect all cylinders. This imbalance could also be caused by a mechanical failure or defect. The results of the compression test and the cylinder leakage test tell us that the cylinder seals well, but do not tell us anything about how well the valves are opening to let the mixture in or the exhaust out. If a push rod were bent, the valve would still close well but would not open as wide as it was designed to. This would limit the amount of mixture that was able to enter or the amount of exhaust that was able to leave the cylinder, depending on whether it was an intake or exhaust valve that was affected. This same condition could exist as a result of a broken rocker arm, a worn camshaft lobe, or a collapsed hydraulic lifter. All these faults would not affect the sealing of the cylinder, but would affect the opening of the valves. To determine the exact cause of the power imbalance, you should test the ignition system, then follow the manufacturer's guidelines for checking these valve-related items.

An engine with good compression, good power balance, and excessive cylinder leakage is typically an evenly worn high-mileage engine. The good results from the compression test are evidence for carbon buildup in the cylinders. The increased compression ratio hides the leakage in the cylinders. The leakage test results verify that there is excessive leakage in the cylinders. Because the engine is worn out evenly, the power balance test results are good. The complaint of the driver of this car would probably be one of poor starting or poor performance. The cure for the problem is to rebuild the engine, which is well beyond the scope of a diagnostic tune-up. However, it may be importnat to locate the source of the compression leak so that the customer can be informed.

While conducting a cylinder leakage test, the intake and exhaust valves must be closed before applying air to the cylinders. If the camshaft timing changed because of a stretched timing chain or belt, the valves would still close and open properly, but they would do so at the wrong time. An engine with this problem would have good cylinder leakage test results because as part of the procedure for conducting the test, you would make sure that the valves were closed. The effects of this problem would be evident in the compression test. The results of the test would be low compression in all cylinders. Similarly, since the camshaft affects all cylinders equally, the results from the power balance test would be good. All the cylinders would be performing just as poorly. An engine with poor compression, good leakage, and power balance usually has a valve timing problem.

The simplest way to check valve timing (Fig. 10–13) is to bring the number 1

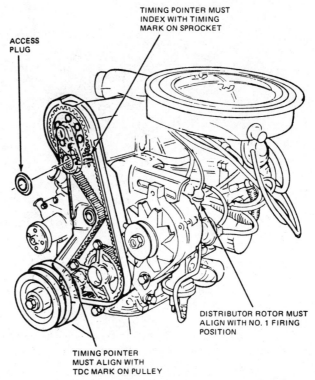

ACCESS
PLUG

TIMING POINTER MUST
INDEX WITH TIMING
MARK ON SPROCKET

DISTRIBUTOR ROTOR MUST
ALIGN WITH NO. 1 FIRING
POSITION

TIMING POINTER
MUST ALIGN WITH
TDC MARK ON PULLEY

Figure 10-13   (Courtesy of Ford Motor Company)

cylinder to TDC on the compression stroke and align the timing marks on the crankshaft with the pointer used to time the ignition. With the marks aligned, remove
the distributor cap. Notice the direction of the distributor rotor. If the rotor is in
the correct location, it will point to the terminal for the number 1 cylinder. If the
valve timing is off, the rotor will not be pointing at number 1. Because the distributor drives off the camshaft, the distributor can be used as a reference for the timing
of the camshaft. If the relationship between the crankshaft and the camshaft is not
in proper time, the valve and ignition timing will be wrong. This simple check can
be verified by removing the valve covers and observing the valves of the number 1
cylinder. Both valves should be closed when the piston is on the compression stroke
at TDC. If the piston is at TDC during the exhaust stroke, both the intake and
exhaust valves are open slightly. Rotating the engine past TDC should cause the
exhaust valve to close and the intake to open wider. Another way to verify the
change in valve timing would be to remove the timing belt or chain cover and check
the alignment marks on the camshaft and crankshaft gears or pulleys.

Using the combination of results from the compression test, cylinder leakage,
and cylinder power balance test, you can identify the systems that should be tested
further. Detailed testing of the fuel and ignition system will be discussed later in
the book. However, the inspection and testing of some of the mechanical parts of
the engine will not. For these items you should check the shop manual for the proper
procedures. Repair of the mechanical parts of the engine is not part of a tune-up.

OIL PRESSURE TEST

Another mechanical test that is commonly used to diagnose engine problems is the
oil pressure test. This test measures the amount of pressure that the engine oil circulates with throughout the engine. Figure 10-14 shows the typical path of oil flow
within an engine. The oil circulation or engine lubricating system is important not

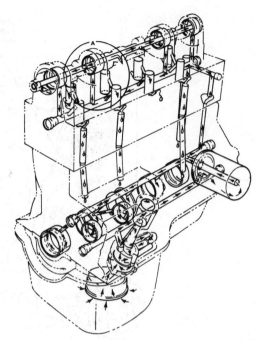

Figure 10-14 An engine's lubrication system with the lubrication of the valve train outlined in the circle marked with an A. (Courtesy of Ford Motor Company)

only to limit the amount of wear on the moving parts of the engine but also has many other important functions. The oil must lubricate the moving parts to limit the amount of power loss due to friction. If the engine's crankshaft is difficult to turn, more of the power produced by the engine will be needed to rotate the crankshaft, leaving less power to move the car. Oil also serves to cool the engine, removing heat from the engine parts as it flows past them. It absorbs the shocks between bearings and other engine parts to minimize engine noise and extend the life of the parts. Engine oil also forms a seal between the piston rings and the cylinder walls by filling the irregular surfaces of the rings and the walls. A certain amount of oil clings to the cylinder walls to stop or limit the amount of leakage past the piston rings. Oil also serves as a cleaning agent, removing varnish and carbon from the engine parts. To perform all these functions, oil must be of high quality and must be supplied in adequate amounts to all the moving engine parts under the correct amount of pressure.

The engine oil pump provides for a continuous supply of oil to the appropriate engine parts and delivers the oil at sufficient pressure to lubricate the parts properly. The oil pump is located on the engine block either outside or inside the crankcase. It picks up the oil from the oil pan through a pickup tube which has a filtering screen on the end of it. The oil is pushed out of the oil pump into a pressure regulator which limits the maximum amount of oil pressure. To provide enough oil to lubricate, clean, cool, and seal the parts of an engine efficiently, the oil must be delivered under pressure. Most shop manuals give specs for the minimum and maximum allowable oil pressure of an engine. The results of an oil pressure test should always be compared to those specs.

An oil pressure test should be performed when there is reason to believe that the oil pressure is not correct or when you want to verify that an engine is worn. Loss of performance, excessive engine noise, and poor starting can all be the result of abnormal oil pressure. Excessive oil pressure can cause the hydraulic lifters to pump up too much and not allow the valves to close completely. Insufficient oil pressure may not be reducing the effects of friction enough, thus causing the crank-

Figure 10-15

shaft to be difficult to turn. This would cause poor starting and poor performance. If an insufficient amount of pressurized oil is delivered to the moving parts, these parts will wear quickly and, while they are wearing, will cause the engine to be noisier.

An oil pressure tester is composed of a gauge to read the pressure, with a high-pressure hose to connect to the appropriate spot on the engine (Fig. 10-15). The scale of the gauge ranges from 0 to 100 psi. The hose is connected to the engine block in a place where oil pressure can be monitored, such as a main oil gallery or the oil filter base plate. The preferred location for a particular engine is given in the shop manual. The engine should be run until the desired test temperature is reached. This temperature is also given in the shop manual. Once the temperature is reached, the engine should be brought to the recommended speed and the gauge should be read. The reading on the gauge should be compared to the specs.

If the oil pressure is too low, the cause could be any of the following: a worn oil pump; excessive wear on the camshaft or crankshaft, causing excessive bearing clearances; a plugged oil pump pickup screen; or a weak or broken pressure relief valve spring. All of these require services that go beyond the scope of a tune-up, and tend to be quite expensive. If the engine's problems are due to any of these conditions, do not attempt to perform a tune-up in an effort to hide or disguise the problem. Inform the customer of what is needed to correct the problem.

If the oil pressure is too high, the probable cause is a stuck oil pressure relief valve. Typically, all that is needed is to remove the valve and clean it. After cleaning the valve with a crocus cloth, the pressure relief spring tension should be checked and replaced if it is not within specs. Crocus cloth is a type of sandpaper that polishes metal surfaces by removing varnish or dirt but does not remove any metal from the surfaces. Normal sandpaper should not be used in place of crocus cloth; removing some metal from the valve will affect the operation of the valve.

SUMMARY

This chapter discussed the basic tests performed to determine the condition of the mechanical parts of the engine. These tests should only be performed in diagnostics when there is reason to believe that the engine may have a problem. Coming to the conclusion that the engine may have a problem results from an understanding of the engine and its systems and from the correct interpretation of results of other systems' tests. Understanding what these tests are testing and what the results are telling you will provide a better understanding of the engine and its systems. As important as this understanding is that you follow the proper procedures. The tests you conduct are only as good as the procedure you use.

REVIEW QUESTIONS

1. What two things are necessary for effective diagnostics?
2. Briefly describe the purpose of the following systems:
   (a). The basic engine.
   (b). The air/fuel system.
   (c). The ignition system.
   (d). The emission control system.
3. What is the purpose of a compression test?
4. What does the compression test actually measure?
5. How does the engine's compression ratio affect the readings from a compression test?
6. How can you calculate the expected readings for a compression test if there are no specifications available?
7. Why should the choke and throttle plates be propped open during a compression test?
8. What is the proper procedure for conducting a compression test?
9. What is indicated by the results of a compression test which show that one or more cylinders have less compression than the others?
10. What can be done to determine whether or not the valves are the cause of low compression readings?
11. What is indicated by a too-high reading during a compression test?
12. What is indicated by two adjacent cylinders having low compression readings?
13. What is wrong with using the readings from a compression test as a final determinant of the condition of an engine?
14. What is the purpose of the cylinder leakage test?
15. How can the cylinder leakage test be used to pinpoint the cause of low compression in an engine?
16. What is the proper procedure for conducting a cylinder leakage test?
17. Why must the cylinder that is being tested during a cylinder leakage test be at top dead center?
18. What is the advantage of a cylinder leakage test over a compression test?
19. What does the cylinder power balance test measure?
20. How can the cylinder power balance test be conducted without an engine analyzer?
21. While conducting a cylinder power balance test on an engine analyzer, the buttons that regulate which cylinder is being canceled are marked according to _____.
22. Why should you not conduct a cylinder power balance test on one cylinder for more than 15 seconds?
23. What is the advantage of conducting a cylinder power balance test over a cylinder leakage or compression test?
24. What is indicated by good results from a compression and cylinder leakage test but poor results from a cylinder power balance test?
25. What is indicated by good results from a power balance test and compression test but poor results from a cylinder leakage test?
26. What is indicated by poor results from a compression test and cylinder leakage test but good results from a cylinder power balance test?
27. What kind of problems can result from having an improper amount of oil pressure?
28. What is the proper procedure for conducting an oil pressure test?

11

ENGINE VACUUM TESTING

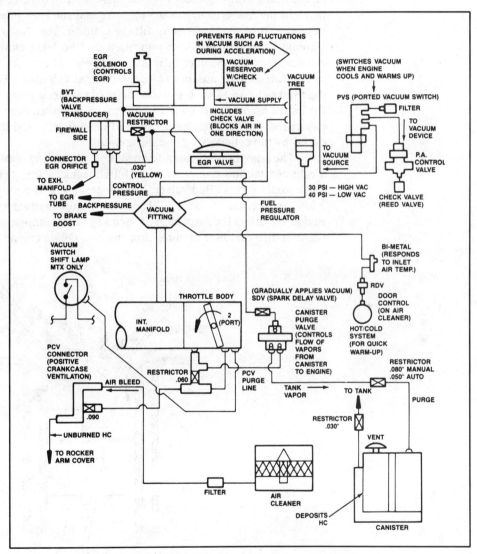

Figure 11-1 (Courtesy of Ford Motor Company)

As the piston moves downward during the intake stroke, it increases the volume of the cylinder. This increased volume causes a decrease in the pressure of the air that occupies the cylinder. The intake stroke does the opposite to air pressure as that done by the compression stroke. The intake stroke increases volume and decreases pressure, while the compression stroke decreases volume and increases pressure. The intake stroke produces a vacuum, while the compression stroke produces high air pressure. Both strokes depend on the sealing of the cylinder to achieve the desired results. Maximum pressure cannot occur if the cylinder is not sealed. Similarly, maximum vacuum cannot be achieved if there are leaks in the cylinder. By changing the volume of a sealed container filled with atmospheric pressure, both strokes change that pressure. At the beginning of both strokes, atmospheric pressure fills the volume of the cylinder. With the movement of the piston, pressure becomes greater or less than atmospheric pressure. Any pressure that is less than atmospheric pressure is considered a vacuum.

A low pressure is formed on each intake stroke of the piston (Fig. 11-2). This low pressure allows atmospheric pressure to enter the cylinder, bringing with it the fuel from the carburetor or the injection system. Overall engine efficiency depends on the filling of the cylinder with the fuel-and-air mixture. The greater the difference between the low pressure in the cylinder and the high pressure of the atmosphere, the more force there will be to fill the cylinder. For the sake of overall engine performance and economy, it is important that no leaks exist to decrease the amount of vacuum in the cylinder or intake manifold. The intake manifold delivers the air or the fuel-and-air mixture to the individual cylinders. The vacuum produced by the intake stroke of each cylinder determines the amount that will be delivered by the intake manifold. If there is a difference in the amount of vacuum formed from one cylinder to another, the amount of air delivered and the amount of power produced by the cylinders will be different.

The amount of vacuum formed depends on how well the cylinder is sealed. If a cylinder had low compression and high leakage, the vacuum formed by that cylinder would be low. Bad valves, leaking gaskets, and poor piston rings would cause a low production of vacuum. The same things that affect the compression of an engine will affect the formation of vacuum. High-compression engines produce more vacuum than do lower-compression engines. On an engine with a compression ratio

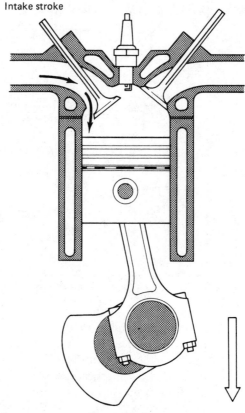

Intake stroke

Figure 11-2

of 8:1, the expected pressure, formed during the compression stroke, is greater than eight times atmospheric pressure or more than 117.6 psi (Fig. 11-3). This occurs because the piston's movement up the cylinder reduces the volume of the cylinder. On the intake stroke of the same engine, the piston increases the volume eight times the original amount, moving from TDC to BDC. This decreases the pressure by more than one-eighth of atmospheric or about 1.8 psi. This pressure is less than atmospheric and is therefore a vacuum. If the pressure formed on the intake stroke was only 13.7 psi, this would still be a vacuum. Atmospheric pressure is normally 14.7 psi. There is a greater difference between the 1.8 psi and atmospheric pressure than the 13.7 psi. The 1.8 psi is therefore a stronger vacuum than the 13.7 psi. A complete vacuum would be formed if the air pressure reached 0 psi. This is never achieved by the engine; in fact, this has never been achieved by anyone or anything.

In cities of high elevation, there is less air above them, and therefore the atmospheric pressure in those cities is lower (Fig. 11-4). Vacuum is still defined as any pressure lower than atmospheric. If pressure of the atmosphere was 12.7 psi, the low pressure formed by the engine above would still be a vacuum. The 1.8 psi is still lower than the pressure of the atmosphere. However, the difference between this lower pressure and atmospheric pressure is not as great as the difference there was with an atmospheric pressure of 14.7, and therefore the vacuum in this high elevated city is not as strong. The strength of the vacuum is proportional to the difference between low and atmospheric pressure.

It is the responsibility of the intake manifold to distribute the incoming air to the appropriate cylinders in response to the vacuum pulses of the cylinders. These vacuum pulses join together at the base of the throttle body assembly, or the carburetor. Present at this point in the intake manifold is a seemingly continuous supply of vacuum. The presence of vacuum at this point keeps the air flowing into the

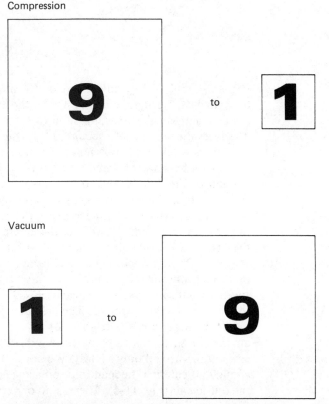

Figure 11-3 During compression, 14.7 psi is compressed to about 135 psi; during the intake stroke, 14.7 psi is reduced to about 1.6 psi.

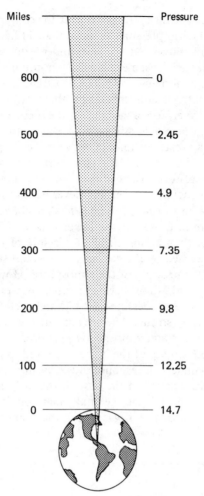

Figure 11-4

engine, so that as the cylinders are on the intake stroke, there is air available to fill the cylinders. During the intake stroke, the intake valve opens. If there is air in the intake manifold, the air will rush in to fill the cylinders. As air enters the cylinder, it reduces the amount of vacuum in the cylinder. The amount of air that enters the cylinder is determined by a number of things. At low engine speeds, the piston is moving rather slow. The amount of time the intake valve is open will determine the amount of time the air has to get into the cylinder. The slower the engine speed, the more time the air has to enter the cylinder. When there is a great difference between pressure in the cylinder and in the atmosphere, the force of the air to enter the cylinder is greater. Therefore, the greater the vacuum produced by the piston, the faster the outside air will attempt to fill the cylinder.

The throttle plates regulate the engine speed by regulating the amount of air that enters the intake manifold for distribution to the cylinders. At engine idle speeds, the throttle plates are nearly closed. This allows only a small amount of air to enter the intake manifold. The cylinders are producing vacuum pulses, and the intake manifold holds nearly all of the vacuum formed by the cylinders. The small amount of air that is entering the manifold through the throttle plates reduces the amount of vacuum only slightly. When the throttle plates are fully opened, there is enough air entering the manifold to balance the pressure between the cylinders and the outside air (Fig. 11-5). There is no or very little vacuum. To form and maintain a vacuum, the low pressure must be contained in a sealed container. The sealing of

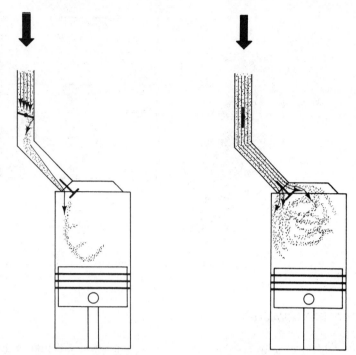

Figure 11-5 The throttle plates control the amount of air that can enter the cylinder; therefore, the throttle plates control the seal of the cylinders.

the intake manifold by the throttle plates seals the cylinders when the intake valves are open. When the throttle plates are open, there is usually a vacuum present under the throttle body and above it. This vacuum above the plates is caused by the venturi effect and the fact that a small column of air is entering through the throttle plates into a larger area: the intake manifold.

Engine vacuum is needed not only for the distribution of the air to the cylinders but is also used to operate other engine systems and some automobile accessories, such as power door locks, power brakes, and air-conditioning ducts. The difference between atmospheric and engine air pressures is used to perform work. By separating the two pressures with a diaphragm or movable piston, the force caused by the difference of pressures causes the piston or diaphragm to move. By attaching a lever or rod to the piston, work can be performed by the force, as shown in Fig. 11-6.

If there is atmospheric pressure (14.7 psi) on one side of the diaphragm and only 9.7 psi on the other, there is 5 psi of pressure available to do work. If the diaphragm is 3 in. in diameter, its surface area is equal to about 7 in.^{2}. There is 5 pounds of force on each square inch of the diaphragm, for a total of about 35 pounds of force on the diaphragm to perform work. This is the force that can be used to operate an accessory. These accessories cannot operate if air is allowed to enter into the low-pressure side, or if there is no atmospheric pressure present on the other side. The pressure differential must be there to perform work.

The feeding of the cylinders also requires the presence of both high and low pressure. Monitoring the production of vacuum will not only allow you to determine how well the accessories or engine systems will work, but will also tell you how well the cylinders work as a vacuum pump. A cylinder's ability to form and hold a vacuum is also its ability to form and hold compressed air. If there are leaks in the cylinder, they will show up on the compression test, the cylinder leakage test, the cylinder power balance test, and the engine vacuum test.

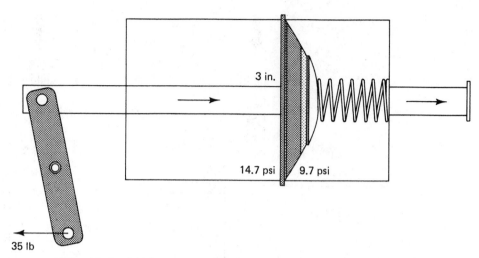

3 in.

14.7 psi   9.7 psi

35 lb

Figure 11-6 A higher pressure from the atmosphere pushing against a diaphragm with a vacuum present on its other side allows work to be performed.

ENGINE CRANKING VACUUM TEST

To conduct an engine vacuum test, all that is needed is a vacuum gauge (Fig. 11-7). This vacuum gauge will give readings of the low pressure present at the point it is attached to the engine. It is calibrated to read inches of mercury (in.Hg). This is the standard used for measuring vacuum. Atmospheric pressure has zero inches of mercury or zero vacuum. Total vacuum, about 30 in.Hg, is equal to 0 psi.

The use of in.Hg as the measurement for vacuum is based on the reaction of mercury to the force of a high and low pressure. To explain this, look at Fig. 11-8, where some mercury, which is a liquid metal, has been placed into a glass tube bent into the shape of a "J." On the short end of the tube is a hose that connects to nothing and is open to the air. Atmospheric pressure will be the only force on it. On the other end, a hose connected to a vacuum source is attached. With no vacuum being supplied by the source, the mercury lies in the bottom of the hook in the tube. As a vacuum is introduced to the tube, the mercury moves up toward

Figure 11-7

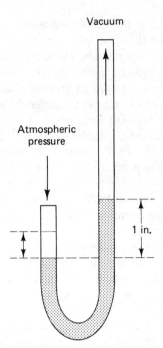

Vacuum

Atmospheric
pressure

1 in.

Figure 11-8 J-tube with mercury in it and subjected to different pressures. The amount of vacuum present is 1 in.Hg.

the low pressure. This is caused by the force of the high pressure in the atmosphere moving in to occupy the lower pressure. When the mercury moves up the tube ½ in. on the low pressure end and ½ in. down on the other end, there is 1 in.Hg of vacuum present from the source. As the pressure gets lower, the mercury will continue to move up the tube. The movement of the mercury is the measurement of vacuum.

Instead of using a glass tube with mercury in it to measure vacuum, vacuum gauges use a Bourdon tube made of thin brass. The tube partially collapses with the presence of a low pressure inside of it. The collapsing is calibrated to equal the movement of the mercury in a J-tube. This collapsing causes the needle of the gauge to move. The typical scale on the gauge gives both a psi and an in.Hg reading, indicating pressures above and below atmospheric pressure.

The vacuum gauge should be connected to the intake manifold below the throttle plates for most testing. A connection here will give you the information you need to determine the condition of the engine and its related systems. When making the connection, it is best not to disconnect any hoses that lead to a system or accessory. Use a "tee" fitting in the main vacuum hose at the manifold to connect the gauge whenever possible. Some engines are equipped with a removable plug in the manifold to allow for the testing of vacuum. Avoid connecting the gauge to an accessory, because the readings at accessories are not always the true reading of the engine's vacuum. True manifold vacuum is best measured at the manifold itself.

Manifold vacuum is measured under two conditions: while the engine is cranking and while the engine is running. The first tests to be conducted are with the engine cranking. To prevent the engine from starting, disconnect the battery lead to the ignition coil. To get a truly accurate reading of manifold vacuum, the throttle plates should be adjusted fully closed, while conducting an engine cranking vacuum test. Crank the engine and observe the reading on the vacuum gauge. There is typically no spec for the reading, but what you should expect is a reading of some vacuum. Some vacuum means at least 1 in.Hg. If the engine produced zero vacuum during cranking, it is very likely that the engine will not run. Some vacuum must be present to draw in the fuel-and-air mixture. If the reading is high enough and the needle is steady, it is an indication that the cylinders are producing the same amount of vacuum. If the needle moves up and down the scale, but spends more time at one reading than the other, it indicates that one or two cylinders are not producing as much vacuum. The pulsating motion of the needle indicates the same condition that the cylinder power balance test did: an unbalanced engine. If the needle seems to spend an even amount of time at each of the readings, it indicates that half of the engine is operating under different conditions than the other half. These needle movements can be more important than the actual reading of the gauge.

While the ignition is disabled, pull out the PCV valve and cover the opening with your thumb. The PCV system is a calibrated vacuum leak. This means that the rest of the engine is set up for the loss of vacuum through the PCV system. If the system is working correctly, plugging this leak should cause the cranking vacuum to increase. By pulling the PCV valve out, with its hose still attached, and putting your thumb over the opening of the valve, as shown in Fig. 11–9, you are plugging this leak. If the vacuum does not increase, while cranking the engine, there is a fault in the PCV system. The problem could be a clogged PCV valve or a hose that has the system plugged already, and what you are doing has no effect on it. Another possibility is that there is a hole or cut in the hose before the PCV valve. This is allowing air to leak in, and by plugging the end of the hose, you are not stopping the vacuum leak.

The term "vacuum leak" needs to be defined before we go further into the discussion of vacuum testing. A vacuum leak is not what it sounds like. A vacuum leak does not allow vacuum to escape. Rather, it allows air to enter and occupy the vacuum that was formed, thereby reducing the amount of vacuum present. Low pressure will never move to a higher pressure. High pressure always moves to a lower pressure.

Figure 11–9 Testing the PCV valve with it plugged and the engine cranked. Vacuum reading should increase with the PCV valve plugged.

ENGINE RUNNING VACUUM TEST

Engine manifold vacuum is also monitored when the engine is running. While running, the engine operates with its various systems. All of these systems can influence the reading you get on the vacuum gauge. Keep in mind that you are looking at the ability of the cylinders of the engine to draw in air, as well as form a vacuum. If there is a lack of air available to the engine, the vacuum readings will be high. If there is an excessive amount of air at the engine, the readings will be low. Readings will also be low if anything causes an extra amount of air to enter the cylinders. A vacuum leak allows air to enter and can affect the operation of the engine, as well as the readings on the gauge. Leaking valves, piston rings, and gaskets are all causes for vacuum leaks. All of these prevent the sealing of the cylinders.

The throttle plates control the amount of air entering the intake manifold. The wider they are open, the lower the vacuum. The faster the piston moves through the intake stroke, the less time there is for air to enter the cylinders, which causes an increase in vacuum. Therefore, if engine speed is increased without increasing the amount that the throttle plates are open, vacuum will increase.

Before taking a vacuum reading with the engine running there are two things you must check. First check the engine's idle speed. To get a true indication of the condition of the engine, the engine must be set to specs. Care must be taken not to upset the present running condition of the engine. Do not attempt to smooth the idle with the mixture screws. Adjust the idle speed as close as possible to the recommendations of the manufacturer. The second item to check is the ignition timing. If you find that the timing is off, do not correct it at this time. To have useful readings, you need only to be aware of the present timing.

When an engine is at idle and the ignition timing is changed, the engine's speed will change. Advancing the timing will tend to increase the idle speed, while retarding the timing will slow down the idle speed. Advanced timing causes combustion to be more efficient. Engine speed will increase with an increase in combustion efficiency. Anything that increases engine speed at a fixed throttle position will increase vacuum, simply because of the increased number of intake strokes at the higher speed. Advanced timing will cause the engine's vacuum to increase with the corresponding increase in engine speed. Retarded timing will cause the vacuum to decrease. Being aware of the present ignition timing will help you decide the expected normal vacuum reading from the engine. If the timing is advanced, expect a slightly higher than normal reading. Expect a slightly lower reading if the timing is retarded.

Before checking the vacuum of the engine at idle speed, bring the car to its normal operating temperature. An engine at operating temperature, correct idle speed, and correct initial ignition timing should have a reading of more than 16 in.Hg. The reading should be steady. There are no specs for the vacuum readings of different engines. The desired reading will vary mostly from engine to engine as the compression ratio differs. The higher the compression ratio, the higher the desired vacuum reading. To formulate your own vacuum spec for the engine, consider the ignition timing and the compression ratio of the engine. If the timing is not correct, add or subtract 1 or 2 in. to the expected reading of more than 16 in.Hg. (Fig. 11–10). If the engine has high compression, add 1 or 2 in. to compensate for the high compression. Use your totals for the desired vacuum spec.

A steady needle on the vacuum gauge indicates that the cylinders of the engine are forming about the same amount of vacuum. If a reading on the gauge is lower than what was expected, it indicates a loss of vacuum. This may be caused by low compression or a vacuum leak somewhere that is common to all cylinders, such as the mounting base of the throttle assembly. Poor piston ring seal, bad valves, and leaking gaskets are all possible causes for a lower reading on the vacuum gauge. To determine whether the low reading is due to a vacuum leak, raise the engine speed and observe the reading at a constant higher speed. If the reading decreases much from the reading taken at idle, the problem is probably in the engine. A slight decrease in vacuum indicates that there is a vacuum leak in the system. This occurs

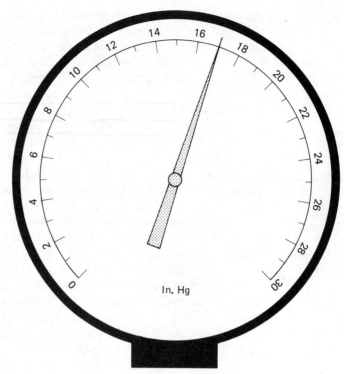

Figure 11-10

because as the throttle opens, the vacuum will normally decrease. If there is a leak in the intake manifold, it will have less effect on the vacuum. To locate the vacuum leak in the manifold, return the engine speed to idle and squirt oil around a variety of sealing surfaces on the manifold. When oil is introduced to the area of leakage, the vacuum gauge will rise temporarily.

Another method of finding the leak is to spray some penetrating oil or carburetor cleaner around the suspected areas. These fluids will burn, and if drawn into the engine through a leak, the engine will tend to speed up. These vacuum leaks are usually located at the sealing surfaces of the manifold, such as the base of the throttle body or where the manifold bolts to the cylinder head (Fig. 11-11). Hoses that are cracked or that do not seal well to a vacuum nipple are also a common cause for vacuum leaks. The more-difficult-to-locate leaks are those caused by a ruptured diaphragm in a vacuum-operated accessory or by excessive throttle shaft wear.

It is very unlikely that all cylinders will have bad valves. If all the cylinders of an engine did have leaking valves, the vacuum reading would be low and steady. Typically, bad valves show up on the gauge as a wavering needle between a higher and a lower reading. Each time the cylinder with the bad valve begins its intake stroke, the needle on the gauge will drop to the lower reading. The more cylinders that have a bad valve, the more often the needle will dip to the lower reading. This is why there would be a steady low reading if all the cylinders had bad valves. The movement of the needle of the vacuum gauge will fluctuate between a high and a low reading anytime there is a difference between the amount of vacuum formed by each cylinder of the engine.

If the engine has a leaking head gasket, only two cylinders will be affected. Therefore, the vacuum gauge needle will fluctuate between a high and a low reading. The gauge will react much the same as if two cylinders had bad valves. The way to identify the problem is to conduct a cylinder leakage test or a compression test. A cylinder power balance test will identify which cylinders are affected.

Figure 11-11   Areas around the base of a carburetor and at the mounting of the intake manifold to the engine are likely spots for vacuum leaks.

By observing the fluctuation of the needle, you can determine how many cylinders are affected. If the engine is four-cylinder, the needle will spend as much time on the high reading as it does on the low if there is a problem with two of the cylinders. If there is one bad cylinder, the needle will spend one-fourth of the time on the lower reading.

Often the needle of the vacuum gauge will waiver quickly between two readings. This is caused by a slight variance in the vacuum formed by the cylinders. Because of this quick flutter, it is difficult to see the needle. By pinching the vacuum gauge hose, you can slow the needle movement and have a better indication of the engine's condition. What should also be observed is the range of the flutter. The larger the range of the flutter, the more variance there is between cylinders. The level at which the range of the flutter occurs should also be observed. If the flutter occurs in the normal area of the scale, it usually indicates that there is a slight imbalance in the fuel-and-air delivery system. Worn and leaking valve guides will also cause the needle to flutter, but the reading will be several inches below normal (Fig. 11-12). This type of problem will not show up in a compression test or a cylinder leakage test. Worn valve guides allow air to enter from the valve covers into the cylinder. This unmetered air will reduce the amount of vaccum in the cylinder.

If there is a sticking valve, the needle of the gauge will drop randomly. The needle will respond to the decrease in vacuum caused by the random sticking of the valve. As the valve sticks open or hesitates before it closes, the vacuum in the cylinder will decrease. It is unlikely that a valve will stick open each time it begins to close, so the needle of the gauge will drop on a random basis. Watching the gauge for a period of time will allow you to observe the occurrence of this condition. This condition causes the engine to have an intermittent miss. Each time the valve sticks, the engine speed will decrease, causing a decrease in vacuum (Fig. 11-13). This type of problem is often difficult to diagnose. The use of the vacuum gauge will aid in this diagnosis.

If the engine has excessive leakage past the piston rings, the vacuum gauge can be used to identify this. In an engine that has good sealing between the piston rings and the cylinder walls, the needle of the vacuum gauge will drop to near zero and rebound to a reading at least 5 in. higher than the reading at idle, when the throttle is quickly opened and closed. This is the result of the pistons moving quickly

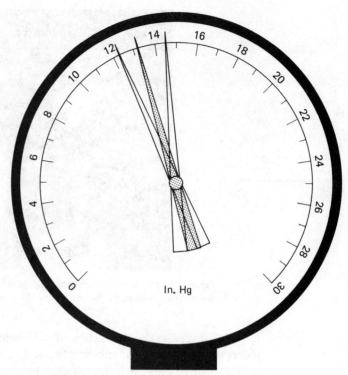

Figure 11–12

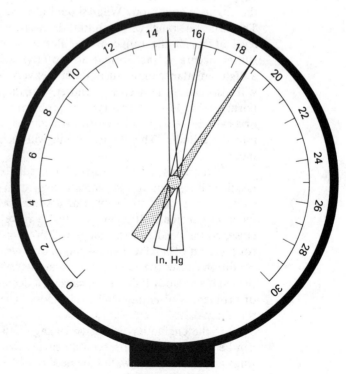

Figure 11–13

through the intake stroke with the throttle closed. The movement of the pistons is faster than at idle, and with the throttle plates closed, it causes a buildup of vacuum in the cylinder. The initial drop to zero, as the throttle is opened, is caused by the large amount of air that is quickly introduced. A cylinder, with excessive leakage around the piston rings, cannot hold the vacuum that is produced with the high piston speeds and the sealing of the throttle plate. Vacuum leaks past the rings and the gauge responds with a rebound reading of less than 5 in. over the idle reading.

While you are quickly opening the throttle and releasing it, watch the vacuum gauge. If the needle vibrates with the increase of speed, it indicates that there is a problem with the valve springs. A weak or broken valve spring will cause the valve to remain open for a time, when it should be closed. The condition of the springs can be checked by holding the engine speed at about 2000 rpm and observing the gauge. If the needle fluctuates with the increase in engine speed, the valve springs are probably weak or worn out. If there is a random drop in the gauge reading as the speed is maintained, it indicates that one or more valve springs are broken.

While the engine is running at 2000 rpm, you can check for an exhaust restriction. An exhaust restriction prevents the exhaust from leaving the cylinders and will reduce the engine's vacuum, because the cylinders will not be empty when the intake stroke begins. A collapsed exhaust pipe or a plugged catalyst can cause this exhaust restriction. To check the flow of the exhaust, observe the vacuum gauge while the engine is running at 2000 rpm. Hold the engine at this speed for at least 15 seconds. If there is a restriction in the exhaust, the vacuum will gradually decrease. If the vacuum stays the same, the exhaust is flowing well. Often, the vacuum reading with a plugged exhaust will drop to a level that is below the idle speed reading.

The proper use of the vacuum gauge can help you decide whether or not you need to conduct a compression or cylinder leakage test. It will also inform you if the cylinders are all forming the same amount of vacuum. This will give you basically the same information as a cylinder power balance test. The vacuum gauge cannot tell you the exact location of the compression losses, but it can tell you if there is a problem and what to do to verify it. Low readings indicate that a problem does exist. Steady needle readings indicate that the problem is a common one, and a fluctuating needle tells you that the problem is not common. The performing of an engine vacuum test is a good way to start a diagnostic tune-up. From the information gathered, you can decide which tests you should conduct to diagnose the problem properly.

ADJUSTMENTS WITH VACUUM GAUGE

The vacuum gauge can also be used to make some adjustments on the engine. These adjustments should be made with the vacuum gauge *only* when the correct equipment is not available. With the vacuum gauge you can adjust both the ignition timing and the idle mixture. The vacuum gauge is especially handy to adjust the mixture of a two- or four-barrel carburetor. With a rich mixture, there is less air to occupy the engine's vacuum; therefore, the vacuum gauge will show a higher-than-normal reading. With a lean mixture, there is more air and the vacuum will be reduced. To adjust a carburetor with a vacuum gauge you simply attempt to achieve a smooth idle, and then turn the mixture adjustment screw in, which leans the mixture until the vacuum begins to drop. On multiple-barrel carburetors, adjust one screw according to the same procedure, and then adjust the other mixture screw to eliminate any fluttering that may occur on the gauge. The flutter of the gauge is caused by the carburetor imbalance. Do not attempt to adjust the carburetor or the ignition timing with a vacuum gauge if the vacuum gauge indicates that there may be another problem.

To adjust the ignition timing with a vacuum gauge, set the engine's idle speed to the recommended speed. Disconnect the vacuum advance and set the engine to the conditions recommended for the adjustment of timing. Loosen the distributor and rotate it in the direction opposite to the one in which it normally rotates. This will advance the timing. As the timing advances, the engine speed will increase. The reading on the vacuum gauge will also increase until there is too much advance. At this point, rotate the distributor in the direction opposite to that in which you were turning it before, until the vacuum has decreased 2 in. from the highest reading achieved by advancing the timing. Readjust the idle speed to specs, and repeat the timing procedure until the idle speed is at the recommended spec and the vacuum reading is about 2 in. below the highest reading. This method will adjust the timing close to the desired specs but should be rechecked as soon as possible with a timing light. Incorrect ignition timing can affect the overall efficiency of the engine.

The use of a vacuum gauge to adjust idle mixture and ignition timing is *not* recommended for engines that are equipped with emission control devices. These engines should always be set to the manufacturers' specifications. Use the two techniques only when it is absolutely necessary.

VACUUM-CONTROLLED ACCESSORIES

Engine vacuum is used to control and operate a number of accessories on the engine, as shown in Fig. 11-14. The malfunctioning of a vacuum-operated device can cause

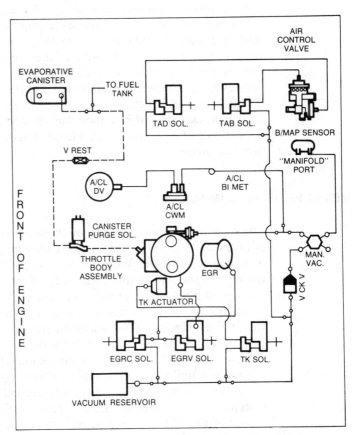

Figure 11-14 (Courtesy of Ford Motor Company)

an engine running problem. These vacuum-operated devices should be checked whenever the conditions suggest that there may be a problem. Usually, these affect the performance of the engine in two ways: they do not operate properly or they leak, which causes a reduction in engine vacuum. For a vacuum-operated device to work, there must be the presence of atmospheric pressure, the presence of a vacuum, and the two must be separated by the use of a movable piston or diaphragm. Two of the most common problems that occur with a vacuum-operated accessory are the lack of vacuum and a loss of a separation between the high and low pressures, because of a leak in the diaphragm or around the piston. At times, the lever or rod that is attached to the piston or diaphragm is not free to move and perform the function for which it was designed.

If the vacuum of the engine is low, we can expect the vacuum-operated accessories not to work to their maximum ability. Low vacuum can also cause an unsafe condition; a good vacuum is essential for the proper operation of power brakes. If the engine has serious problems, do not expect the vacuum-controlled or vacuum-operated accessories to function normally. Before checking the operation of the accessories, check the source of this needed vacuum: the engine.

These vacuum-operated accessories are best tested without the vacuum supply of the engine. If the engine's vacuum is not used, engine controls can also be checked during times they normally do not operate. To test these systems, use a hand-operated vacuum pump (Fig. 11–15) or an electric vacuum pump such as those included in some engine analyzers. These vacuum pumps can supply the needed vacuum to systems in order to test them. Each system will be tested for leaks in the vacuum chamber and for the mechanical action that should result. With the engine running, tests can be performed to see if the system is causing the effect for which it was designed. Vacuum components can be checked for leaks and mechanical action with the engine off. Any vacuum component can be checked with a vacuum pump. To test the operation of a vacuum-operated accessory, you must have an understanding of what should happen when a vacuum is supplied to it. Following are the test procedures for the more common vacuum devices. Others can be tested

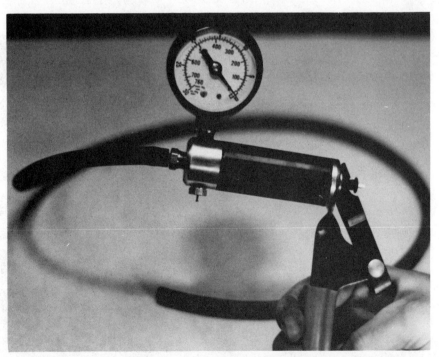

Figure 11–15 Hand-held vacuum pump.

in similar ways. Your understanding of the components will dictate the way to test them.

The purpose of the air intake control valve is to draw in air from around the exhaust, to provide for warm air to the engine when the engine is cold. This warmer air improves the operation of the engine when it is cold. Once the engine warms up, the valve switches, to allow fresh air from the outside to enter the air cleaner assembly. If the valve does not make this switch, warm air will always be delivered to the engine. This is not desirable for complete combustion. Similarly, if the valve does not switch to draw air from around the exhaust, the engine will have poor performance when it is cold. The air intake control valve is controlled by vacuum. A vacuum signal is delivered to the valve and opens it. The vacuum signal is controlled by a thermostatic switch. The vacuum is delivered to the air control valve only when a particular temperature, usually around 100°F, is reached. The operation of the valve can be tested by hooking the vacuum pump to the vacuum fitting on the valve, as shown in Fig. 11-16. Using the pump handle, attempt to form a vacuum of 18 in.Hg. This should cause the door or flap of the valve to open. The valve should also hold the vacuum without having to continue pumping the handle of the tester. If you are unable to form 18 in. of vacuum, or if the valve will not hold the vacuum, the diaphragm of the valve leaks and the valve should be replaced.

Another component that can be tested with the vacuum pump is the distributor advance unit. This unit should be tested with the engine running and with the engine turned off. With the engine off, connect the vacuum pump to the fitting on the vacuum advance unit (Fig. 11-17). Apply 18 to 20 in. of vacuum to the advance unit. The unit should be able to hold this amount of vacuum for quite awhile. If the vacuum cannot be held by the unit, it must be replaced. If the advance unit holds the vacuum, find the vacuum advance specs in your shop manual. The information given will include at least two readings. These will tell you how much vacuum will cause a particular amount of advance. With this information, plug the vacuum advance hose and start the engine with the vacuum pump still connected to the distributor. Hook up a timing light and check the engine's initial timing. With no vacuum pumped up in the vacuum pump, the timing light will only show the

Figure 11-16 Vacuum pump connected to air intake control valve for testing.

Figure 11-17 Testing the vacuum advance unit with a vacuum pump.

initial advance. Pump the vacuum pump to form the amount of vacuum given first in the specs. The timing should have increased to the amount given in the specs. If it does not increase the correct amount, the advance unit should be replaced. If the unit causes the correct amount of advance at the first vacuum level, the unit should be tested at the other vacuum levels.

If the distributor vacuum advance unit is not working properly, overall engine driveability and economy will suffer. The required changes in ignition timing for changes in engine load will not occur. If the advance unit does not hold the applied vacuum, not only will it not function, but the leak will result in an improper amount of air in the cylinders, as well as incorrect ignition timing.

Some distributor advance units and other engine components have their vacuum lines interrupted by a delay or switching valve (Fig. 11-18). These valves can also be checked with the vacuum pump. The purpose of a delay valve is to slow the vacuum signal to a particular component. When vacuum is supplied to a delay valve, it will hold the vacuum and then slowly lose it. To test a delay valve, apply a vacuum

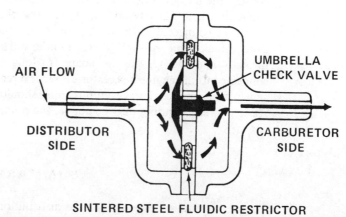

Figure 11-18 Spark delay valve. (Courtesy of Ford Motor Company)

to one side of the valve. The valve should hold the vacuum and slowly lose it. If it does not hold the vacuum, it is leaking, or it is not delaying the vacuum signal. In either case, the delay valve should be replaced. If the delay valve never loses the vacuum and holds the vacuum for a long period of time, the valve should be replaced. A malfunctioning delay valve can cause a number of engine problems. The exact problems will depend on the systems controlled.

Most switching vacuum valves are controlled by temperature. These valves can be the source of vacuum leaks or may be allowing the vacuum to switch at the incorrect times. To test these, identify the conditions of the switching and duplicate those conditions with the vacuum pump. By removing the temperature-controlled switches from the engine, you can change their temperature. By holding the valve under cold water or by placing on it a plastic bag filled with ice, you can test its cold operation. Using warm water or a heat gun, you can test the warm operation.

Faulty EGR valves can cause the engine to have a rough idle, poor performance, and poor fuel economy. The opening and closing of the EGR valve is controlled by vacuum. If the valve is open when it should not be, the engine will be receiving the exhaust into the cylinders, and this will affect the mixture in the cylinders. Some EGR valves can be tested for vacuum leaks and to see if they open and close properly. To test for vacuum leaks, apply a vacuum to it with the engine off. The valve should be able to hold the vacuum at least momentarily. When the vacuum is applied to the valve, the valve stem should move. This is the opening of the valve. If the valve holds a vacuum and opens with the vacuum, the valve closing should be tested. To do so, start the engine and run the engine at an idle speed. Normally, the EGR valve is closed when the engine is at idle. Apply a vacuum to the valve. The opening of the valve should cause the engine to run rough. If this does not happen, the valve was probably open before the vacuum was applied or the EGR passage is plugged. The EGR valve should be suspect if there is little or no change in the running of the engine when a vacuum is applied to it. If the problem is found to be the EGR valve, it should be replaced.

A vacuum pump can be used to test a number of other systems on the engine, such as the power valve in the carburetor, vacuum amplifiers, vacuum lines, nonvented gas caps, and air-conditioning control units. Through an understanding of how these components operate, a testing method can be devised for these, as well as any other vacuum-controlled or vacuum-operated device. A vacuum leak in any vacuum system will affect the operation of that system and may cause the engine to run abnormally.

Some gasoline and most diesel automobile engines are equipped with an engine-driven vacuum pump (Fig. 11–19). This pump is used to operate the accessories of the vehicle. The reason for the addition of the pump on the engine is because these engines do not produce a sufficient amount of vacuum to operate its accessories continuously and safely. To test the vacuum pump, a vacuum gauge should be connected to the outlet port of the pump. With the engine running, the gauge should read 16+ in.Hg. upon initial starting and the reading should increase through time. Do not run the engine for long periods without the vacuum output connected to the engine's accessories. Once a reading of 16+ is given on the gauge, shut off the engine. The vacuum reading should decrease very slowly. If the gauge readings were anything but the above, the pump probably needs to be replaced.

SUMMARY

Engine vacuum is necessary for the distribution of the air to the cylinders and is used to operate and control a number of systems on the engine. The testing of these vacuum-operated devices can help you locate the cause of some engine problems.

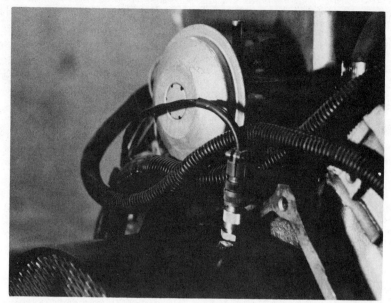

Figure 11-19 Vacuum pump driven by the engine as used on some gasoline and most diesel engines.

The monitoring of the vacuum formed by the engine is quick way to begin engine diagnosis. With the engine vacuum test, you can define the problem and decide what tests chould be conducted to identify the problem. It takes only minutes to conduct the engine vacuum test. This test is a quick but precise way to start your engine troubleshooting.

REVIEW QUESTIONS

1. How and when is vacuum formed by the engine?
2. Why will a cylinder that has low compression also produce a lower amount of vacuum?
3. How does a vacuum test determine how well a cylinder is sealed?
4. What is a vacuum leak?
5. What two things determine how fast the outside air will enter and fill a cylinder?
6. How do the throttle plates regulate engine speed?
7. How is engine vacuum used to allow other engine systems to perform work?
8. What is the relationship between an engine's ability to form a vacuum and its compression?
9. In what units is vacuum measured?
10. While conducting an engine vacuum test, why should the vacuum gauge be connected to the intake manifold below the throttle plates?
11. What is indicated by a zero reading on a vacuum gauge during an engine cranking vacuum test?
12. What is indicated by a steady needle on the vacuum gauge during a vacuum test?
13. What is the proper procedure for conducting a test of the PCV valve during an engine cranking vacuum test?
14. Why should the engine's idle speed and ignition timing be checked prior to conducting an engine running vacuum test?
15. How will the vacuum gauge readings be affected by an engine's ignition timing being overly advanced?
16. How does the engine's compression ratio affect the vacuum readings?

17. What tests can be done to determine if the engine has an internal or external vacuum leak?

18. What is usually indicated by a low but steady reading on a vacuum gauge?

19. How will the vacuum gauge respond to one bad valve in one cylinder?

20. If the individual cylinders of an engine are receiving different amounts of an air-and-fuel mixture, how will the vacuum gauge respond?

21. How can the cause of an external vacuum leak be found?

22. How will the vacuum gauge react to a sticking valve?

23. How will the vacuum gauge react to a broken valve spring?

24. How can the vacuum gauge be used to determine how well the piston rings are sealing against the walls of the cylinders?

25. How can the vacuum gauge be used to determine if a catalyst is plugged or if there is a restricted exhaust?

26. What are the similiarities between a cylinder power balance test and an engine vacuum test?

27. How can a vacuum gauge be used to adjust a two- or four-barrel carburetor?

28. A vacuum gauge can be used as a temporary means to adjust ignition timing. What is the proper procedure for doing this?

29. What are the two consequences of having a vacuum leak in a vacuum-operated accessory?

30. What is the purpose of the air intake control valve?

31. A distributor vacuum advance unit should be tested in two ways. What are they and how do you test them?

32. If you suspect that an engine has a leaking EGR valve, how do you test the valve using a vacuum pump?

12

EXHAUST QUALITY TESTING

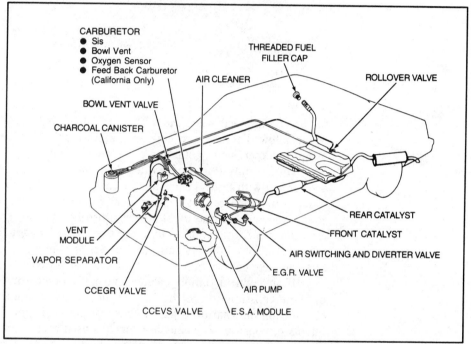

CARBURETOR
● Sis
● Bowl Vent
● Oxygen Sensor
● Feed Back Carburetor
(California Only)

THREADED FUEL
FILLER CAP

ROLLOVER VALVE

AIR CLEANER

BOWL VENT VALVE

CHARCOAL CANISTER

VENT
MODULE

VAPOR SEPARATOR

CCEGR VALVE

CCEVS VALVE

E.S.A. MODULE

AIR PUMP

E.G.R. VALVE

AIR SWITCHING AND DIVERTER VALVE

FRONT CATALYST

REAR CATALYST

Figure 12-1 (Courtesy of Chrysler Corporation)

During the past 20 years, the concern over the quality of the exhaust from cars has caused automobile manufacturers to add and modify many things, attempting to reduce the amount of pollutants released by the exhaust. The basic design of engines has been modified to make them run more efficiently. Systems have been added to prevent the formation of the pollutants during the combustion process. These have resulted from the fact that an internal combustion engine will release some pollutants unless the engine is run at ideal conditions. Yet to be developed is a car with an engine that operates under ideal conditions. These conditions seem to exist only on paper. The air is not pure oxygen and the fuel is not without some contaminants. Temperatures and pressures change from hour to hour (Fig. 12-2). These are all problems with the environment in which the engine must operate.

The engine itself cannot operate in an ideal way either. Because of the environment, the engine does not operate in the same conditions every time it runs. Fine adjustments should constantly be made to compensate for these environmental changes, to allow the engine to have total combustion. If these adjustments were made, the engine still requires the correct amount of fuel to be mixed with the correct amount of air in a totally sealed container. This mixture must be compressed the right amount to allow a sudden and precise amount of heat, to cause the immediate expansion of the gases in the container. This heat increase must occur at

Figure 12–2   Weather conditions change from hour to hour as the barometric pressure changes.

the precise moment when the mixture is fully compressed. These exact conditions would allow the engine to emit safe gases from the exhaust, because the engine would be experiencing total and complete combustion.

Total combustion occurs when the right amount of fuel and air are combined and ignited with heat. The results of complete combustion are harmless gases and water. An engine cannot achieve total combustion because of the environmental conditions and because it cannot be controlled precisely enough to allow for the exact needs of total combustion. Advances in computer technology allow for improved control of the systems, and the result is more complete combustion. Total and complete combustion has still not been achieved.

If an engine is running, some degree of combustion is occurring. The purpose of some emission control devices on engines is to convert the products of poor combustion into harmless gases. These products of poor combustion are the pollutants of concern. The emission controls used today cannot eliminate all the pollutants from the exhaust, but they do reduce them quite a bit. Any engine operating with all its systems in good condition and in good condition itself will emit a certain amount of pollutants. As the condition of the engine becomes poorer, the level of pollutants from the exhaust will increase. The minimum level of pollutants will be emitted only when the engine and its systems are in a fine state of tune. Certain problems will cause particular pollutants to increase in the exhaust. If the engine is equipped with emission controls and if any of these malfunction, the exhaust will have an increase in the pollutant that the device was to control.

The reduction of exhaust emissions was mandated and is governed by the federal government. With the government mandates came standards for the maximum allowable pollutant levels in the exhaust. These levels had to be met by automobile manufacturers before they could release their cars for sale. The engines of these cars were built and modified to meet these standards. After the car was sold, it was subjected to inspection of exhaust quality by agencies of some states. For the car to meet the standards after the car had been driven for some time, it had to be kept in good condition and in good tune. These exhaust quality inspections were often used to identify the problems of the engine or diagnose engine problems.

Testing the exhaust quality is a means to examine the end results of combustion. If the pollutant content is excessive, there is evidence of a problem. The determination of what is excessive is based on a comparison of the standards for that model and year of engine. Although the standards refer to emission controls and mandates from the government, the standards can be used as specs for the quality of the exhaust. When a technician is testing the exhaust, the primary concern should be diagnostics, not compliance with the standards!

With the establishment of the standards, certain pollutants were identified: hydrocarbons (HC), carbon monoxide (CO), and oxides of nitrogen (NO_x). Exhaust analyzers typically measure the content of two of these: HC and CO. These will be the focus in our study of exhaust testing. NO_x is difficult to measure, and although it is of great concern, it is ignored somewhat in engine diagnostics. The standards for the levels of HC and CO vary from year to year. This is the result of the legislation reducing the maximum levels of the pollutants on a yearly basis. For diagnostic purposes, certain years and their standards can be grouped together. This allows for diagnostic guidelines that are easier to understand and remember.

HYDROCARBONS

The measurement of HC is expressed in parts per million (ppm) (Fig. 12-3). This measurement is based on the amount of HC that is present in a sample of exhaust that has a total of 1 million parts. The result of total combustion would be 0 ppm of HC. HC emissions are unburned fuel. If the engine burns all the fuel it receives, it is quite efficient. Amounts of HC that are greater than the allowable emission standard levels indicate that the engine is not as efficient as it should be. Combustion is not as complete as it was designed to be. The lower the amount of HC in the exhaust, the more complete the combustion is in the engine. If the engine is equipped with an air injection system and/or a catalytic converter, the amount of HC in the exhaust not only determines the degree of combustion but also the effectiveness of the emission controls. For diagnostic purposes, the concern should

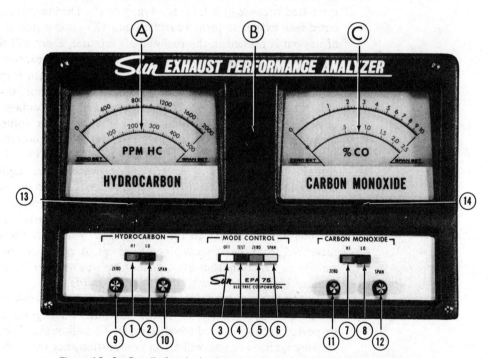

Figure 12-3 Detail of typical exhaust analyzer: A, hydrocarbon meter; B, low flow/check filter indicator; C, carbon monoxide meter; 1 and 2, scale selectors for HC meter; 3 to 6, controls for the different modes of operation of the analyzer; 7 and 8, scale selectors for the CO meter; 9 to 12, meter calibration knobs; 13 and 14, mechanical adjustments for the meters. (Courtesy of Sun Electric Corporation)

be on the amount of HC released by the engine, not just the amount of HC that leaves the exhaust pipe of the car. If the emission controls are working properly but the engine is running poorly, the HC levels from the exhaust may be within the standards. This would falsely indicate that there is no problem with the combustion of the engine. If an exhaust sample is obtained before the exhaust quality is altered by the emission controls and the engine is running poorly, the level of HC would exceed the standards.

The HC standards for 1967 and older vehicles fall into two categories: 900 ppm or less for all engines equipped with a distributor vacuum advance that senses intake manifold vacuum and 500 ppm for all other engines, including those equipped with a vacuum advance sensed by ported or venturi vacuum. Vehicles produced from 1968 to 1970 should have no more than 400 ppm HC in all states but California, where the standard is set at 275 ppm. Because of the number of vehicles and the presence of heavy smog, California has standards for lower levels of pollutants than in the other 49 states. The maximum allowable limit of HC for vehicles produced from 1971 to 1974 is 300 ppm. Most of these vehicles were equipped with an air injection system to reduce the HC level. In 1975, manufacturers added catalyst converters to the vehicles. This was an attempt to meet the standard of 200 ppm. Engines with converters easily met this standard; in fact, engines equipped with converters from 1975 to the present should have 100 ppm or less. Use this spec for diagnostics on engines equipped with a converter.

CARBON MONOXIDE

The other major pollutant of the internal combustion engine is a product of combustion. HC is emitted from the exhaust as a result of incomplete combustion. CO is emitted because of a lack of oxygen or air. During combustion, the HC is combined with oxygen to form a harmless gas, CO_2, and water. If an insufficient amount of oxygen is present in the fuel-and-air mixture, there will not be enough to form CO_2 and the pollutant CO is the result. One part of carbon (C) is mixed with two parts of oxygen (O) to form CO_2. If not enough oxygen is present, the result is the formation of CO. This is a deadly gas which is odorless, tasteless, and invisible.

The measurement of CO is expressed in a percentage (Fig. 12–4). This percentage is based on the volume of CO in a particular volume of exhaust. If there is enough oxygen to combine with all the available carbon molecules in the mixture, the exhaust will have no CO and a large quantity of CO_2. The air in the atmosphere contains a low amount of oxygen and in order for an engine to come close to receiving the proper amount of oxygen, it needs to receive as much air as possible. Induction systems have a difficult time suppling the correct amount of air and fuel to provide for the ideal fuel-and-air mixture. Air injection systems supply an additional amount of air to the exhaust and will lower the amount of CO that is measured from the exhaust pipe of the car. A rich fuel-and-air mixture is best described as a mixture that has a lack of air. An overly rich mixture will result in the exhaust levels of CO being above the standards.

The CO standards for vehicles produced before 1968 are set at 1 to 5%. There are no allowance differences in the standards for engines equipped with manifold or ported vacuum-sensed advance units. This allowance is concerned only with HC because ignition timing will affect combustion, not the fuel-and-air mixture. Combustion must take place for CO to be formed. The CO guidelines for 1968 to 1970 vehicles is 0 to 4%. These levels were reduced to 0 to 3% for vehicles produced from 1971 to 1974. Engines equipped with a catalytic converter from 1975 to present have a standard of 0.2% CO. Engines without a catalyst of the same years have a standard of 0.5 to 1.5% CO. Since most vehicles have been equipped with a con-

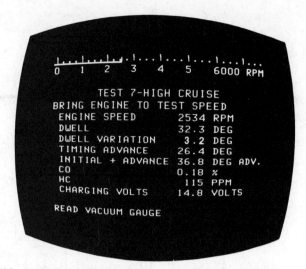

Figure 12-4 Some engine analyzers show the readings of HC and CO in a digital readout, as shown here. (Courtesy of Sun Electric Corporation)

verter since 1975, use the 0.2% spec as a diagnostic spec when measuring CO from the tailpipe of the car.

EMISSION STANDARDS

These standards for CO and HC refer to readings taken at engine idle speeds. The exhaust quality of an engine should be lower than these standards at higher engine speeds. This is especially true for the HC standards. An engine should run more efficiently at highway speeds and under no-load conditions. Table 12-1 presents a summary of the standards or guidelines to use in engine diagnostics. It is important to compare the readings of the engine's exhaust to these specs at both idle and high speeds. This is necessary to check those systems that are unique to particular speeds.

Most vehicles are equipped with an emission control and compliance decal or tag in the engine compartment (Fig. 12-5). This decal lists the standards that the engine meets. Often, this also contains the specs for exhaust quality. It should be noted that these standards are given in different units of measurement than those given in this text and those used on most exhaust test equipment. In 1970, there was a change in the manner of measuring the pollutant quantity of the engine's

TABLE 12-1 GUIDELINES FOR ENGINE DIAGNOSTICS

Year	At idle speeds		At 2000 to 2500 RPM	
	Max. HC (ppm)	CO (%)	Max. HC (ppm)	CO (%)
1967 or older with manifold dist. vacuum	900	1–5	300	1–2
1967 or older without manifold dist. vacuum	500	1–5	300	1–2
1968 to 1970	400	0–4	300	1–2
1971 to 1974	300	0–3	200	0.5–1.5
1975 and newer with converter	100	0.2	100	0.2
1975 and newer without converter	200	0.5–1.5	150	0.2–0.5

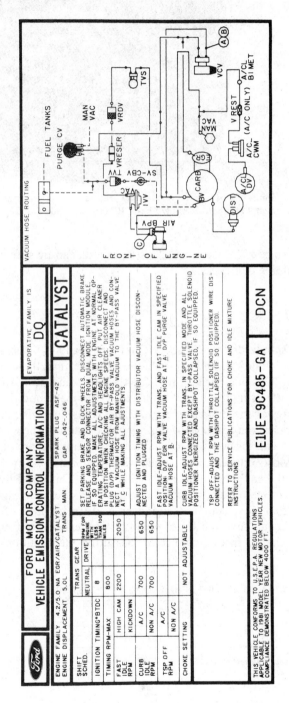

Figure 12-5 (Courtesy of Ford Motor Company)

exhaust. The original measurements were based on parts per million for HC and in percentage for CO. The new units of measurement are based on grams per mile. Most diagnostic equipment measures the HC and CO in ppm and percent. Do not let the decal confuse you. The guidelines given here are for diagnostics, not for the certification of exhaust quality. However, if the vehicle is within the specs given in this text, the engine should meet the federal standards.

EXHAUST GAS ANALYZER

An exhaust gas analyzer is a diagnostic tool that can be used to identify problem areas and to pinpoint malfunctioning systems. The analyzer in Fig. 12–6 is equipped with meters that display the HC and CO content of the exhaust. The exhaust is the product of combustion, and by monitoring the results of combustion, much can be discovered about the engine and its systems.

Most exhaust analyzers are infrared analyzers. The exhaust gas is analyzed by a beam of infrared light as the exhaust passes through it (Fig. 12–7). The light beam then passes by a series of optical filters that convert the amount of light passing through the exhaust into electrical signals. These electrical signals operate the meters. Infrared light is used to analyze the exhaust because it has the ability to penetrate. The amount of light that penetrates the exhaust determines the content. Exhaust analyzers are calibrated to allow for separate readings of the HC and CO content.

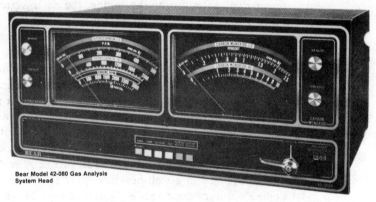

Bear Model 42-080 Gas Analysis
System Head

Figure 12–6   (Courtesy of Bear Automotive)

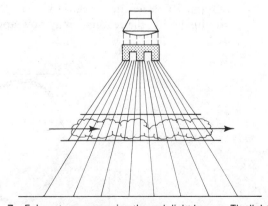

Figure 12–7   Exhaust gases passing through light beams. The light that shines on a series of filters determines the reading on the scales.

Some exhaust gas analyzers have a third or fourth meter. These analyzers also determine the amounts of oxygen and CO_2 present in the exhaust. These gases are harmless and are of no concern to environmentalists. However, by monitoring these gases, the effectiveness of emission controls can be determined. More precise diagnostics are also possible from comparison of the amounts of pollutants to the amounts of oxygen and carbon dioxide. The use of these diagnostic readings will be discussed later in this chapter. Since the more commonly used gas analyzers measure the amounts of HC and CO, these will be the initial focus of the use of the analyzer as a diagnostic tool.

Before using an exhaust analyzer, it must be turned on for a short period. This is necessary to allow the light to warm up. After about 10 to 30 minutes, the analyzer must be calibrated. There are many manufacturers of exhaust analyzers and each has its own method of calibration. Consult the operator's guide for the analyzer you are going to use to find the specific way to calibrate that analyzer. Most analyzers have some means to adjust the zero setting on the meter. Because the quality of air present in the analyzer will vary, the meters need to be adjusted to the air in which it is operating. Most analyzers will calibrate the air around them as zero readings in both CO and HC. Doing so, the exhaust will be compared to the air in the shop or in the environment of the tester. There is also an adjustment in the calibration for the maximum range of the meters. Once calibrated, the analyzer is ready to use.

The analyzer is equipped with an exhaust probe fastened to the end of a long tube (Fig. 12-8). This probe should be inserted into the exhaust pipe. If the engine is equipped with an air injection system, this should be disconnected. To do so, locate the diverter valve. Disconnect and plug the vacuum line that goes to the valve. This is done to prevent the air injection system from giving false information about the engine. After testing the engine, the air injection system can be reconnected to test the effectiveness of the system on the exhaust quality.

Some Chryslers and other engines have an access plug before the catalytic converter for the insertion of the exhaust probe. Inserting the probe before the converter and disconnecting the air injection system allows the analyzer to measure the contents of the engine's exhaust without alterations in these emission controls. If the car you are testing is equipped with such a plug, use it. After the engine is tested, the emissions from the tailpipe can be measured. Care should be taken on any engine equipped with a catalytic converter not to allow the engine to run for long periods of time. The longer the engine runs, the hotter the catalyst gets. The heat can cause damage to the exhaust probe. A secondary combustion process may also take place in a well-heated converter, which can drop the readings from the engine's true exhaust reading significantly. The biggest problem with this is that after the engine has been running for awhile with the exhaust analyzer hooked, the CO and HC levels may begin to drop. If the original readings were above the allowable limits, the readings may now appear to be lower, as if the problem cured itself.

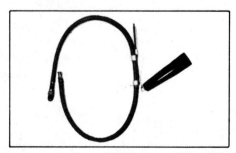

Figure 12-8 Exhaust probe assembly. (Courtesy of Sun Electric Corporation)

The engine should be started and run at an idle speed. The readings on the meters should be compared to the guidelines, according to the year in which the engine was produced. Any reading higher than the spec should be noted. For example, if the spec for the engine is 200 ppm HC and the reading at idle is 300 ppm, there is evidence of high HC. HC, you may recall, is unburned fuel and it is in the exhaust as a result of poor combustion.

Not only should you be concerned with high readings but also with readings that are within the specs but seem to be unbalanced. If the HC reading is at the low end of the allowable range and the CO is at the top of its range, there is an imbalance. This should be regarded as high CO, which is caused by a lack of air. This type of imbalance may be caused by an improper fuel-and-air mixture, a slightly rich one.

The presence of HC in the exhaust is evidence of poor combustion. The higher the level of HC, the worse the problem is. Not enough air in the mixture will cause CO. CO is a product of combustion. When you use the readings of the exhaust analyzer as a diagnostic tool, you should keep the readings of the two meters separate. CO is formed by combustion and HC is the result of a lack of combustion. If a cylinder of an engine is not firing, the exhaust will contain an excessive amount of HC and a low amount of CO (Fig. 12-9). The fuel that is delivered to that

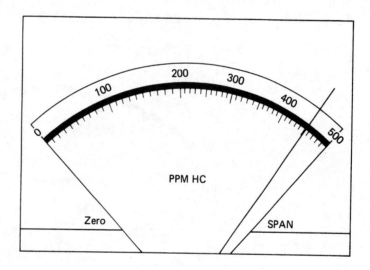

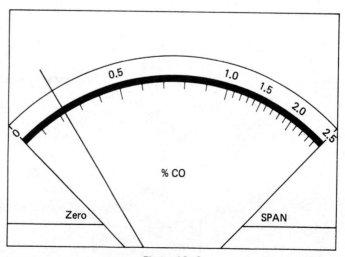

Figure 12-9

cylinder will leave the engine through the exhaust. This is the cause for the high reading in HC. This cylinder will also affect the CO level because combustion is not taking place. The total content of the exhaust will be minus the amount of CO that would be normally produced by the cylinder.

The probable causes for a cylinder not firing would be systems or components that are not common to all cylinders. Examples of these types of problems include an open plug wire, bad valves in a cylinder, and a fouled or broken spark plug. If all the cylinders were firing and the readings were high HC and low CO, the most probable cause would be an excessive amount of air. Too much air in the mixture would not support complete combustion but would allow for the production of CO_2. The abundance of air would limit the production of CO which forms because of a lack of air.

From these readings on the exhaust analyzer, two areas need to be tested further. Conducting a cylinder power balance test will identify whether or not there is a cylinder that is not firing. If all cylinders are firing, the area of concern is the excessive air that is being delivered to the cylinder. This can be caused by a lean mixture or a vacuum leak. Testing the engine with a vacuum gauge should identify the problem as either a leak or a problem with the mixture.

If the results of an exhaust test show that there is high HC and a normal amount of CO (Fig. 12–10), it indicates that there is a combustion problem but that

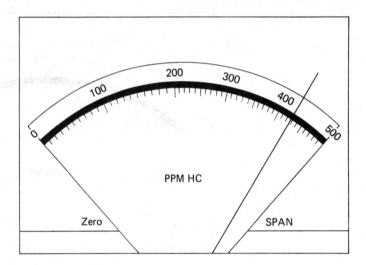

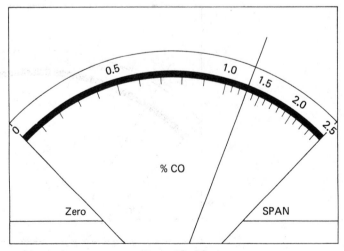

Figure 12–10

some combustion is taking place. The normal amount of CO suggests that all of cylinders are producing some CO. The problem causing the high HC is probably common to all cylinders and is having a slight affect on combustion. If all cylinders are releasing slightly more HC than what is normal, the total of all the cylinders in the exhaust pipe will be great. The probable causes for high HC and normal CO would be low compression, retarded or advanced ignition timing, or a problem in the primary ignition.

An extremely rich mixture can cause the exhaust to contain both high HC and CO (Fig. 12-11). Too much fuel in the mixture can cause this in two ways: the amount of fuel may be too great to support combustion, or the excess fuel may create an excessive amount of carbon in the cylinders and on the spark plugs which cannot effectively ignite the mixture. If the amount of fuel is too great to support combustion, the engine will tend to quit after running a short while at idle. An extremely rich mixture can be caused by a number of carburetor problems or fuel system problems. Diagnostics should continue with an inspection of the spark plugs and then a visual inspection of the fuel system and carburetor. Testing the exhaust at a higher engine speed will allow for the separation of the circuits of the carburetor and fuel system. If the levels of HC and CO decrease at the higher speeds, this indicates that the problem is in the idle circuit. The problem cannot be any system that operates at all engine speeds, such as the fuel pump.

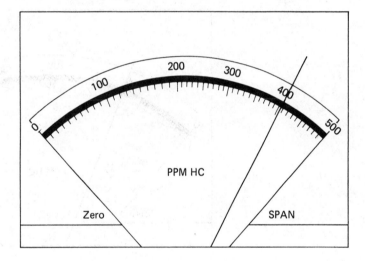

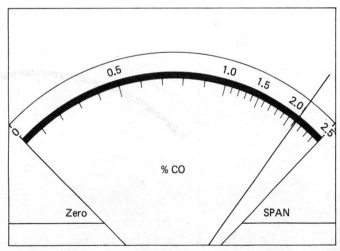

Figure 12-11

A lack of air or a rich mixture is shown on the meters in Fig. 12–12 as normal HC and high CO readings. The engine is able to burn the extra amount of fuel, and therefore the HC level is within standards. There is a rich mixture because of a lack of air or because of extra fuel. The probable causes for this condition are a plugged PCV system, a dirty air filter, a choke stuck closed, a high float level, or a leaking power valve in the carburetor. Diagnostics should continue with a visual inspection of the air filter and choke. The PCV system should be tested either with a vacuum gauge or with the exhaust analyzer. The method for doing so will be discussed later in this chapter. If the problem ends up being in the carburetor, it must be taken apart and rebuilt to correct.

The key to effective diagnostics with an exhaust gas analyzer is to determine carefully the cause of the readings. Whenever HC is high, the cause is poor combustion. High CO is always caused by a lack of air or a rich mixture. These facts will lead to the area that should be tested next. The exhaust analyzer can be used for pinpointing problems once there is an understanding of its readings.

The analyzer can be used to determine whether or not a vacuum leak is the cause of a high HC reading. It can also be used as a means to identify weak or dead cylinders. There are many systems that can be checked with the analyzer. What follows are the test procedures for many of these systems and the testing for vacuum leaks and dead cylinders.

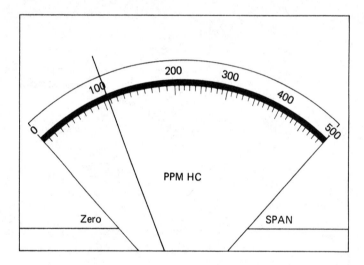

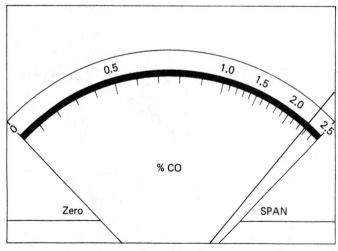

Figure 12-12

Vacuum leaks and overly lean mixtures are the common cause for readings of high HC and low CO. If the engine has a vacuum leak, the idle-mixture screws can be turned in without a reaction to the CO meter. You should be aware that it takes some time, about 7 seconds, for the exhaust analyzer to reflect any change. This is the amount of time needed for the exhaust to exit the engine, enter the analyzer, and be measured. If the mixture screw can be turned without causing a change in the CO meter, unmetered outside air is diluting the mixture and a vacuum leak is evident.

To locate the vacuum leak, inspect all vacuum hoses and fuel system bolts for tightness. Then remove the air cleaner assembly and plug the vacuum lines leading to it. Partially cover the air horn of the carburetor or throttle body. Note the readings on the analyzer. If the CO and HC readings increased, the problem is not a vacuum leak but is in the fuel system. If the CO and the engine speed increased, a vacuum leak exists. To find the cause of the vacuum leak, squirt oil around the throttle assembly base and watch the CO meter. If the CO reading goes up, the oil has sealed the leak and the location of the leak is where the oil was squirted. All likely points for a vacuum leak should be checked in the same way. This should be continued until all vacuum leaks are found.

If no leaks are found, disconnect the vacuum lines to each vacuum accessory. Plug the vacuum port from where they were disconnected. If the CO increases with a particular system disconnected, the problem is in that system. A vacuum hose or component may be leaking.

To determine whether or not all the cylinders are firing, the exhaust analyzer can be used. Note the readings of the CO and HC meters when the engine is at idle. Using an insulated pair of pliers, remove one spark plug wire and ground it. If that cylinder was firing, the HC reading will increase and the CO decrease. If the cylinder was not firing, there will be no change in the readings on either gauge. Do not keep the wires off for an extended period of time if the engine is equipped with a catalyst converter. The raw fuel (HC) that would be present in the exhaust could ruin the converter. This procedure can be done on all cylinders one at a time, until all are checked.

PINPOINT TESTING WITH AN EXHAUST GAS ANALYZER

The accelerator pump of a carburetor can be tested with an exhaust analyzer. To do this, run the engine at idle and note the readings on both gauges. Now, quickly open the throttle and allow it to close. The HC and CO readings should climb quickly to a high reading and then drop back to the idle speed readings. If the accelerator pump is operating properly, the CO will increase at least 3% on engines that are not equipped with a catalytic converter, and 2% on engines equipped with one.

If the reading of the CO meter does not increase, the pump is not operating. If the reading increases, but not to a high level, this can be caused by a worn accelerator pump assembly, an improperly adjusted pump linkage, or an improperly seated check ball in the accelerator pump. If the reading decreases before it increases, it indicates that a large volume of air entered the cylinders prior to the squirting of the fuel. This is caused by poor accelerator pump action due to a poorly adjusted pump actuator linkage or worn pump linkages. Any of these problems would be evident by the engine's inability to accelerate smoothly.

The PCV system is a calibrated vacuum leak (Fig. 12–13). The engine is normally designed and adjusted to run with this source of unmetered air. If this leak is removed, the engine will emit an excess of CO in its exhaust. To test the PCV system, run the engine at idle speed and note the CO reading. Remove the PCV

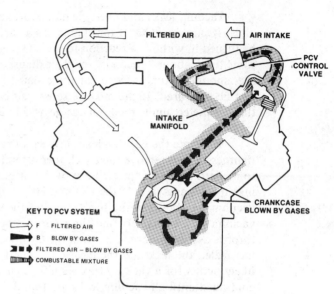

Figure 12-13 (Courtesy of General Motors Corporation)

valve from the engine and plug the end of it with a rag. If the PCV system has been working properly, the CO reading will increase when the valve is plugged. If there is no increase in the reading, the system is plugged. To determine where the blockage is, remove the PCV valve from the hose and plug the end of the hose. If the reading increases, the valve is plugged. If there is no change in the reading, remove the hose from the manifold and plug the fitting on the manifold. An increase now in the CO indicates that the hose is plugged and should be cleaned or replaced.

The exhaust analyzer can be used to test for fuel and exhaust leaks. Fuel leaks will be spotted by a reading on the HC meter. The HC and CO meters will be used to find an exhaust leak. To find the leaks, take the probe of the analyzer and scan the car at the most likely spots, such as wherever two pipes join together. Keep in mind that it takes approximately 7 seconds for the fumes to be analyzed; therefore, do not scan the vehicle too fast! The leak will cause the meters to show a reading.

The analyzer can also be used to test for oil contamination. Engine oil needs to be acid and dirt free. If the oil becomes contaminated, it will have CO content. Since oil is a hydrocarbon, the HC meter should be ignored in this test unless the HC reading is very high, which indicates the presence of raw gas in the oil. To test the oil, place the exhaust probe into the dipstick tube with the engine off. If the reading is 0.3% or more, the engine oil and filter should be replaced.

ADJUSTMENTS WITH ANALYZER

The idle mixture can be tested and adjusted with an exhaust analyzer. Before doing so, the rest of the engine must be in good condition. Prior to starting the adjustment sequence, the engine should be run at a high speed (1500 to 2000 rpm) for at least 30 seconds, to clear out the exhaust and converter. A tachometer should be connected to the engine and the analyzer calibrated. Return the engine to its recommended idle speed and take the readings from both meters. Turn the idle-mixture screw out to richen the mixture. Reset the idle speed and turn the idle mixture screw in a little at a time, until the engine's exhaust is within the standards. If the engine begins to run rough while attempting to bring the engine to the CO specs, stop! Because of the secondary combustion that takes place in the exhaust on an engine equipped with an air injection system and/or a catalytic converter, there is a 1%

inaccuracy in the CO readings taken at the tailpipe. Add 1% to the CO spec and readjust to meet this altered spec. You will notice that as the CO drops and the engine begins to run rough, the HC begins to increase. This is the result of a too-lean mixture. Adjusting for the additional 1% CO will compensate for that overly lean condition.

FOUR-GAS ANALYZER

Newer exhaust analyzers are equipped to measure the carbon dioxide, oxygen, HC, and CO content of the exhaust (Fig. 12-14). These additional readings are valuable for diagnostics on engines equipped with catalytic converters. A converter, when functioning properly, will decrease the CO and HC readings at the tailpipe of an engine's exhaust. Often, it is difficult to determine if the readings are the result of the engine or the converter.

During the combustion process, oxygen is combined with fuel (HC) to form water and carbon dioxide (CO_2). This is the end result of total combustion. The more CO_2 that is formed, the less CO will be formed. By monitoring the amount of CO_2 in the exhaust we are able to obtain an indication of the efficiency of combustion. As combustion efficiency increases, the amount of CO_2 increases and CO levels decrease, as will the level of oxygen (O_2). High amounts of oxygen with low amounts of CO_2 in the exhaust indicate that the engine is experiencing poor combustion. Normally, there would be a high HC reading with poor combustion, but the catalytic converter may clean the HC from the exhaust, and this will not be measured by the analyzer. The CO_2 and O_2 readings will not be greatly changed by the converter.

A well-tuned engine with an air/fuel ratio of about 14.5:1 will have exhaust gases consisting of approximately 2.5% oxygen and 14.5% carbon dioxide. As the air-and-fuel mixture becomes rich, the amount of oxygen in the exhaust will increase and the CO_2 will decrease. With an air/fuel ratio of 10.5:1, the CO_2 level will drop

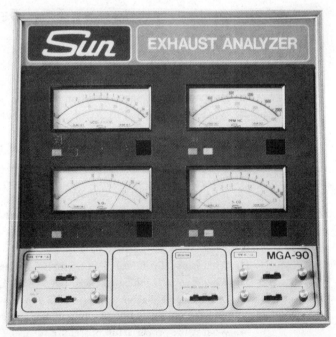

Figure 12-14 Four-gas exhaust analyzer. (Courtesy of Sun Electric Corporation)

to about 8.0%. With this rich a mixture, the CO level would be at approximately 10%. By monitoring the CO_2 levels, a truer estimation can be made of the air/fuel ratio. The ideal air/fuel ratio for a particular engine will allow the engine to emit a high level of CO_2. If the air/fuel mixture is adjustable on a particular engine, it should be adjusted to have the highest level of CO_2 that it can produce.

Because an overly lean mixture causes poor combustion, if the mixture is set too lean, the level of CO_2 will be lower than desirable and the level of oxygen will increase. When the engine is producing its maximum amount of CO_2, the amount of O_2 it emits will be low.

Four-gas analyzers can be used for the same purposes as the two-gas analyzers but will tend to give more information about the engine than will the two-gas types. The efficiency of combustion and the proper air/fuel ratio can be determined with a four-gas analyzer. This information can be an aid to the diagnostics of all engines, whether or not they are equipped with an air injection system or catalytic converter. The four-gas analyzer allows a better look at the end results of combustion.

SUMMARY

Exhaust gas analyzers display the results of the combustion that is taking place within the engine. Monitoring the exhaust can identify the inefficiencies of the engine and its systems. The desired gas in the exhaust is CO_2, while excess amounts of CO, HC, and O_2 indicate that there are problems. Understanding what happens during combustion and what is indicated by the results of combustion will enable you to reduce considerable time from your diagnostic survey of an engine problem. Effective diagnostics occurs only when the problem is isolated after it has been defined. The measuring and monitoring of the engine's exhaust can help you determine what system should be tested to identify and solve the problem.

REVIEW QUESTIONS

1. Why is it currently impossible to produce an engine that produces no pollutants or releases none in its exhaust?
2. What are the results of complete and total combustion?
3. How does the engine's condition determine the quality of its exhaust?
4. What should be used to determine whether the amount of pollutants in an engine's exhaust is excessive?
5. What are the three major pollutants emitted by a gasoline internal combustion engine?
6. In what units of measurement do most exhaust analyzers measure amounts of HC?
7. What two major emission control systems were designed to reduce the amounts of HC from the exhaust of an engine?
8. Why do the levels of HC increase as the efficiency of combustion decreases?
9. What major pollutant is the result of combustion, *not* the result of a lack of combustion?
10. What pollutant is caused by a lack of air?
11. What is usually indicated by a low HC reading?
12. What is usually indicated by a high CO reading?
13. During engine diagnosis, why should you use the emissions specs given in the text rather than the specs given on the automobile's emission decal?
14. What is the proper setup procedure for using an exhaust analyzer?
15. Why should an engine's exhaust quality be tested at idle and high speeds?

16. How can a PCV system be tested with an exhaust analyzer?

17. How will a vacuum leak affect the readings on an exhaust analyzer?

18. How will a dirty air filter affect the readings on an exhaust analyzer?

19. How will the readings on the exhaust analyzer be affected by a "dead" cylinder?

20. How can an accelerator pump be tested with an exhaust analyzer?

21. How can an exhaust analyzer be used to locate the source of a fuel leak?

22. How can an exhaust analyzer be used to locate the source of an exhaust leak?

23. What is the proper procedure for adjusting a carburetor with an exhaust analyzer?

24. Newer exhaust analyzers measure four gases. What are these gases?

25. What is indicated by the presence of high amounts of oxygen in the exhaust?

26. What is indicated by the presence of high amounts of carbon dioxide in the exhaust?

27. As the air-and-fuel mixture being delivered to the engine becomes more rich, what happens to the amount of oxygen present in the exhaust?

28. What is the advantage of using a four-gas analyzer over a two-gas analyzer during engine diagnosis?

13

IGNITION SYSTEM TESTING

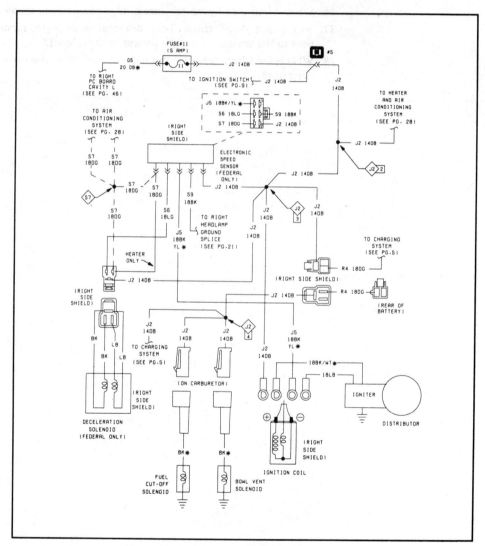

Figure 13-1 (Courtesy of Chrysler Corporation)

The ignition system is responsible for supplying heat needed for combustion. It must supply the correct amount of heat at the correct time. To accomplish this task, the ignition system has two basic circuits and is timed to operate with the movement of the pistons of the engine. If a component of either system fails or if the timing is no longer correct, the engine will not run efficiently. If the other tests conducted on the engine indicate that there may be a problem in the ignition system, each component that could be a probable cause should be tested. The testing of these components can be done as a unit or as individual parts. This chapter covers the testing methods for each component of the ignition system.

The three basic faults possible in an electrical circuit are a short, an open, and high resistance. A short causes lower resistance and higher current flow in the circuit, whereas an open causes such a great increase in resistance that current *cannot* flow in the incomplete circuit. High resistance or resistance higher than normal decreases current flow. These electrical faults can be present in the ignition system as well as in any electrical circuit. From an understanding of Ohm's law, we know that a change in resistance will cause a change in the current flow and in the voltage dropped across a component or load (Fig. 13–2).

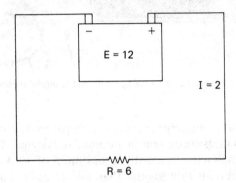

R = 6   Figure 13–2   Ohm's law: $E = IR$

Testing of the ignition system includes resistance checks of the starting, charging, primary ignition, and secondary ignition circuits. Some of these resistance checks are taken while the engine is operating, measuring current and voltage, and others are conducted when the engine is not running, measuring resistance.

To test each component of a conventional or electronic ignition system properly, different types of test equipment are necessary. This equipment will test the electrical properties of the component. When the engine is running, the voltage and amperage will be measured and compared to the component's specs.

Measurements of resistance are taken when the engine is off and there is no electrical activity in the circuit. The amount of resistance in a circuit determines the amount of voltage dropped in the circuit and the amount of current that will flow through the circuit. Although it may be necessary at times to do so, measuring the resistance of the ignition system when the engine is off will not give a precise picture of what is happening in the circuit when it is in operation. The heat caused by current flow can affect the actual operating resistance; therefore, it is better to measure the current and voltage in a circuit while it is operating, to determine its resistance. The measurement of current and voltage drops will indicate the amount of resistance in the circuit. Excessive resistance will cause decreased current flow and increased voltage drops. Increased current flow and lower voltage drops can be caused only by a decrease in resistance.

VOLT-OHMMETER

The basic instrument for testing voltage, current, and resistance is the multimeter. There are two basic types of multimeters: an analog meter and a digital meter (Fig. 13–3). A digital meter displays the measurements as a digital readout and usually has a high operating resistance or impedance. These are usually required for use on computer systems, where the high impedance allows only a small amount of current from the meter to flow through the component being tested. The use of this meter is similar to the use of an analog meter; the major difference is in the reading of

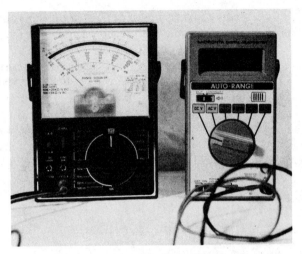

Figure 13-3 Analog meter (left); digital meter (right).

the measurements. A multimeter is often referred to as a volt-ohmmeter because it is seldom used to measure current in automotive systems. The scale of current measurements on these meters is usually a maximum of 10 A. Most electrical systems on a car have a current flow greater than this amount. The resistance and voltage scales measure amounts that include the smallest to the largest amounts found on the car, and are used most often.

These meters are equipped with a selection control which allows the technician to choose the electrical property to be measured. Also on this control is a selection for the scale to be used. There are different scales to select from for voltage, current, and resistance readings. The many scales and ranges of the scales make this meter difficult to read initially, but also give this test instrument much flexibility. Certain abbreviations are used to define individual scales. For current, the prefix "m" is used to represent milli, which stands for 0.001 or one-thousandth of an ampere; 0.5 mA equals 0.0005 A. For resistance, two prefixes are used, "k" and "M". The prefix k represents 1000; 33 kΩ equals 33,000 Ω. The prefix M represents 1,000,000; 5 MΩ equals 5,000,000 Ω.

The meters are also equipped with two leads that plug into the meter: a red lead for the positive probe and a black lead for the negative probe. For accurate measurements, the red probe should always be placed at the most positive point in the circuit and the black lead at the least positive. Proper placement of the probes in a circuit are usually given in the test procedures for the component, but when they are not given, connect the red lead to the positive side or terminal of the component. The positive end is the side that is wired closest to the battery's positive post.

The meter is hooked into the circuit that is being tested in one of two ways: in series and in parallel. When the meter is hooked in series, as it must be to read current, the leads are attached to the circuit as if the meter was part of the circuit (Fig. 13-4). The leads of the meter are attached so that all current flowing in the circuit must pass through the meter. The usual hook-up for a series connection is between a connector and the component. To do this, the connector is unplugged from the component and the red lead of the meter is attached to the connector and the black lead to the component. This hook-up will give the current flowing at that point of the circuit. This hook-up can also be used to measure voltage, although it is not commonly used.

The parallel hook-up is used for measuring resistance and voltage (Fig. 13-5). To measure voltage the circuit must be activated. The red probe is placed at the

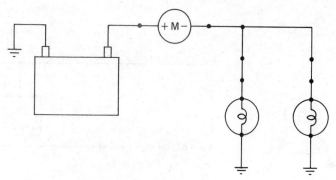

Figure 13–4 Meter hooked in series.

point of the desired measurement and the black lead is placed at the point to which the positive is being compared. This hook-up allows for comparison of one point to another. While measuring resistance, the reading will be the resistance of everything between the red lead and the black lead.

All resistance checks with an ohmmeter should be done with the power supply disconnected from the circuit. The component or circuit tested should be isolated as much as possible. To measure the resistance of a component, remove the component from the circuit and hook the red lead to the terminal that was the most positive and the black lead to the least positive, or to the desired location if you wish to measure partial resistance of the component. (This hook-up is usually explained in the test procedures for components when it is applicable.)

To measure the amount of voltage present at a particular point (Fig. 13–6), attach the red lead to that point in the circuit and the black probe to the ground or negative post of the battery. Because the car uses its chassis as a ground for many components, the black lead can be attached to any clean spot on the frame or body of the car that is grounded. Obviously, body panels that are made of rubber or plastic are not good grounds because they are not conductors of electricity. The ground or negative post of the battery has zero volts present on it and the voltmeter compares the reading or the amount of voltage present at the red lead to the ground voltage, and displays the difference. If a voltmeter is hooked up across the battery with the red lead attached to the positive post and the black lead to the negative post, the meter would display the potential difference between the two points, which would be the voltage of the battery.

If the red lead of a voltmeter is hooked to the positive side of a load and the black to the negative side, as shown in Fig. 13–7, the reading on the meter would be the voltage drop of the load. The meter would compare the voltage present before the load at the red lead to the voltage left after the load or the black lead. The meter is measuring the potential difference between the two points. If the load is the only resistance in the circuit, the meter will display full battery voltage. When there is

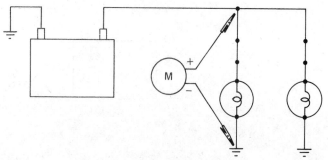

Figure 13–5 Meter hooked in parallel.

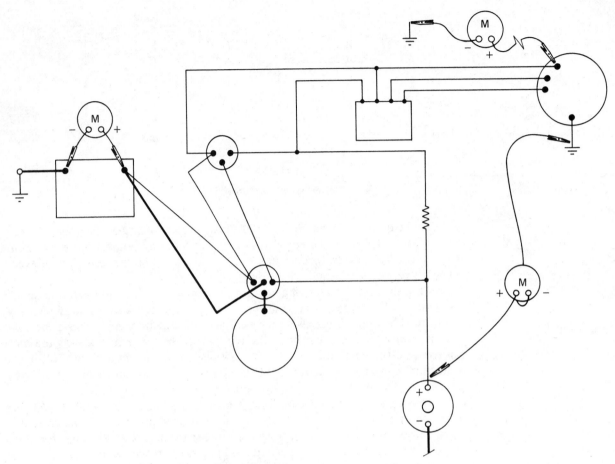

Figure 13-6 Measuring available voltage at a variety of spots in a circuit.

more than one load in a circuit, the parallel hook-up across each load will measure the voltage drop across each of the loads. Remember, the sum of the voltage drops in a circuit is equal to the voltage of the source.

When current passes through a conductor, a magnetic field is present around that conductor. The strength of the magnetic field is determined by the current that flows in the conductor. Many current-measuring devices use this principle to measure current. These test instruments are equipped with an inductive pickup which converts the strength of the magnetic field to a reading of current flow. The inductive pickup is like a clamp that fits around the wire and eliminates the need to hook the ammeter in series with the circuit. This type of ammeter is the one most commonly used on automotive equipment and is capable of measuring the large amounts of current needed by most of the systems.

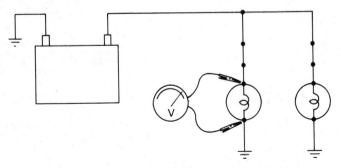

Figure 13-7 Measuring voltage drop across a component.

Testers that incorporate the features of the volt-ohm tester and the inductive amperage tester are made by many manufacturers. Most of these testers are specifically designed to test the battery, the starting system, and the charging system. These components and systems are the source of power for the ignition system and therefore must operate correctly for the ignition system to operate properly. Our testing of the ignition system will begin with the testing of the battery and the starting and charging systems with the use of both the multimeter and the special testers.

BATTERY TESTS

The first item to test is the battery. A totally charged battery will have 12.6 V in it. This is the result of the six cells of the battery having a maximum charge of 2.1 V each. The battery must be able to maintain 9.6 V in it while the starter motor is turning the engine. Batteries must be able to supply the starter with the electrical power it needs and still have enough power left over to allow the ignition system to work. The charging system recharges the battery while the engine is running. A poor starting problem can be caused by a battery, the starting or ignition system, or by problems with the engine or its systems. It is best not to assume that the battery is all right when you are beginning your diagnostics.

Batteries are usually rated in one of two ways: cold-cranking amperes and amperehour rate. These are ratings used by battery manufacturers to state the capability of a battery to maintain a particular current flow for a period of time. Particular ratings of batteries are recommended for certain applications because different engine starters have different power needs and the battery must be able to meet these needs and still have power available for the ignition system. These ratings are used for specs when testing a battery with a battery tester (Fig. 13-8).

To test a battery with a multimeter, hook the meter across the two terminals of the battery. If the battery has 12.6 V, it is fully charged. The battery should have at least 12 V across it before any further testing is done on the battery or the starting system. Using the multimeter to test the battery during starter load may also be a test of the starting system, so care must be taken not to jump to any conclusions. The use of a multimeter to test a battery is not the preferred method of testing. Once the meter is hooked up (Fig. 13-9) and the battery voltage noted, the ignition should be disabled. This is done by disconnecting the battery lead to the ignition

Figure 13-8 (Courtesy of Sun Electric Corporation)

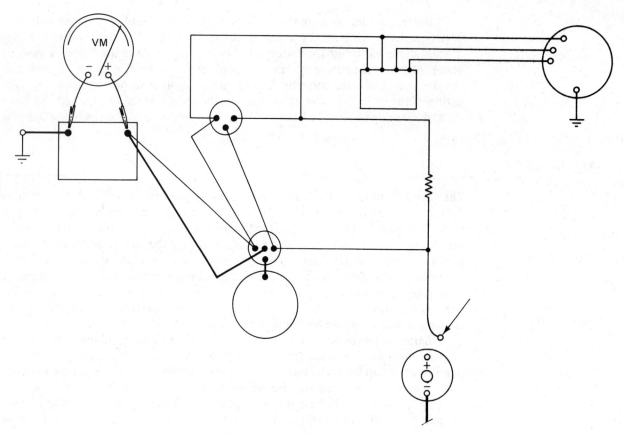

Figure 13-9 Testing the starting system with the ignition system disconnected and with a multimeter.

coil. The engine is then turned over with the starter. This cranking of the engine should continue for at least 15 seconds without stopping. At the end of this time period with the engine still cranking, the meter should be read. If the voltage of the battery is above 9.6 V, the battery did not drop below the minimum allowable and can be considered good. If the voltage was below 9.6 V, this indicates that the battery did not maintain the proper level and can be considered weak. The drop of voltage at the battery can also be the result of the starter pulling more current than it should, so care must be taken before replacing the battery in an attempt to correct the problem.

To separate the battery from the starter, when there is an excessive drop in battery voltage, the cause of the drop must be identified. If the battery had more than 12 V, prior to the operation of the starter, the most likely problem is high resistance in the starter or the battery cables that supply the power to the starter. If the cables are corroded or loose, the extra resistance will decrease the amount of voltage available for the starter and will cause the starter to draw more current than normal from the battery. To test the cables, connect the voltmeter across each cable individually, as shown in Fig. 13-10. With the starter activated, the cables should not have more than 0.1 V dropped across them. If the cables check out, the probable cause of the decreased voltage at the battery is the starter or solenoid. If the cables have excessive resistance, they need to be cleaned, tightened, or replaced. When the corrections are completed, the original test should be repeated to see if the condition still exists.

To test the battery with a battery tester, follow the recommended procedure of the tester's manufacturer. Most testers require that you determine from the bat-

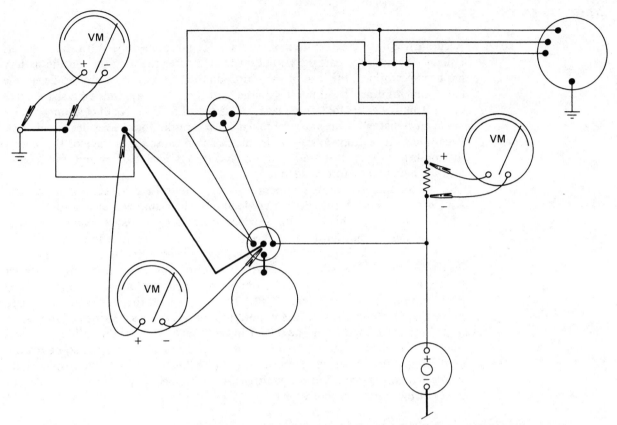

Figure 13-10 Measuring voltage drops of the starting and ignition systems; to measure the voltage drops, the system must be activated.

tery's rating the load that should be placed on the battery for testing. If the battery is rated in amperehours, the recommended load is three times the amperehour rating. A battery that has a 50-Ah rating will have a testing load of 150 A. If the battery is rated in cold-cranking amperes, the test load is one-half of that rating. A battery with a cold-cranking rating of 400 A has a test load valve of 200 A.

The battery performance test requires that a load be applied to the battery for about 15 seconds. This load is determined by the rating of the battery. A battery should be able to maintain over 9.6 V while this load is being applied. The battery testers are equipped with a control by which you can adjust this load amount. This adjustment causes a certain amount of current to flow from the battery to the tester. The load of the tester is a carbon pile which is a special heavy-duty adjustable resistor. Once this current flows through the tester, the technician looks only at the voltmeter on the tester. If battery voltage drops below the 9.6-V level, the battery is not good. This is a correct conclusion because the only component on the car that is being tested is the battery. The tester hooks up directly to the battery posts for the voltmeter, and an inductive pickup is used to measure the current that flows from the battery through the tester. The load of the tester is the only load in the circuit.

The amount of voltage present at the battery after the load has been applied determines the condition of the battery. If the battery voltage dropped very little during the test, it is in good condition. If the battery dropped to less than 10 V, but still had more than 9.6 V, it is not as strong as it should be and may need further testing to determine whether or not it will last; however, the battery is in good enough shape to continue with the testing of the starting system.

STARTER MOTOR TESTING

Accurate testing of the operation of the starter can be done only with a starter tester. Usually, this tester is part of the battery tester. After the load test on the battery has been performed, the load of the starter on the battery is measured to determine the amount of drain the starter has on the battery. If the starter has excessive current draw, it must be repaired or replaced. This test should be conducted only with a satisfactory battery. Care must be taken when interpreting the results of this test. Excessive voltage drops in the starter circuit could cause the voltage of the battery to drop below the spec of 9.6 V. The voltage drop of each battery and starter cable must be measured before condemning a starter.

To conduct the test, the amperage pickup should be attached around the negative battery cable. During the battery load test, the clamp was around the tester's lead to the battery. This hook-up provided for reading current through the tester. When testing the current draw of the starter, the ampere lead should be attached to the car's electrical system (Fig. 13–11). Doing so enables the tester to measure the current draw of the starter. Most test procedures consist of operating the starter for approximately 10 seconds and monitoring the current draw of the starter and the voltage of the battery (Fig. 13–12). The readings on the meters are compared to the manufacturer's specs. It is important that all other accessories of the car be turned off while conducting the test. This includes having the doors closed to prevent the operation of the dome light in the car. The ignition must also be disconnected to prevent the engine from starting as well as to eliminate the current drain of the ignition system. These conditions must be met to measure accurately the effect of the starter on the battery.

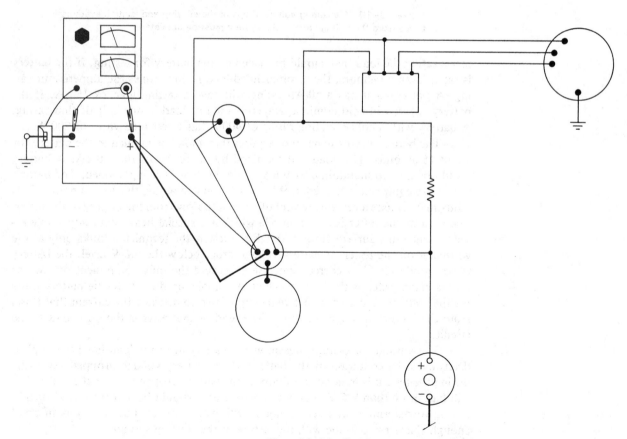

Figure 13–11 Starting system tester hook-up.

Figure 13–12 Ampere and volt readings during starter current draw test.

Normally, the spec for starter current draw on a four-cylinder engine is approximately 150 A. Six-cylinder engines draw slightly more, about 175 A, and eight-cylinder engines draw approximately 200 A. Higher-compression engines will normally have a higher current draw. Most higher-compression engines are older models. Specs, when available, should always be used. Often listed in the specs of an engine are the "free-running current draw" specs. These specs are not to be used when conducting a starter current draw test with the starter installed in the car. These specs are for testing the starter after it is removed from the car. Use the starter current draw specs for comparison to the measurements taken in this test.

If the readings of the voltmeter or the ammeter are not within the specs, the voltage drops of the battery and starter cables should be tested. Excessive current draw by the starter will cause the battery voltage to drop below the spec of 9.6 V, and this indicates that the starter has a problem. The repair or replacement of a starter is not part of a normal tune-up but is often the solution for a hard-starting engine. The excessive drain of the starter on the battery will decrease the amount of voltage available to the ignition system, which will decrease the chances of starting the engine.

ALTERNATOR TESTING

The charging system of an engine is designed to replenish the amount of current available from the battery after the engine is started. The alternator, which is the charging unit of the system, is designed to deliver a certain amount of current and voltage back to the battery, to recharge it. Alternators are designed for a variety of current outputs. The outputs are determined by the manufacturers to meet the demands of the electrical loads of the car. If the charging system does not adequately recharge the battery, the electrical energy available for the accessories or the ignition system will not be adequate. The voltage output of the alternator is regulated by a voltage regulator. This regulator limits the peak output voltage of the alternator. The normal range for the voltage output of the alternator is 14 to 15 V. If the voltage from the alternator is too great, it will tend to heat up the fluid in the battery, which will tend to ruin it. Testing the charging system includes testing of both the maximum current output and the regulator's limiting capability. The alternator will only put out as much current and voltage as the battery needs; therefore, when testing

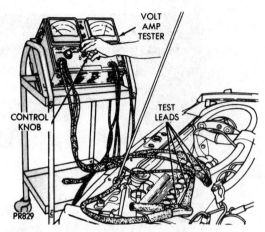

Figure 13-13 Putting a load on the system. (Courtesy of Chrysler Corporation)

the alternator, the load on the battery must be great. This allows the alternator to have a maximum output and it can then be accurately tested.

To test the charging system, the tester hook-ups are the same as for the starter tester (Fig. 13-13). The ignition is not disconnected because the engine must run in order for the alternator to work. Specs on the charging system will include a test speed for that the engine. To conduct the test, the procedures given by the tester's manufacturer should be followed. Most require the use of the load control knob to cause the necessary drain on the battery. With the engine running at the specified speed, the load control is increased to cause a heavy drain on the battery. The control should be increased until a maximum reading is present on the ammeter; however, the voltage of the battery should never be allowed to drop below 12 V. The amperage reading should be compared to the specs for that alternator. After the maximum current output of the alternator has been measured, the load control should quickly be decreased to zero. The voltmeter should be observed and this reading should be within the range of the specs for the voltage regulator. If the voltage is greater than the regulator's specs, the regulator needs to be replaced or there may be a short in the lead from the alternator to the regulator.

This test can be conducted with a multimeter. Although the current output of the alternator cannot be measured with the meter, the charging voltage can. The alternator should be able to maintain its specified charging voltage rate regardless of the load put on the system. The load that is put on the battery while testing it with a multimeter is from the operation of the accessories of the car. The lights, radio, and other accessories should be turned on with the engine running at the recommended speed. The voltage from the alternator can be measured at the battery with the meter hooked across the battery terminals. If the voltage is low, the most probable cause is the alternator, unless there was no increase over the amount of voltage present at the battery before the engine was started. In this case the regulator is probably at fault. If the voltage is greater than the specs, the regulator circuit is again probably at fault.

PRIMARY CIRCUIT: BREAKER-POINT IGNITION SYSTEM

The battery and its charging system are part of the primary ignition system. The rest of the primary circuit can be checked by visual inspection or through the use of the multimeter. There are also many testers available for the testing of the individual systems and components of the primary circuit. The use of these specialized testers will not be covered in this text; rather, the use of the basic test methods will. There are many types of these testers and they have step-by-step procedures to follow. Both point-type and electronic ignitions have components that should and can be tested through these basic means. The ballast resistor, breaker points, condenser, and primary wiring are the components of the point-type primary circuit that often need to be checked during diagnostics (Fig. 13–14).

The purpose of the ballast resistor (Fig. 13–15) is to ensure the proper amount of current flow for the primary. The ballast resistor is bypassed during the operation of the starter. This allows for the ignition system to receive full battery voltage and current during the starting of the engine. The ballast resistor can be a resistor mounted on the engine, ignition coil, or firewall. It can also be a resistance wire that connects the ignition coil to the ignition switch. This resistance causes a decrease in the amount of voltage at the positive terminal of the coil. To test the ballast resistor and the ballast bypass system, an ohmmeter or voltmeter can be used. The resistance specs of the resistor are usually available in a complete list of tune-up specs. Care should be taken when testing a ballast resistor because its resistance value will change with an increase in temperature. The hotter the resistor is, the more resistance it will have. It is important to point out that some electronic ignition systems are equipped with a ballast resistor. The most common application is on Chrysler products. Testing these resistors is no different from testing the resistors in a point-type ignition circuit.

By disconnecting the wires leading to the resistor or the ends of the resistance wire, an ohmmeter can be connected across the resistor and a reading taken. The reading should be compared to the specs for the resistor. A higher-than-normal resistance reading will reduce the current too much and will affect the chances of coil saturation. Damage to the points, caused by too much current flow, can be the result of too little resistance in the ballast resistor.

A defective ballast resistor circuit can also cause the engine to quit once the ignition switch is changed from the start to the run position. The ballast bypass system is responsible for the bypassing of the resistor during engine cranking. Once the engine is running, the ballast becomes part of the primary circuit. If the ballast is opened or shorted, the coil would not receive the current it needs.

The ballast resistor can also be checked with a voltmeter. During this test, the amount of voltage available to the coil is measured. With the red probe of the meter

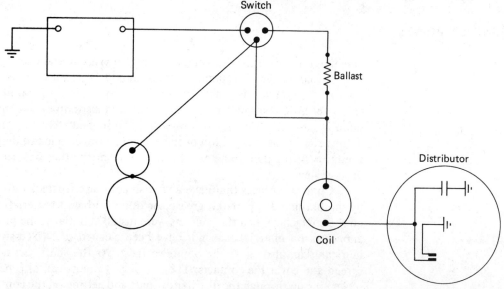

Figure 13–14

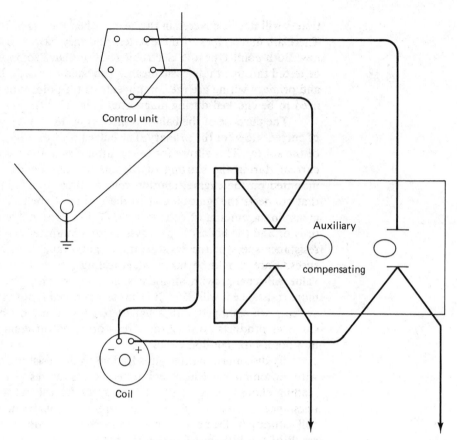

Figure 13–15 Chrysler's dual-ballast resistor.

hooked to the positive post of the coil, a reading is taken. This reading should be less than battery voltage if the resistor is allowing for the delivery of decreased voltage to the coil. The reading of voltage should be compared to the specs for the engine. If the reading is the same as the battery voltage, this indicates that the ballast is not in the circuit. This can be caused by faulty wiring or a defect in the bypass circuit. Full battery power present at the points will cause them to burn.

IGNITION POINTS

The ignition's breaker points can be inspected visually and checked for conductivity with an ohmmeter. The point surfaces should be clean and smooth. The normal color of the contacts is a dull gray. It is important that the point contacts be clean and offer little resistance to current flow. It is also important that the surfaces be smooth, so that when they separate they can immediately stop the flow of current in the primary. Many ignition problems can be traced back to defective points. The condition of the points can also be used to identify other problems with the engine or its systems.

Contact surfaces that have a whitish color are frosted. This frosting is caused by the burning away of oil, grease, or other oil-based materials that were present on the contacts. When the contacts were installed initially, the protective coating of anticorrosion material may not have been cleaned off. Excessive distributor cam lubricant can also cause the points to frost. As the shaft turns, the lubricant can spread and cover the contacts. The frosting of points can also be caused by engine oil leaking up through the distributor shaft and getting on the contacts. The presence of oil inside the distributor would be another indication of this problem. The solution to this problem is to replace the seal in the distributor housing or on the shaft. The distributor seal may have been ruined by a defective PCV system. If the pressure in the crankcase is not vented properly, the pressure will force oil out of the engine any place it can. The distributor shaft seal is a likely place for these oil leaks.

If the contacts are black in color, this indicates that the current flowing through them has been low. This can be the result of many problems, including the ballast resistor. Although the presence of this color indicates that there is another problem, the contacts should be replaced.

Another possible cause for the low current flow in the primary is a poor distributor ground. This can be tested with an ohmmeter. If the negative lead of the coil is disconnected at the coil and the red lead of the meter connected to the lead and the black lead to a good ground, the meter should read zero resistance when the points are closed. This would indicate that the circuit from the coil to the ground is good. If the reading indicates some resistance, the red lead should be moved to touch the end of the wire at the points. If the reading is now zero, the cause of the resistance is the wire. If the readings still indicate that there is resistance, the red lead should be attached to the ground at the points. If the resistance reads zero, the points are defective. If the reading still shows resistance, the problem is at the ground for the breaker plate of the distributor. Any resistance in the circuit will cause the points to take on a blackish appearance after running, due to the low current flow.

If the contacts are blue-black in color, it indicates that there has been high current flow across them. This can be caused by a defective ballast bypass circuit or by a problem in the charging system. Both of these circuits and the tests for them have been discussed previously. Irregular surfaces on the contacts will cause either high resistance or will allow for point arcing when they open. In either case, the points should be replaced and the cause identified and corrected. These irregular surfaces are caused by the burning or pitting of the points. Points can burn and become disfigured if the dwell is set too long. This causes the point gap to be too narrow and allows for point arcing.

Point arcing can also cause point pitting. The pitting results from some arcing taking place in a small area of the contact to the other (Fig. 13-16). Metal is transferred from one contact to the other, by the arcing. Pitted and burned points cause problems with the opening and closing of the points. With the irregular surfaces, the points will not be able to close fully. This will cause a resistance to current flow. Current will only have a good path across the part of the surfaces where the contacts meet. There will be slight air gaps in the other areas. Any projections on the surface of the points will increase the chance of point arcing when the points open. The points must cause an immediate interruption of current flow, when they open, to

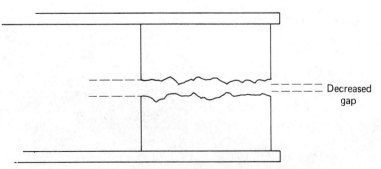

Decreased gap

Figure 13-16 Pitted points and the potential for point arcing.

provide for the total collapsing of the field in the coil. Not only can incorrect dwell cause arcing across the points but so will a defective condenser.

IGNITION CONDENSER

The primary purpose of the condenser is to prevent arcing across the contacts when they first open. To prevent this arcing, the condenser must offer less resistance to current flow than the air gap of the points. It must also be able to store that current long enough to allow the field in the coil to collapse. To measure the ability of a condenser to do these things, special testers are available to test condensers. Many of the engine analyzers have a condenser test mode as part of the equipment.

The condensers are tested for leakage, capacity, and series resistance. The leakage test measures the condenser's ability to hold the current long enough to allow for the induction of the secondary voltage in the coil. The capacity test measures the amount of current the condenser can store, and the series resistance test measures the resistance through the condenser. These tests cannot be performed without a special tester.

PRIMARY CIRCUIT: ELECTRONIC IGNITION SYSTEMS

Because of the different systems used by the various automobile manufacturers for electronic ignitions, the test procedures for the components of the primary circuit (Fig. 13-17) cannot be easily divided. Detailed testing of these components will be presented according to the type of system. The complete testing procedure for each and every type will not be given in this text. Much of this information is given in the shop manuals and those procedures should be followed.

What is discussed in this chapter are the common systems and the common tests that can be conducted. A visual inspection of all wiring and components is part of a maintenance and diagnostic tune-up. Individual manufacturers have special testers for their ignition system, but there are also appropriate volt-ohmmeter checks that can be made to test the components of electronic ignition systems.

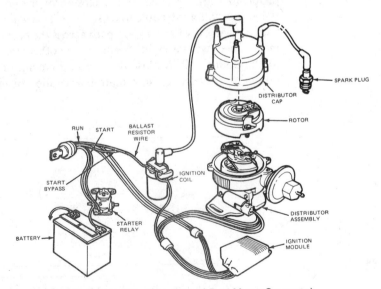

Figure 13-17 (Courtesy of Ford Motor Company)

To determine whether or not a problem exists in the primary of an electronic ignition system, pull a plug wire off a plug. Install a test plug (Fig. 13-18) in the end of the plug wire and attach its ground wire to a good ground. Hold the test plug about 3/8 in. away from a good ground at the engine. Next, crank the engine with the starter to see if a good spark is produced. If the spark is able to jump the gap to the ground, the ignition is working fine. If the spark cannot jump the gap, remove the coil wire from the distributor cap and hold it the same distance from a ground. If the spark is good, the problem is in the secondary at the distributor cap, rotor, or plug wires. If the spark did not jump the gap, there is a problem with the primary ignition. General Motors engines equipped with an HEI system, with the coil built into the distributor cap (Fig. 13-19), cannot have the coil wire removed, since there is none.

To make a quick check of the secondary of the HEI system, connect an ohmmeter to the distributor cap "tach" terminal and the carbon button inside the distributor cap. The meter readings should be between 6000 and 30,000 Ω, with the meter set to the ×1000 scale. This spec also applies to all GM HEI coils that are mounted outside the distributor cap and the coils of GM cars produced in 1974 and 1975. Later models should have an infinite reading between these two points. If the

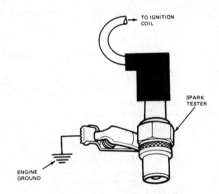

Figure 13-18 (Courtesy of Ford Motor Company)

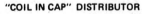

"COIL IN CAP" DISTRIBUTOR

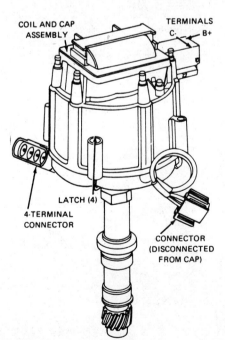

Figure 13-19 (Courtesy of Pontiac Motor Division, General Motors Corporation)

readings are not within these specs, the coil needs to be replaced. If the readings were good and there was no spark when the test plug was held close to ground, the problem is in the secondary circuit.

CHRYSLER ELECTRONIC IGNITION

To test the primary circuit of the Chrysler electronic ignition (Fig. 13-20), first check the gap between the reluctor and the pickup coil. This gap should be 0.008 in. and checked with a nonmagnetic feeler gauge. If the gap is incorrect, correct it by moving the pickup coil. Next, make sure that the ignition switch is "off." Disconnect the multiwired connector at the ignition control module. Doing so will expose the female terminals used for testing (Fig. 13-21).

Turn the ignition switch to the "on" position and connect the black lead of a voltmeter to a good ground. The red or positive lead should be connected to the number 1 terminal inside the connector. The voltage reading should be within 1 V of the battery's voltage. If the reading is less than that amount, there is excessive resistance in that circuit.

Next, move the red lead to the number 2 terminal and take a reading. The reading should also be within 1 V of the battery's voltage. If the reading is not within this spec, there is excessive resistance somewhere in the circuit. In this circuit

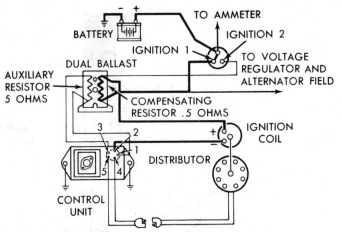

Figure 13-20   (Courtesy of Chrysler Corporation)

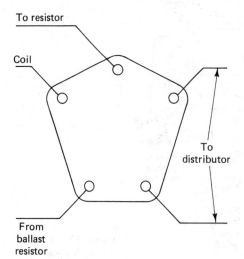

Figure 13-21   Ignition module connector.

are the ignition coil, related wires and connectors, ballast resistor, and ignition switch. These components should be tested to identify the cause for the high resistance.

The next test is at the number 3 terminal. Only vehicles produced before 1980 will have this terminal in the connector. Therefore, this test is conducted on 1979 and earlier models only. Again the reading should be within 1 V of the battery's voltage. Lower readings indicate high resistance in the circuit, which includes the ballast resistor and ignition switch. These components should be checked before the test of the primary circuit is continued.

To test the ballast resistor, set the ohmmeter to the X1 scale and disconnect the resistor from the circuit. Before disconnecting any wires, the ignition switch should be turned off. Many Chrysler products are equipped with a dual-ballast resistor. One of the resistors in this assembly is used to control the voltage going to the control module and is called the auxiliary resistor. This resistor has a normal value of 4.75 to 5.75 Ω. The other resistor is the primary ballast resistor and has a value of 0.5 or 1.2 Ω. The correct spec is usually stamped on the brass tab terminal of the resistor. With the ohmmeter connected across the terminals of the resistor, one at a time, the readings are compared to the specs. A reading that is not within the specs on either resistor indicates that the resistor should be replaced. On 1980 and newer models, a single-ballast resistor is used. The auxiliary resistor has been removed from the circuit. All single resistors have a value of 1.2 Ω.

With the ignition switch off, the pickup coil can be tested through terminals 4 and 5 of the module connector (Fig. 13–22). With the ohmmeter set on the $\times 100$ scale, connect one lead to the number 4 terminal and the other to the number 5 terminal. If the pickup and its wires are good, the ohmmeter should read between 150 and 900 Ω. If the reading is lower or higher than the spec, disconnect the wiring harness at the distributor and connect the meter to the two terminals at the distributor. If the reading is now within specs, the wiring harness between the distributor and the module connector has a problem and should be repaired or replaced. If the reading is still not within specs, the pickup coil is defective and should be replaced.

To further test the wiring to the pickup coil, manually move the vacuum advance arm and observe the ohmmeter. If the readings fluctuate or change with the movement, the wiring has a fault and should be replaced. To test for a possible short to ground at the pickup coil, connect one lead of the ohmmeter to either the number 4 or 5 terminal and the other lead to a ground. The reading should be infinite, showing that there is no path from the pickup coil to ground. If there is a reading, either the wiring harness is shorted and needs to be repaired or the pickup coil is defective and needs to be replaced.

In 1980, Chrysler introduced a Hall-effect type of electronic ignition on their front-wheel-drive cars. This type of ignition requires a change in the testing of the primary ignition components. To test this circuit, disconnect the connector to the ignition module with the ignition off. With the ignition turned back on, connect the voltmeter to a ground and to the number 2 terminal of the connector, as shown in Fig. 13–23. The voltmeter should be read with this connection, then with connections to terminals 1 and 5. These readings should all be within 1 V of the battery's voltage. If the readings are not within this range, the appropriate circuit should be tested and repaired.

To test the pickup of this system, disconnect the coil wire from the distributor cap and hold it $\frac{3}{16}$ in. from a good ground. Crank the engine over and check the spark. If there is no spark, disconnect the three-pin connector at the distributor and with a small jumper wire use it to connect the female terminals numbered 2 and 3. If the coil wire now sparks, the pickup coil in the distributor was not working and should be replaced. If no spark occurs, check the wiring from the module to the distributor.

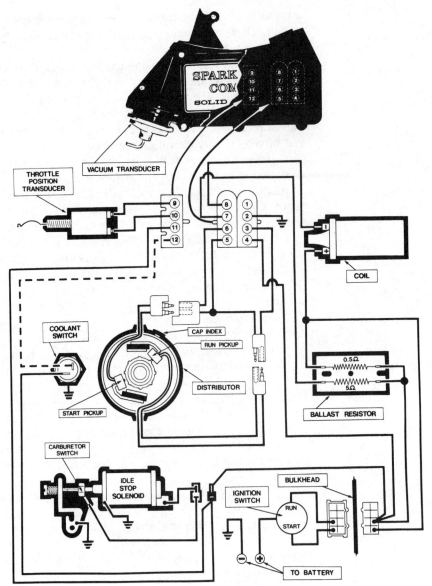

Figure 13-22 (Courtesy of Chrysler Corporation)

FORD ELECTRONIC IGNITION

To test the magnetic pickup assembly on the Ford's Solid State and Dura-Spark systems (Fig. 13-24), set the scale on the ohmmeter at ×100. With the ignition turned off, disconnect the four-wire connector at the control module. Connect one lead of the meter to the orange wire terminal and the other to the purple wire terminal. On 1978 and older-model cars, the reading should be between 400 and 800 Ω. On 1979 and newer models, the reading should be 400 to 1000 Ω. If the reading is not within these ranges, the test should be repeated at the connector at the distributor. This allows for a check of the wiring harness from the module to the distributor. If the readings are still not within the specs, the pickup should be replaced.

The connector at the distributor is a three-wire connector (Fig. 13-25). Two of the wires are the orange and purple wires for the pickup. The other wire is black and is the ground for the circuit. To test the ground of the circuit, connect one lead

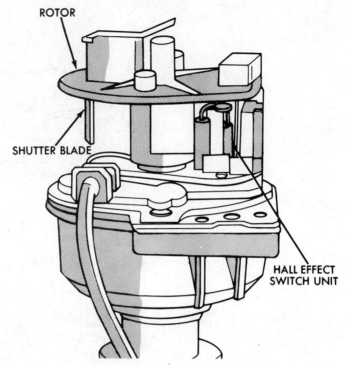

ROTOR

SHUTTER BLADE

HALL EFFECT
SWITCH UNIT

Figure 13-23   (Courtesy of Chrysler Corporation)

of the ohmmeter to the black wire terminal and the other to a good ground. If the
circuit has a good ground, little resistance will be measured. If there is a measurable
amount of resistance, the ground path should be checked. Most Ford distributors
have a ground tab located at the connector as it enters the distributor body. This
tab is held in place with a small screw. If the screw has become loose, the ground
will not be good. This is one item that should be checked in your diagnosis of a
poor ignition ground.

The pickup coil may be also checked for output. Connect a voltmeter to the

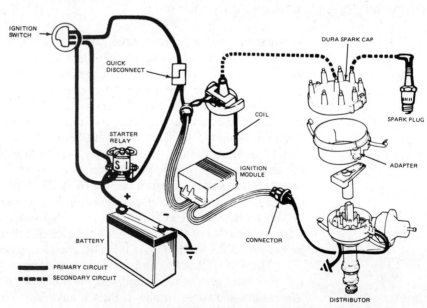

IGNITION
SWITCH

QUICK
DISCONNECT

DURA SPARK CAP

COIL

SPARK PLUG

STARTER
RELAY

ADAPTER

IGNITION
MODULE

BATTERY

CONNECTOR

DISTRIBUTOR

▬▬▬   PRIMARY CIRCUIT
▪▪▪▪▪   SECONDARY CIRCUIT

Figure 13-24   (Courtesy of Ford Motor Company)

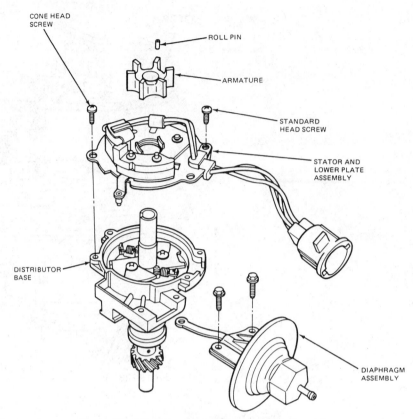

CONE HEAD
SCREW

ROLL PIN

ARMATURE

STANDARD
HEAD SCREW

STATOR AND
LOWER PLATE
ASSEMBLY

DISTRIBUTOR
BASE

DIAPHRAGM
ASSEMBLY

Figure 13-25   (Courtesy of Ford Motor Company)

wire terminals of the pickup coil. Switch the voltmeter to its lowest scale and rotate the distributor shaft. The movement of the shaft and the action of the pickup will cause the needle of an analog meter to deflect and cause inconsistent readings on a digital meter. This technique can be used on all pickup coils, not just Ford products.

GENERAL MOTORS ELECTRONIC IGNITION

The first test on the primary circuit of GM's HEI system is to verify that battery voltage is available at the battery terminal of the distributor. HEI systems do not use a ballast resistor and battery voltage is delivered to the distributor during both the starting and the running of the engine. To perform this test, connect the red lead of a voltmeter to the battery terminal at the distributor. To do so, the connector should be disconnected. Whenever disconnecting the wires on engines equipped with electronic ignition, the ignition should be off. When the ignition switch is in the "start" position and the "on" position, the reading should be equal to the voltage of the battery. If the reading is lower, there is resistance in the circuit from the battery to the switch, or from the switch to the distributor.

To test the pickup coil, set the ohmmeter on the ×100 scale. Disconnect the pickup leads from the control module located inside the distributor (Fig. 13-26). Connect the leads of the ohmmeter to the pickup coil terminals. The reading should be between 500 and 1500 Ω. If it is not, the pickup coil should be replaced.

The pickup should also be tested for a short to ground. Connect one lead of the ohmmeter to a good ground and the other to one of the leads of the pickup. The reading should be infinite. If there is any reading other than infinite, the pickup coil should be replaced. If the pickup tests out all right, keep the ohmmeter hooked

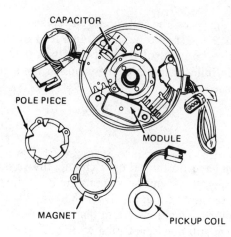

CAPACITOR

POLE PIECE

MODULE

MAGNET

PICKUP COIL

Figure 13-26 (Courtesy of Chevrolet Motor Division, General Motors Corporation)

up. Manually move the vacuum advance arm and watch the meter. If the reading changes with the movement of the advance arm, the wires are contacting a ground and should be repaired or replaced.

Many imported cars have ignition systems similar to those of domestic cars. The procedures for testing the individual components are given in shop manuals. Given in these procedures are the specs for resistance and, in some cases, the voltage specs for checking the pickup and the primary wiring. For all electronic ignition systems, domestic or foreign, if all checks of the primary show no evidence of a problem and there is no spark available from the coil, the control module should be replaced with a known good one and the system retested. If the replacement module corrected the problem, the cause of the problem was the module.

IGNITION COIL TESTING

Whether the engine is equipped with an electronic or a conventional ignition system, the ignition coil can be checked with an ohmmeter. Specs are given for the normal resistance across the primary and secondary windings of the coil. To measure the resistance of the windings, connect the ohmmeter across the windings with the coil removed from the engine, as shown in Fig. 13-27. With one lead of the meter connected to the positive terminal of the coil and the other to the negative or distributor terminal of the coil, the resistance of the primary winding can be measured. To measure the resistance of the secondary winding, connect the ohmmeter from the positive terminal to the secondary terminal or the center tower. If any of these read-

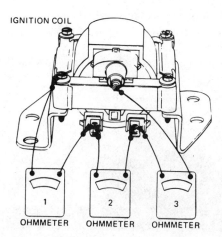

IGNITION COIL

1   2   3

OHMMETER   OHMMETER   OHMMETER

Figure 13-27 (Courtesy of Cadillac Motor Car Division, General Motors Corporation)

ings are not within the specs, the coil should be replaced. If the readings are within the specs, one final check should be made. One lead of the meter should be connected to the positive terminal of the coil and the other to a good ground. The reading should be infinite. If it is not, the windings are shorted to ground and the coil should be replaced.

SECONDARY CIRCUIT

With the tests completed on the primary circuit, the secondary should be tested. The secondary circuits of all ignition systems are basically the same, and therefore the procedures for testing them apply to all engines. There is one minor exception to this, the distributorless ignition system that General Motors introduced in 1984. However, this system still uses spark plug wires and spark plugs.

A visual inspection of the spark plugs can tell you quite a bit about them and the engine. The insulator nose of the plug and the ground electrode reach into the combustion chamber and give helpful hints about the combustion process. To be able to interpret the information on the plug into some useful diagnostic hints, you must know what a normal spark plug (Fig. 13–28) looks like after it has been used for many miles in an engine.

The color of the insulator nose of a normal spark plug is a light brown to a whitish gray. These colors indicate that the plug is in the correct heat range and that the engine is healthy. The corners of the center and ground electrode are square and there is a minimum amount of deposits, or buildup of materials, on the plug. A slight amount of oil present on the threads of the plug can be the result of a new engine that is breaking in or a slightly worn engine with leaks in the valve seals or guides. A fluffy whitish-gray deposit on the ground electrode and the insulator tip is normal for emission-controlled engines that have lean mixtures and use unleaded fuel. Soft white with a light tan color on the plugs indicates that the engine is normal and that the engine is used mostly at medium or high speeds. This results from heat, which has less time at higher speeds to leave the plug totally. A soft light brown color is normal for an engine that uses leaded fuels and is not equipped with emission controls. These engines also tend to run a little richer. The additional fuel causes the plugs to have a darker tan color than the plugs of a lean engine.

A normal worn-out plug will show a color that is in the same range as a normal plug but will have the corners of the electrodes rounded off (Fig. 13–29). The gap of the plug will also have been increased. It is normal for a plug on today's cars to increase its gap by 0.001 in. every 5000 to 10,000 miles. Sharp square corners on the electrodes are needed for efficient sparking. A worn-out plug should be replaced with a new one of the same heat range.

If the firing end of the insulator is broken and there is evidence of metal transfer from the center electrode to the ground electrode, engine detonation has probably caused the damage. When detonation occurs, part of the air-and-fuel mixture begins to burn after ignition has taken place. This additional burning is the result

Figure 13–28 (Courtesy of Champion Spark Plug Company)

Figure 13-29 (Courtesy of Champion Spark Plug Company)

of the increased heat and pressure present in the combustion chamber after ignition. The explosion that takes place as a result of detonation puts an extreme amount of pressure on the engine and the spark plug. This pressure causes the damage to the plug. Detonation can be caused by overadvanced ignition timing, a defective EGR system, or the use of fuel with too low an octane rating.

If the insulator nose is a clean white in color and has deposits on it of the same color, the plug is blistered. The blistering of a plug is caused by high temperatures in the combustion chamber. The blistering can also be the result of the plug being in the wrong heat range. High combustion-chamber temperatures can be the result of overadvanced timing, a lean mixture, a vacuum leak, or an overheated engine. Extreme heat conditions are indicated by a plug that is white and blistered. The electrodes will also show signs of extreme wear. An engine that has this type of plug will also have experienced preignition as a result of the extreme heat. The cause for the heat should be identified and repaired, and then a new set of spark plugs should be installed.

Too cold a heat range of plug will be indicated by the presence of soft sooty-black deposits on the electrodes and insulator. These deposits offer extra resistance to the flow of current across the gap of the plug. Excessive amounts of these deposits can cause the plug to foul out. This prevents the plug from firing. This carbon fouling can be caused by too cold a plug. If the plug is in the correct heat range, the problem can be an overly rich mixture, a lack of air, a weak ignition system, low compression, retarded timing, or low engine temperatures. The cause of this condition should be repaired and a new set of plugs installed.

If the insulator and electrode are covered with an oily black deposit, the plug is oil fouled. Oil fouling is the result of an excessive amount of oil entering the combustion chamber. This can be caused by worn piston rings, valve seals, or guides. A plugged PCV system may also cause this condition. A buildup of crankcase fumes can force oil past the piston rings or valve guides, and allow oil to enter into the chamber. A plug with too cold a heat range can also allow oil fouling. The heat that should be retained by the plug should burn off any oil left on the plug after firing. If the plug does not retain an ample amount of heat, oil will collect on the plug and cause it to be fouled.

At times, the gap between the electrodes of the plug becomes filled with the deposits that collect on the plug. When this occurs, the plug will not fire the mixture because the plug has its gap bridged. These deposits can also collect between the center electrode and the metal body of the plug. If this occurs, the plug is core bridged. Although there is still a gap between the center and ground electrodes, the current will flow through the deposits and not across the gap. This is a path of less resistance for the current to ground.

If the end of the plug looks like it was hit with a hammer, it probably has been damaged by some mechanical means. If the piston or valve has experienced some damage or has broken, some pieces of the metal may be present in the combustion chamber. Upon the compression stroke, this metal can be pushed toward the plug and cause damage to it. This damage can also be caused by the presence of some object in the combustion chamber. A bolt or washer may have entered the chamber through the carburetor or another place and is beating around on the inside of the chamber. To correct this problem, the engine must be disassembled to remove the object or to correct the engine damage, and new plugs installed.

Careful inspection of the plugs, upon removal, can indicate quite a bit about the engine (Fig. 13–30). To aid in your inspection, always place the plugs in order so that you can tell what cylinders they came from. This will help identify the cylinders that have a problem. This is especially handy on an eight-cylinder engine if four of the plugs were darker brown than the others. This indicates that four cylinders are receiving a richer mixture than the others.

The spark plug wires can be inspected visually, tested with an ohmmeter (Fig. 13–31), and tested on an engine analyzer. The use of the engine analyzer is covered in the next two chapters. What follows here are the procedures for testing the wires with an ohmmeter. The wires should be inspected visually for cracks and corrosion. To test the wires with an ohmmeter, connect the leads of the meter to the ends of each individual wire. If the wires are nonresistor types, the readings expected will be very low. Most engines are equipped with resistor-type wires and the desired range of resistance is about 10,000 Ω or 10 kΩ per foot. If the reading is greater than this amount, the wires should be replaced. Readings lower than this amount

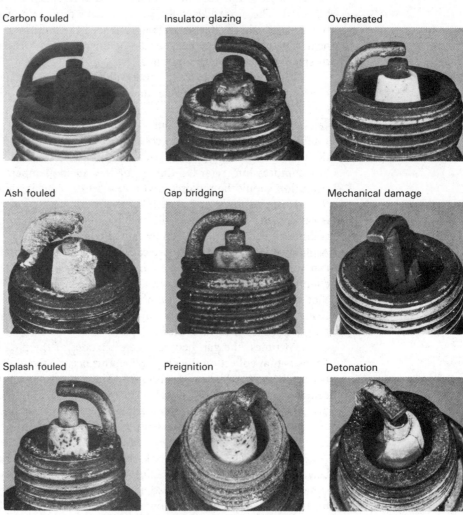

Carbon fouled   Insulator glazing   Overheated

Ash fouled   Gap bridging   Mechanical damage

Splash fouled   Preignition   Detonation

Figure 13–30 (Courtesy of Champion Spark Plug Company)

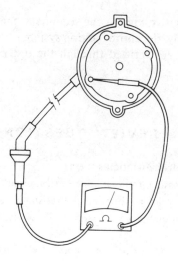

Figure 13-31 Testing spark plug wire resistance with an ohmmeter. (Courtesy of Ford Motor Company)

should not be of concern unless there is evidence of low firing voltages due to low resistances in the secondary. Because the wires meet this spec does not definitely mean that the wires are good. The flow of voltage and current in them can raise the resistance to the point where they present a problem. Testing the wires in operation on the engine is accomplished with an engine analyzer or scope.

The rotor and distributor cap should be inspected carefully for corrosion and cracks. On some electronic ignition systems, particularly GM HEI systems, there is a problem with high voltage burning small holes through the rotor as the voltage seeks a ground (Fig. 13-32). To see these holes, a careful look must be made. Do not assume that if a distributor cap or rotor is clean, it is in good shape. A small amount of corrosion is normal and can be cleaned off. Heavy deposits of corrosion and cracks or holes in the cap or rotor require replacement of the cap and rotor.

Figure 13-32 HEI distributor cap and rotor; note the small burn mark on the rotor.

SUMMARY

Detailed testing of the secondary system is better done with an engine analyzer. With this instrument, the electrical activity in the circuit can be monitored while the engine is running. The high voltage of the secondary can cause resistance to change, causing problems. Ideally, the entire ignition system is tested with an engine analyzer and then the problems are verified through the test procedures outlined and discussed in this chapter.

REVIEW QUESTIONS

1. What is the purpose of the ignition system?
2. What are the three possible faults in an electrical circuit?
3. What should be done prior to conducting a resistance test with an ohmmeter?
4. When referring to amounts of current, "m" is used to indicate that the current is _____.
5. When referring to amounts of resistance, "k" and "M" are used to indicate _____.
6. What is the general rule for connecting a meter to a circuit?
7. What are the two ways to hook up a meter to a circuit?
8. How is a meter hooked up to measure the voltage drop of a load?
9. Some ammeters are equipped with an inductive pickup. How do these pickups work?
10. How many volts will a fully charged battery have?
11. What are the two ways that batteries are typically rated?
12. How are battery cables tested with a voltmeter?
13. How does the engine's compression ratio affect the readings of starter current draw?
14. How can a ballast resistor be tested?
15. What can cause the surfaces of the breaker points to appear frosted?
16. Burned points are usually caused by _____.
17. What is the proper way to test an ignition condenser?
18. When using an ignition condenser tester, what three properties of the condenser should be measured?
19. What can you quickly do to determine if the problem in the ignition system is in the primary or secondary circuit?
20. What is the proper procedure for checking the reluctor air gap on a Chrysler electronic ignition distributor?
21. What is the procedure for testing the Hall-effect switch on ignition systems that use one?
22. Why is a visual inspection an important part of testing an electronic ignition system?
23. Where are the leads of an ohmmeter placed to test the resistance of the primary winding of an ignition coil?
24. If the engine is operating normally, what color will the insulator of the spark plugs be?
25. What will be the appearance of the spark plugs be if the engine has been running excessively lean?
26. Why is it a good practice when removing spark plugs to set them aside in order?
27. How can spark plug wires be tested?
28. What can result from excessive resistance in a spark plug wire?
29. What should be noted when visually inspecting a distributor cap or rotor?
30. What can result from corrosion being present at any connector in an ignition system?

14

BASIC OSCILLOSCOPES

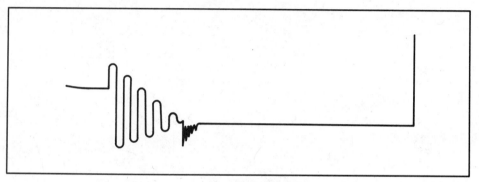

Figure 14–1

One of the most common pieces of test equipment used for electrical and electronic diagnostics is the oscilloscope. An oscilloscope can be likened to a television set; however, the screen shows only a line of light, called a *trace,* instead of a picture. When connected into an electrical circuit, an oscilloscope will display the electrical activity at that point in the circuit (Fig. 14–2). The trace displayed on the screen of an oscilloscope indicates the voltage levels present at that point in the circuit over a period of time.

While using an oscilloscope for diagnostics, a technician observes these voltage levels and the times at which they occur. The operation of the ignition system causes increases and decreases in voltage. These variations must occur at precisely the right time. Therefore, an automotive technician can use an oscilloscope to monitor the actions of the ignition system in order to diagnose that system. The oscilloscope, or "scope" as it will be referred to from here on, will display anything that may interfere with the normal ignition process, as well as displaying the normal process.

An understanding of the trace is imperative when using the scope as a diagnostic tool, as is the understanding of the operation of the ignition system. The scope screen, like a television screen, is the front of a *cathode ray tube* (CRT). The trace that is displayed is the result of a stream of electrons being shot to the inside of the CRT screen (Fig. 14–3). The inside surface of the CRT is specially coated, causing it to glow at points where the particles hit the surface. These particles are shot from a gun located at the rear of the CRT.

The stream of particles shot is regulated by the input into the scope and will be bent or deflected in its path to the screen by deflection plates located in the CRT (Fig. 14–4). The vertical deflection plates bend the stream up and down to change the height of the trace. These plates are controlled by the voltage that is present in the circuit tested, and will cause the stream to rise as voltage increases. The horizontal deflection plates bend the stream from side to side, causing the trace to move from left to right. The speed of this horizontal movement is regulated by the input. On some scopes this can be controlled by the technician. The speed of this move-

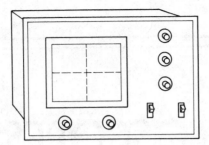

Figure 14-2

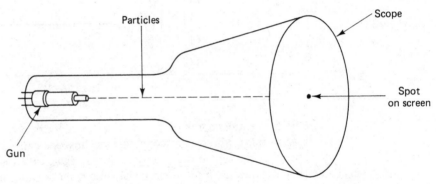

Figure 14-3

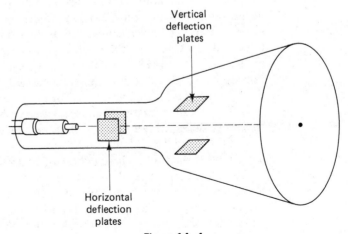

Figure 14-4

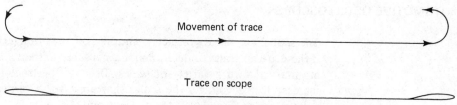

Movement of trace

Trace on scope

Figure 14-5

ment is often referred to as the *sweep rate*. The sweep rate on most automotive scopes is controlled by the speed of the engine.

The trace is continuous; although it appears to end at the right of the screen, it actually continues itself at the left (Fig. 14-5). The movement right to left is so quick that it can hardly be seen. Because the trace is continuous, it is capable of responding to any and all electrical changes that might occur. This capability makes the scope an invaluable diagnostic tool!

If a scope were connected to a 12-V battery, the trace would be a flat line going across the screen as shown in Fig. 14-6, indicating a constant 12 V over that period of time. The trace would remain flat until something caused the voltage to change.

When connected to a 12-V ac power source, the trace would be a perfect *sine wave* (Fig. 14-7), indicating a peak of exactly 6 V negative and 6 V positive. The increase from negative to positive voltages would be displayed in the trace. The time displayed as it goes positive would be equal to the time displayed as it goes negative. This sine wave would remain constant until the peak voltage is changed. At that time, if the change affected only the negative voltage, the trace would no longer be a sine wave but would still reflect the electrical activity in that circuit (Fig. 14-8).

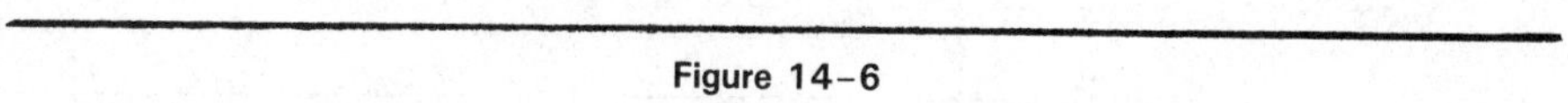

Figure 14-6

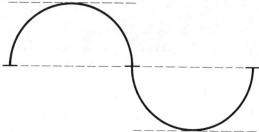

Figure 14-7

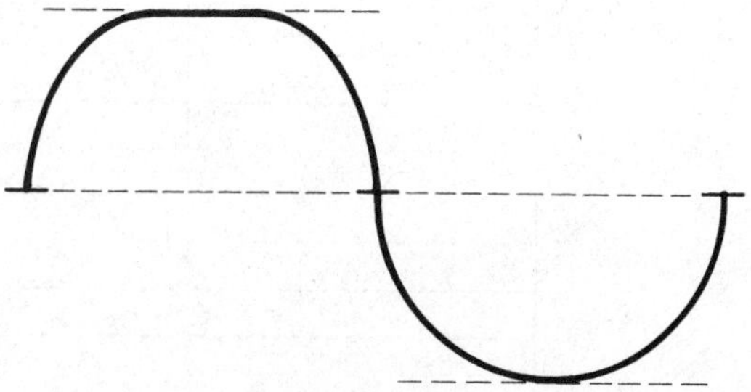

Figure 14-8

AUTOMOTIVE OSCILLOSCOPES

The scope is a visual voltmeter, and like other voltmeters, the readings are useless if there are no scales to determine the measured amounts. Automotive scope screens are marked with a variety of scales. Because the trace is moved upward with an increase in the voltage, the scales that are on the vertical sides of the screen are voltage scales (Fig. 14–9). At the bottom of the scale is the zero line. At the top will be the maximum amount measurable within that scale. You can select the range of the scales by one of the controls on the scope console. Typically, the screen will have two different scales marked off on the scope screen. On the left side of the screen, the scale will range from 0 to 20 or 25 k, and on the right side it will range from 0 to 40 or 50 kV. These two scales are usually referred to as the low and high voltage scales.

The scope monitors voltage over time, and the travel of the trace across the screen represents time. On the scope, time is measured in degrees of distributor rotation. At the bottom of the scope screen there are graduations of time, scaled from 0 to 90 degrees for four-cylinder engines, 0 to 60 degrees for six-cylinder engines, and 0 to 45 degrees for eight-cylinder engines (Fig. 14–10). You should refer to the appropriate scale for the type of engine being worked on. Degrees of distributor rotation are used to measure time, because the engine's crankshaft will rotate a different number of degrees in a second depending on engine speeds.

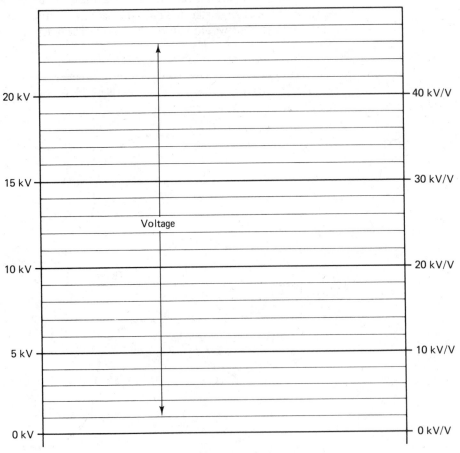

Figure 14–9

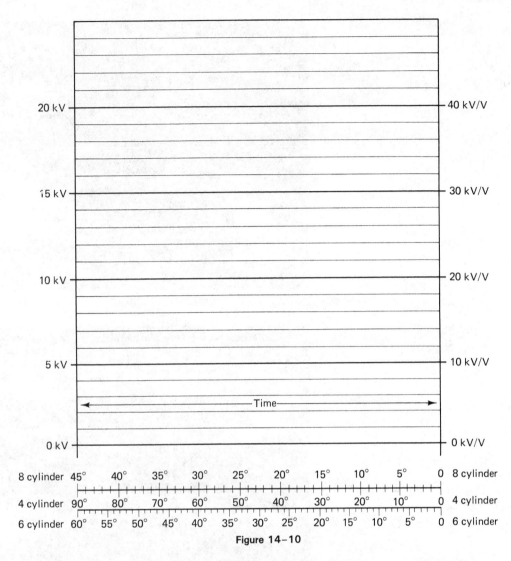

Figure 14–10

To measure the timing of electrical events, some engine analyzers are equipped with a millisecond scale. A *millisecond* is equal to one-thousandth of a second. Scopes with this scale will have a time selection control for the scale to be used. Typically, you will be able to choose either a 5-millisecond or a 25-millisecond scale. The choice of scales should be based on the event to be measured. The scope can only display events that occur over a certain period of time. By using the 5-millisecond scale, only the events that occur during that time will be displayed. The graduations for the millisecond scales appear in the center of the scope screen.

Many of the newer computer-based engine analyzers are equipped with millisecond scales that do not have fixed ranges (Fig. 14–11). Usually, when using this time reference, the screen will initially display 10 divisions; each division is equal to 1 millisecond.

The number of millisecond divisions can be increased or decreased. Increasing the number of divisions will compress the displayed pattern (Fig. 14–12). The first portion of the pattern will be lengthened as the number of divisions is reduced. The capability of changing the time reference allows for very detailed studies of electrical events.

The scope can be used to monitor and diagnose many systems. The most common systems tested with a scope are the ignition, charging, and fuel injection systems. These systems have changes of voltage occurring in their normal operation.

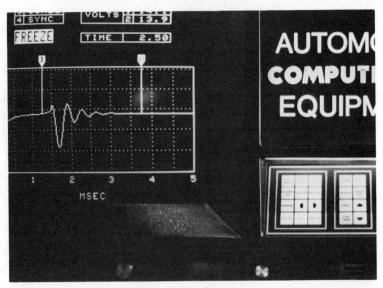

Figure 14-11 This pattern is displayed on a 5-millisecond sweep.

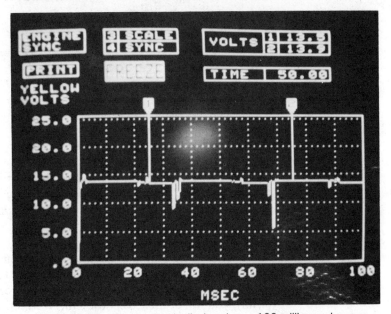

Figure 14-12 This pattern is displayed on a 100-millisecond sweep.

The key to successful diagnosis with a scope is knowing what causes each change in a normal pattern displayed by trace (Fig. 14–13). A technician must be able to notice all abnormalities in a given pattern and then further test the components in the circuit that could cause these abnormalities.

IGNITION SCOPE PATTERNS

The ignition system is composed of two circuits: a low-voltage or primary circuit and a high-voltage or secondary circuit. The scope can display either circuit for individual cylinders or all cylinders of the engine. Through the controls of the scope console, the desired pattern can be chosen. One of the controls will be that of pattern selection, which allows the choice of primary or secondary patterns. The pri-

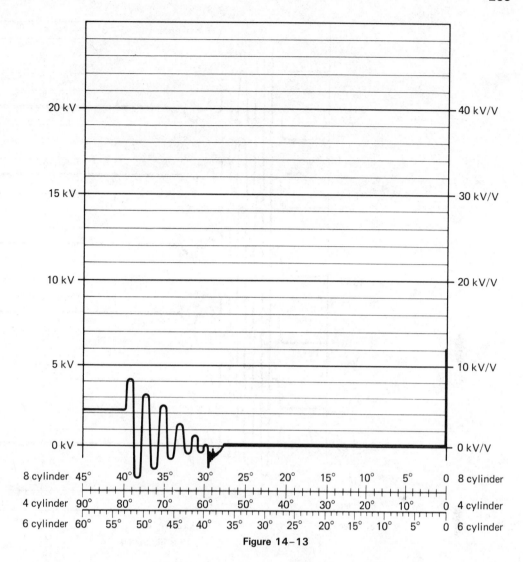

Figure 14–13

mary pattern will display the electrical activity of the primary circuit: the breaker points, condenser, coil, and so on. The secondary will display the activity of the spark plugs, coil, cables, and so on.

Another control on the console will be a position selector. With this control, choices of one or all cylinders can be made and can normally be used with either primary or secondary patterns. The positions available are called: raster, superimposed, and display or parade. The *raster position* (Fig. 14–14) displays the trace of individual cylinders, stacked one above another on the screen. For a six-cylinder engine the screen will show six separate traces, representing the six different cylinders. The order of the cylinders on the screen will be the firing order. The first cylinder is the bottom trace and the last cylinder in the firing order is the top trace.

If it is desirable to review all cylinders as a single pattern, the position selector would be changed to the *superimposed* position (Fig. 14–15). In this position the traces for each cylinder appear on top of one another. If the activity of each cylinder is the same, the superimposed pattern will appear to be one trace. This position allows for a quick comparison of cylinders. If a variance is found, the individual cylinder causing this variation can be identified by switching back to the raster position.

The *display* or *parade pattern* displays the trace of each cylinder, side by side,

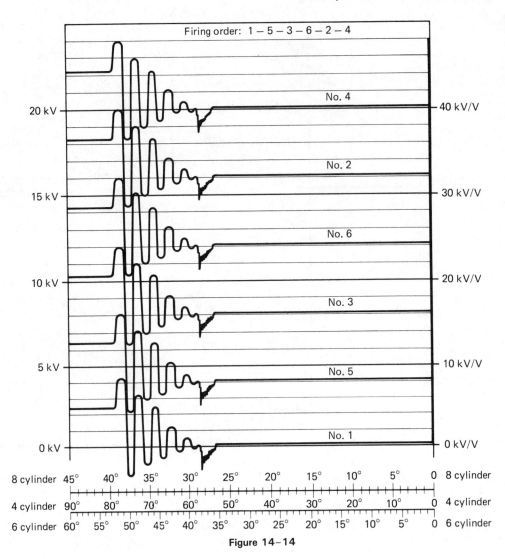

Figure 14–14

across the screen (Fig. 14–16). The main concern in viewing a display pattern is in comparing the peak voltages of each cylinder as the trace goes across the screen. The arrangement of the cylinders will be according to the engine's firing order; therefore, individual cylinders can be easily identified.

Most scope consoles also have controls for proper positioning of the waveform within the printed scales on the screen. There is a vertical control that raises or lowers the total trace on the screen. This allows you to align zero voltage traces to the zero voltage line, ensuring that all other voltage readings can be precisely made.

There is also a horizontal control to move the waveform from right to left on the screen. To complement this control, there is a pattern length, which can reduce or increase the length of the pattern. These two controls allow for the proper placement of the pattern: between the two extremes of the time scales on the screen of the scope (Fig. 14–17). Proper positioning on the degrees scale allows for accurate measurements of time.

In addition to being able to measure voltage over a period of time, the scope also displays the polarity of the voltage. All voltages displayed above the zero line are negative voltages and all below are positive. It is important to note that all traces moving up are going negative and those moving down are going less negative or positive (Fig. 14–18).

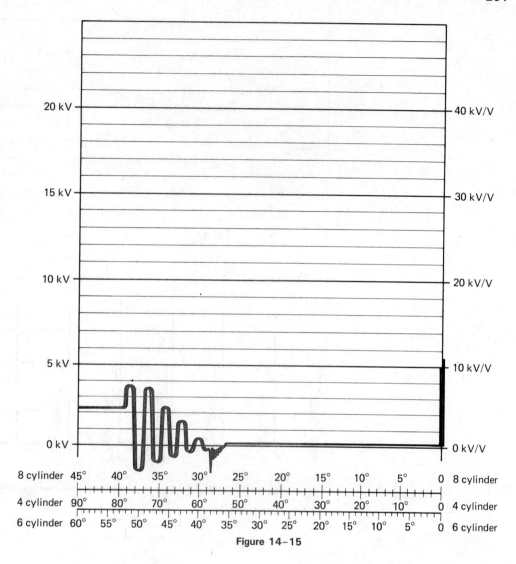

Figure 14-15

There are many manufacturers of automotive scopes and each has its own set of controls. Most will have the ones just mentioned. To define additional controls of a particular scope, it is best to refer to the manufacturer's manual for that scope. As there are variances between manufacturers on console controls, there are also differences in the connections of the scopes to the engines. While the specific hook-up sequences of the different scope manufacturers are not covered in this text, the basic connections to the ignition systems for all scopes are.

An automotive scope requires input from three sources: the primary circuit, the secondary circuit, and a time reference. The primary pattern pickup is connected in parallel with the primary ignition wire at the distributor side of the coil. The secondary pattern pickup is connected in parallel with the coil wire. The firing number 1 cylinder is the time reference; therefore, the time pickup or "pattern trigger" is connected in parallel with the number 1 spark plug. A typical scope hook-up is shown in Fig. 14-19.

When using a scope to diagnose an ignition system, voltage is observed over a period of time. Each part of the waveform represents a specific event in the ignition system. An understanding of the ignition process leads to an understanding of scope patterns.

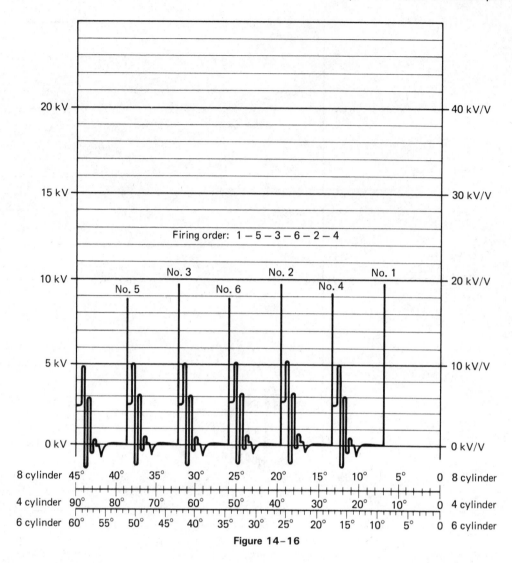

Firing order: 1 — 5 — 3 — 6 — 2 — 4

Figure 14-16

PRIMARY PATTERN

The ignition process begins with the contact points closed, which allows battery voltage to cause current flow through the primary windings of the ignition coil. The points remain closed for a period of time. During this time battery voltage is flowing through the primary circuit. On the scope, the trace would display a line moving downward, indicating a change from zero volts to a positive voltage when the points close, causing current to flow. The trace would then move in a straight line across the screen, indicating a constant voltage over time (Fig. 14-20).

When the points open, current flow is interrupted and the trace moves upward far above the zero line. This interruption of current flow causes the magnetic field around the primary winding to collapse across itself and the secondary winding, inducing a high voltage in both. The high voltage induced in the primary—approximately 250 V—goes to the condenser, which causes the voltage to move back and forth between ground and the coil, through its charging and discharging. The trace follows this movement of voltage by displaying the oscillations of voltage going negative to positive, and gradually diminishing as the amount of voltage decreases (Fig. 14-21).

The condenser action continues until the firing of the spark plug is completed

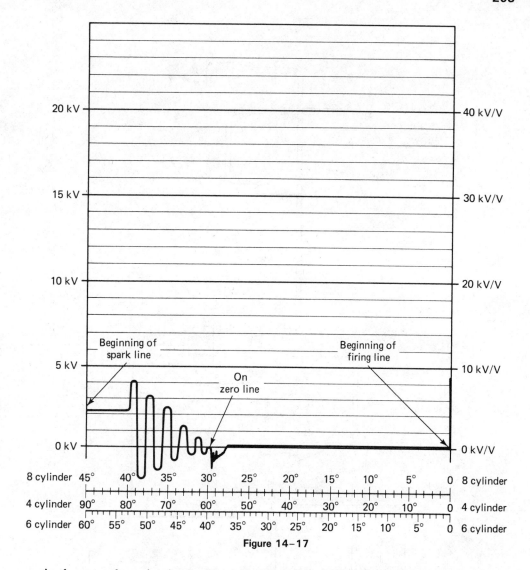

Figure 14–17

in the secondary circuit. At this time all voltages left in the coil are dissipated and the coil returns to a zero potential, ready for another cycle. The dissipation of the voltage in the coil is displayed on the scope again, as an oscillation of gradually diminishing voltages over time (Fig. 14–22). At the point of time where no voltage is present in the coil, the trace returns to the zero line, where it moves across the screen until the points close and a new cycle for the firing of the next cylinder begins.

The waveform constructed by the foregoing events is the primary pattern, as it displays the events of the primary ignition circuit (Fig. 14–23). The pattern is

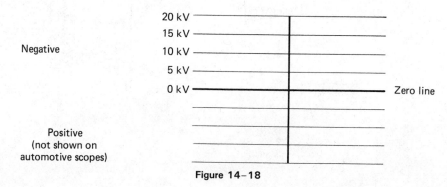

Figure 14–18

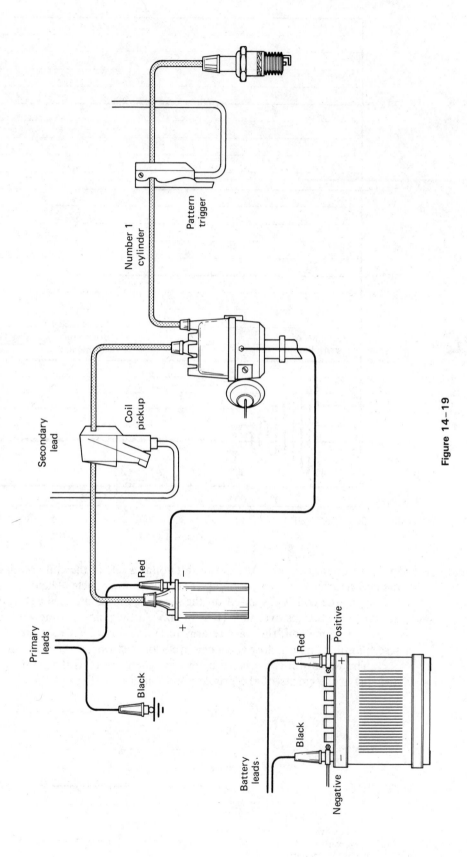

Number 1
cylinder

Pattern
trigger

Coil
pickup

Secondary
lead

Red

Primary
leads

Black

Battery
leads.

Red

Positive

Black

Negative

+

—

+

—

Figure 14–19

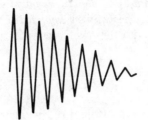

Figure 14-20

Figure 14-21

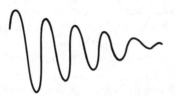

Figure 14-22

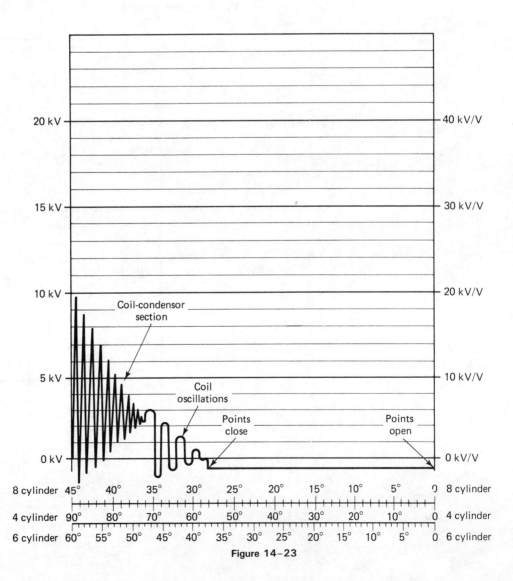

Figure 14-23

commonly divided into three sections: the firing, intermediate, and dwell sections. The firing section displays the condenser action, the intermediate displays the excess voltage being dissipated, and the dwell section begins with the point-close signal and ends with the point-open signal. Because scopes use the firing of a cylinder for a reference of time, the primary waveform that is displayed begins with the firing section and ends with the dwell section.

SECONDARY PATTERN

The secondary pattern displays the events of the secondary ignition circuit. It also begins at the left with the firing of the spark plug. When the magnetic field in the primary winding collapses, it induces a high voltage in the secondary winding. This high negative voltage seeks to complete a path for current flow through the secondary circuit. This voltage pushes its way through all the resistances in that circuit until it bridges the spark plug gap and completes its path. The amount of voltage needed to overtake the resistance found in the ignition coil wire, rotor to distributor cap gap, spark plug wire, spark plug, and gap is represented on the scope as a spike that goes up. The height of this spike is a measurement of the voltage needed to overcome the resistance in the secondary. It is called the *firing line* (Fig. 14–24).

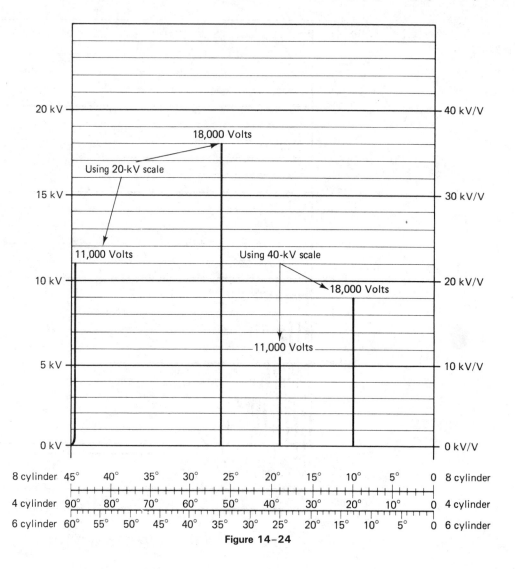

Figure 14–24

Once the voltage has completed its path, current flow is now established, as the voltage bridges the gap in the form of a spark. The voltage necessary to maintain this current flow or spark is approximately one-fourth of the amount that was needed to cause current flow. The spark continues for a period of time until the rotor moves away from the terminal inside the distributor cap and the ignition coil can no longer supply enough voltage to keep the current flowing across the gap of the spark plug. During this time, the voltage remained constant; therefore, the trace on the scope shows the beginning of the spark as a decrease in voltage and moves across the screen at that level until the circuit becomes incomplete. At this time, the excess voltage is dissipated in the coil. This horizontal line, formed on the scope by the amount of voltage needed to maintain the spark, is called the *spark line* (Fig. 14-25).

Following the spark line are oscillations, indicating gradually diminishing voltages, as the excess voltage in the secondary winding is being dissipated within the coil (Fig. 14-26). This section on the scope is similar to the intermediate section of the primary. At the point when there is no voltage left in the winding, the trace returns to the zero line and moves across the screen until the points close and a new cycle begins.

Although the breaker points are part of the primary circuit, the closing and

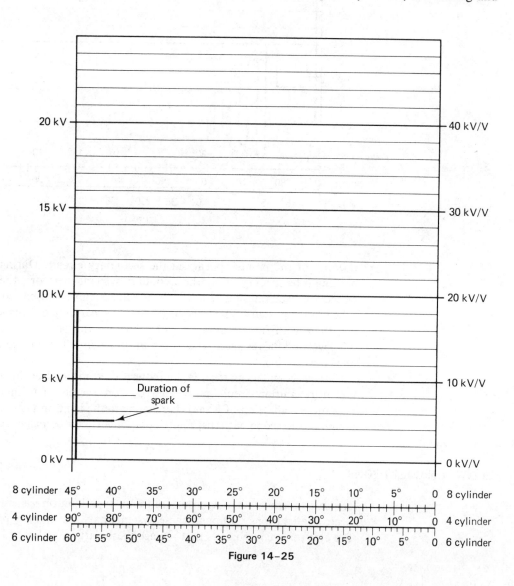

Figure 14-25

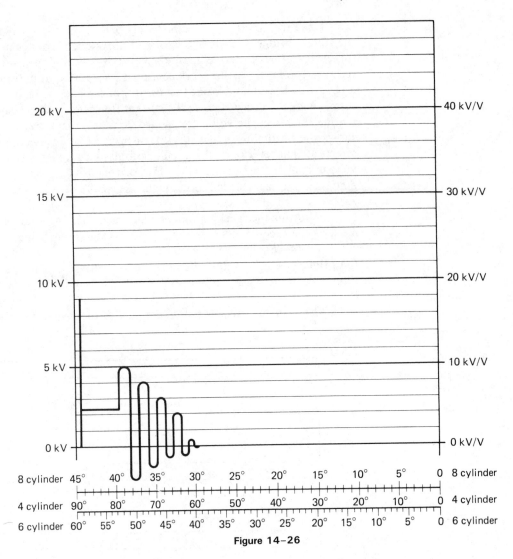

Figure 14-26

opening of the points appears in the secondary circuit. During the dwell period, nothing is happening within the secondary system; therefore, the scope displays the effects of dwell on the secondary. When the circuit is closed, current flow is established in the primary. The trace of this event shows a line going straight downward, and then, with fine oscillations, going back to the zero line. The trace remains at this level and moves across the screen until the circuit opens, causing another firing line (Fig. 14-27).

The secondary pattern is the pattern used most often for diagnostics. Nearly the entire ignition circuit can be viewed, from the points to the plugs (Fig. 14-28). By a mere realization of what is happening in the ignition system, and relating these events to the scope patterns, many components can be inspected and tested.

ELECTRONIC IGNITIONS

The same method for diagnostics on point-type ignition systems can be applied to electronic ignitions, although these systems have no breaker points or condenser. The electronic components that replace these parts can also be inspected and tested

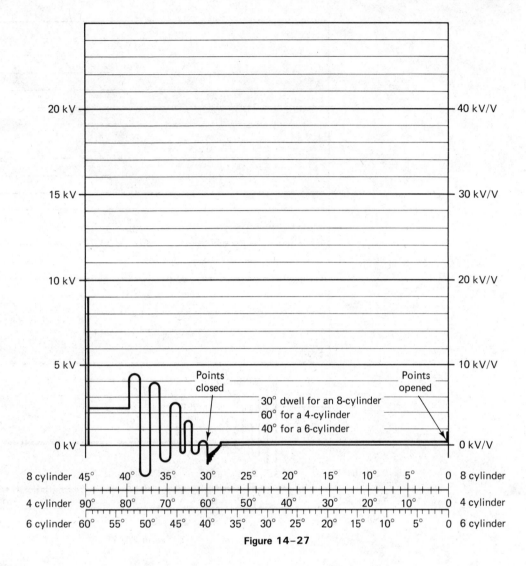

Figure 14-27

on the scope. These electronic components function only in the primary ignition system, and therefore have little or no effect on the appearance of secondary ignition patterns. The only exception to this is GM's Distributorless ignition, which currently cannot be tested with a scope.

The normal primary patterns for electronic ignitions vary according to the systems used. Chrysler's electronic ignition primary pattern (Fig. 14–29) begins at the left, with coil oscillations that appear to be lower than those of a conventional ignition primary pattern. These coil oscillations are followed by the trace going straight down to below the zero line. This trace continues downward until the main switching transistor turns on, allowing for current flow through the coil's primary windings. At this point the trace moves across the screen until the transistor is turned off by the signal received from the ignition module from the pickup unit. The trace now moves upward as the magnetic field in the primary windings collapses, starting a new primary pattern for the next cylinder.

The secondary pattern (Fig. 14–30) for the Chrysler electronic ignition begins at the left with the spark line, which ends with the main transistor being turned on. This leads into the secondary coil oscillations, which continue until the voltage reaches a zero value. The trace continues across the scope at this zero level until

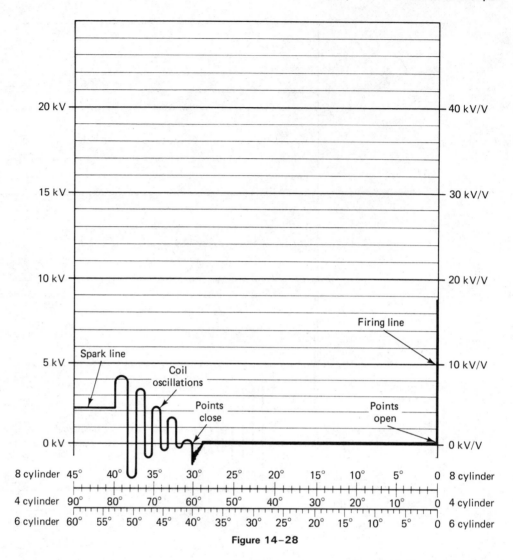

Figure 14–28

the transistor is turned off, when secondary voltage heads out to bridge the spark plug gap. This causes the trace to form a normal firing line, thereby starting the beginning of another cycle or the scope pattern of the next cylinder.

The primary scope pattern on the Ford Solid State Ignition system is nearly identical to that of point-type ignition systems (Fig. 14–31). The firing section begins the pattern at the left. The height of the coil oscillations might not appear as high as those of a point-type pattern, but there may be more. This is also true of the intermediate section, which follows. When the trace from this section reaches a zero value, the trace drops down, as if the points closed. Actually, what happens is that the ignition module, triggered by the magnetic pickup, is allowing primary current flow. The trace then moves across the screen and displays a dwell section. This trace is very similar to that of a breaker-point ignition system, except that it is longer and has a slight curve to it. When primary current flow is interrupted by the ignition module, the trace moves into the next firing section.

The secondary pattern (Fig. 14–32) of Ford's Solid State Ignition begins with a spark line that is identical to that of a point-type system. The coil section that follows is, again, comprised of oscillations that are not as high but are more nu-

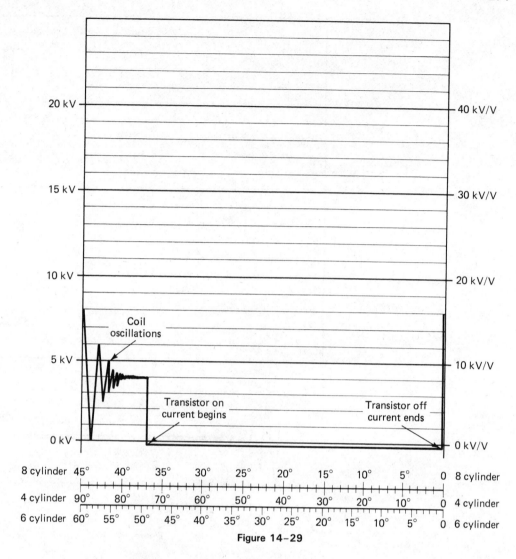

Figure 14-29

merous. The dwell section is longer and curved, just as it was in the primary. This section ends with a firing line that is no different from that of a point-type system.

The General Motors electronic ignition system has the most unique looking primary (Fig. 14-33) and secondary (Fig. 14-34) scope patterns. There is much similarity between the primary and secondary patterns. Both have waveforms which begin with a trace that looks like a spark line moving into the coil oscillations, which appear to be fewer and not as high as those of a point-type system. When the voltage level in the coil reaches a zero value, the trace moves to the right until the ignition module is triggered, to allow primary current flow. This triggering appears in the primary as in most primary patterns: as a straight drop-off of the trace as it goes into the dwell section. The secondary trace shows the same effect of the primary on the secondary as in breaker-point ignitions: displaying small oscillations as primary current flow begins.

The dwell section appears to be shorter than normal and increases in length with increases in speed. An increase in voltage is shown near the end of the dwell section. This increase levels off and the trace continues until primary current is turned off, which causes the firing line to appear.

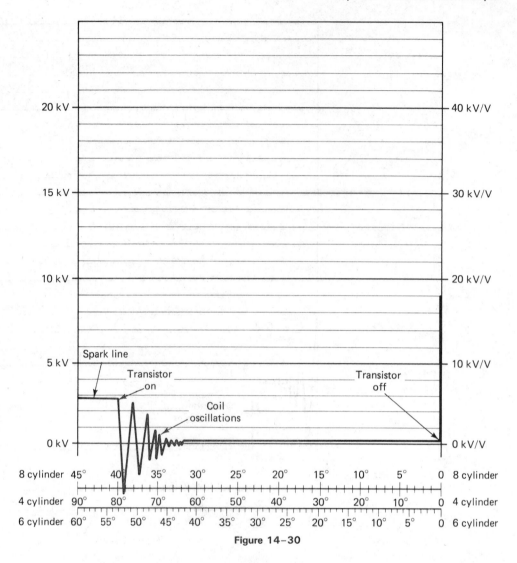

Figure 14-30

The uneven firing system used by GM on some V-6 engines creates unique-looking patterns in both raster and superimposed. Because the cylinders alternately fire every 150 and 90 degrees, the "on" signal for primary current is staggered to represent this unevenness (Fig. 14-35). Knowing that this is a normal pattern for this type of engine can be a great guide in using the scope for diagnosis.

Although there are differences between types of electronic ignitions, an understanding of how the systems work is the key to using the scope as an effective diagnostic tool.

CHARGING SYSTEMS

The scope can also be used to test and diagnose charging systems. The alternator sends a current back to the battery. This current is pulsating dc. The resultant trace of this charge is a series of small equal-size curves. The presence of this trace (Fig. 14-36) on the output of the alternator indicates that the rectifier and related components are operating properly.

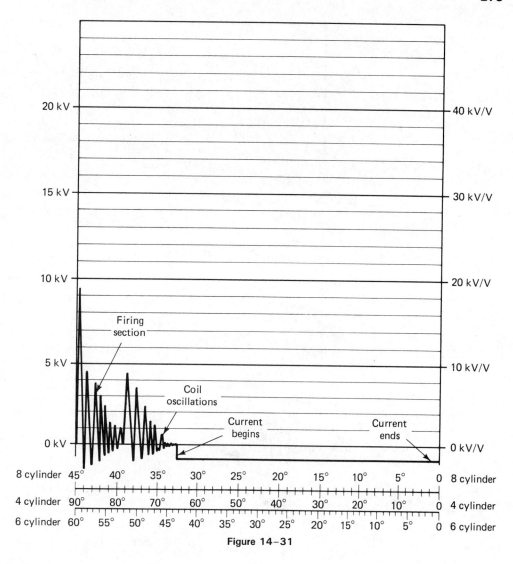

Figure 14–31

ELECTRONIC FUEL INJECTION

Electronic fuel injection works much like the firing of a spark plug. Voltage surges to the injectors, causing them to activate and inject fuel. These voltage surges can be likened to the firing line, as the trace on a scope will show up as a spike. The injectors are activated for a period of time. The resultant trace of this duration of time is much like the spark line of the secondary ignition pattern (Fig. 14–37). The absence of the injector firing line or spark line can be the clue a technician needs to determine why a car has a miss or some other performance problem.

SUMMARY

Automotive systems that function normally with sudden changes in voltage can be monitored, and thereby diagnosed, with an oscilloscope. A thorough understanding of the normal patterns and the events that cause the patterns is imperative to using

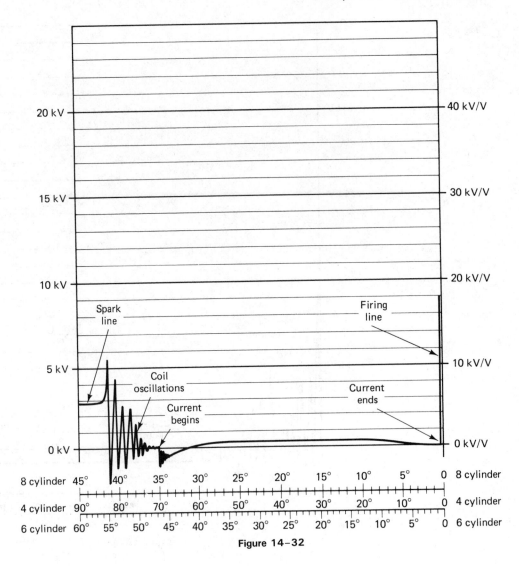

Figure 14–32

scope patterns for diagnostics. In the next chapter, ignition and alternator scope patterns with common problems will be examined and explained in detail.

TERMS TO KNOW

Trace   Superimposed
CRT   Display
Sweep rate   Parade
Sine wave   Firing line
Millisecond   Spark line
Raster

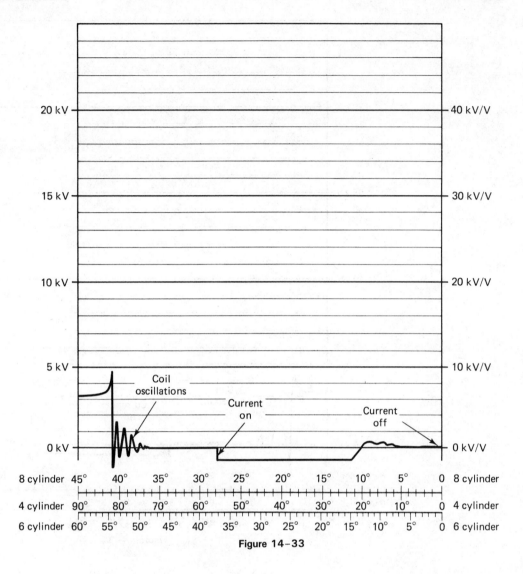

Figure 14–33

REVIEW QUESTIONS

1. What do the traces on the screen of an oscilloscope indicate?
2. Why is an oscilloscope an important diagnostic tool?
3. On an automotive scope, what scales are usually on the vertical sides of the screen?
4. On a scope, how is time measured?
5. Why is time usually measured in degrees of distributor rotation?
6. Where does the pattern for the number 1 cylinder appear in a raster pattern?
7. When is the superimposed pattern used?
8. When is the display pattern used?
9. In what order are the cylinders displayed in the display pattern?
10. How can the voltage's polarity be determined by a scope?
11. All automotive scopes require inputs from three sources. What are they?
12. Why must you have an understanding of the ignition system before you can use the scope as an effective diagnostic tool?

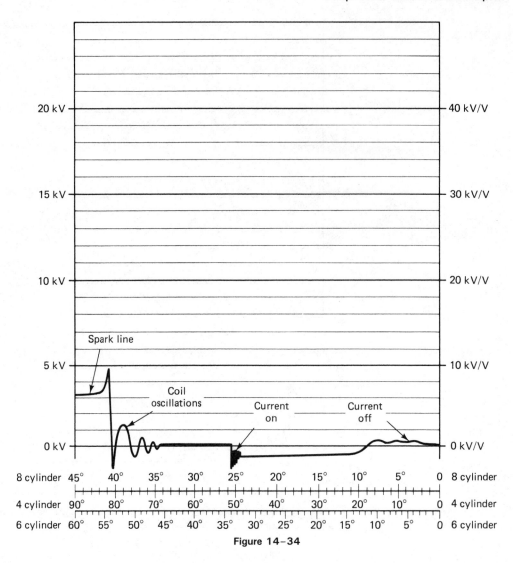

Figure 14–34

13. What events make up the primary waveform?
14. What are the three sections into which a primary pattern is broken?
15. What is indicated by the height of a firing line?
16. What event took place in the primary which caused the firing line?
17. What is taking place in the primary during the spark line?
18. What is taking place in the secondary during the spark line?
19. In the secondary pattern, the spark line is followed by a series of oscillations. What do these represent?
20. What scope pattern is most often used for diagnostics?
21. Why is the use of the scope in diagnostics of electronic ignitions only slightly more difficult than for conventional systems?
22. What does the normal output signal from an alternator represent?
23. Compare the pattern caused by electronic fuel injection to the normal firing and spark lines of an ignition system.

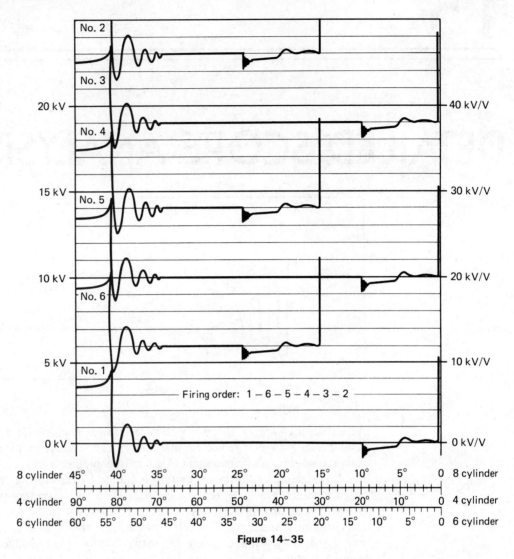

Figure 14-35

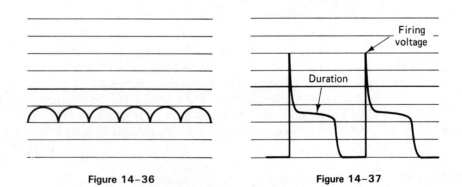

Figure 14-36   Figure 14-37

15

DETAILED SCOPE ANALYSIS

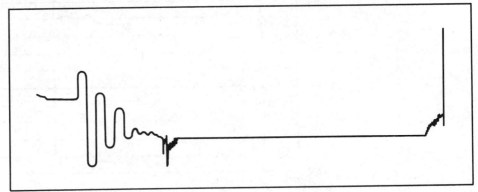

Figure 15-1

As stated in Chapter 14, the key to successful diagnosis with a scope is knowing what causes each change in a normal pattern. The change in a waveform is usually caused by a particular component or group of components (Fig. 15-2). For example, the dwell section of a primary pattern is caused by the closing and opening of the contact points. Similarly, the coil oscillations in the secondary pattern are caused by the coil dissipating its excess voltage. This voltage is the energy left over after the firing of a spark plug. In other words, it is the voltage produced by the coil minus the firing voltage. There are many factors that can affect the amount of voltage produced, and many factors that can affect the amount of voltage necessary to cause and maintain the spark across the spark plug gap. Therefore, there are many factors that affect the amount of unused voltage or the look of the coil oscillations on the scope.

The effects of one component on another must also be kept in mind while analyzing a scope pattern. The firing line of the secondary, which is best viewed in the parade or display pattern, represents the amount of voltage necessary to overcome the resistance in the secondary circuit. This resistance can be present anywhere in the circuit, from the ignition coil wire to the spark plug gap. Resistances are present either by design or by fault. Until the plug gap is bridged, there is no current flow, and with no current flow there are no voltage drops through the resistances; instead, the amount of voltage increases with an increase of resistance.

FIRING LINE

To shock the air-and-fuel mixture adequately, some amount of resistance must be present to cause enough voltage to ignite the compressed mixture properly. The ideal amount of voltage delivered to each cylinder is 8 to 11 kV. This amount allows for enough shock and also allows for enough coil reserve to be available when the demand for more electrical energy is there and/or the coil output is low. At high speeds

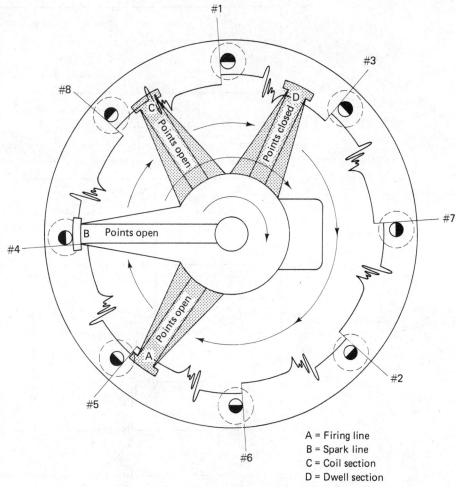

A = Firing line
B = Spark line
C = Coil section
D = Dwell section

Figure 15-2

the time for primary current flow is decreased, and the coil is less likely to become saturated, which would cause a lower coil output. If the output is less than what is needed to fire a cylinder, the cylinder will not ignite.

The first step in diagnosing any ignition system with a scope is to observe the heights of the firing line of all cylinders (Fig. 15-3). The amount of voltage measured is also the amount of resistance in the secondary system. A reading of less than 8 kV indicates that the resistance is too low. If the voltage is more than 11 kV, the resistance is too high. In either case, the firing line is out of the normal range and a problem exists. The severity of the problem will usually be indicated by how far out of range the voltage is. The general area of the problem must be determined. By observing the required voltage for all cylinders, one can make a general but important assumption. If all the cylinders require a lower or higher amount of voltage than the amount that is considered normal, it is likely that the problem is something common to all cylinders. One bad spark plug will not affect the operation of all the other cylinders and therefore is an uncommon problem. Reviewing the components that can cause an increase or a decrease in required voltage, and determining if the problem is a common or uncommon one, can lead to the correct general area of the system.

By identifying all the possible areas of resistance in the secondary circuit, all the possible causes of an abnormal firing line are examined. It is possible that the secondary cable, from the ignition coil tower to the distributor cap, has the wrong

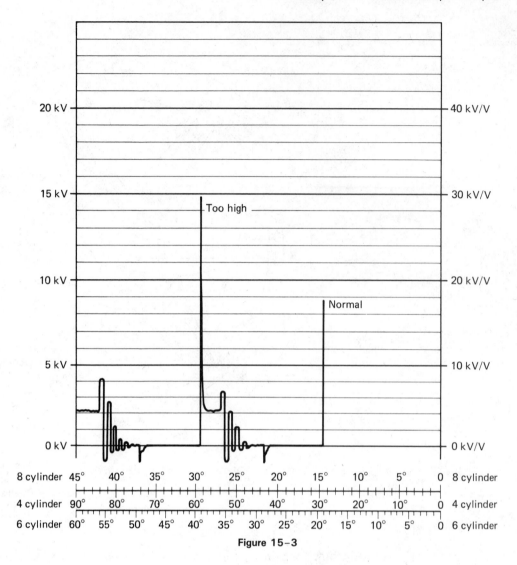

Figure 15-3

resistance or that its ends are corroded. The distributor center button could be damaged, causing a change in resistance. The rotor-to-distributor cap air gap may be too wide or too narrow. The rotor itself may have some resistance. Some manufacturers even put in a resistor to connect the center contact of the rotor to its tip. All these components and problems would appear as a common problem, as they would affect the firing line and actual firing of all cylinders (Fig. 15–4).

The castle pins inside the distributor cap and their connection to the spark plug cable tower may have resistance. Spark plug cables and their ends may be the source of high or low voltage requirements. The connection between the cable terminal and the center electrode of a spark plug can have an improper resistance. The air gap of the spark is surely a source of resistance, but more than the distance of the air gap determines the amount. The quality of air within the gap affects this the most. How well the mixture is compressed and the air/fuel ratio are the major concerns here. The higher the compression in a cylinder, the more pressure there is against the arc, and therefore there is more resistance. Fuel conducts electricity better than air, so the leaner the mixture, the higher the resistance.

Problems between the distributor cap and the individual cylinders are likely to show up on the scope as uncommon problems (Fig. 15-5). Air/fuel ratio problems are an exception to this. A carburetor has a common function to provide all cyl-

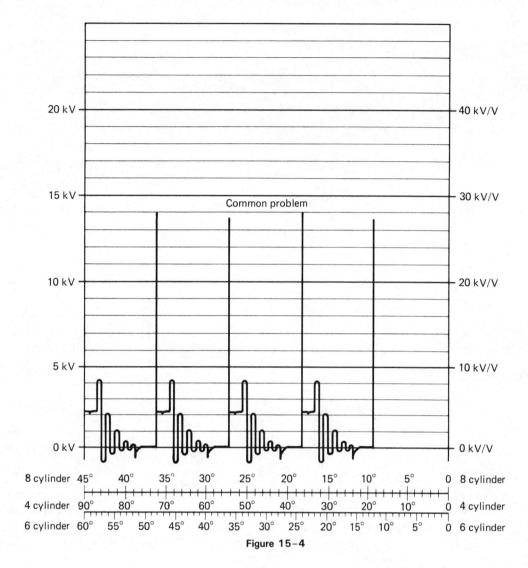

Figure 15-4

inders with a fuel-and-air mixture; therefore, a carburetion problem is a common problem. On some eight-cylinder engines, where one-half of the carburetor feeds one-half of the cylinders, a problem here will be displayed as common to half of the cylinders. This is not to say that whenever four of eight cylinders show a lower or higher voltage requirement, a carburetion problem is evident. Because of intake manifold designs and the vacuum pulses of an eight-cylinder engine, one-half of the carburetor usually feeds cylinders 1, 4, 6, and 7 and the other half feeds cylinders 2, 3, 5, and 8. If a carburetion problem exists in half of the carburetor, a set of four cylinders will be affected. The scope would display this with lower or higher firing voltages on every other cylinder. V-6 engines with an imbalanced carburetor will also display different firing line heights on every other cylinder.

The heights of the firing lines of all cylinders should be the same. This indicates that each cylinder is firing with the same voltage, and that each is firing in approximately the same pressure and air-and-fuel mixture. This equality is not always obtainable, but the maximum allowable variance between cylinders is 3 kV and all must be within the normal range 8 to 11 kV.

To find the cause of abnormal firing lines, the major components of the secondary circuit must be systematically bypassed (Fig. 15-6). A firing line can be divided into four general sections: (1) the amount of voltage necessary to overcome

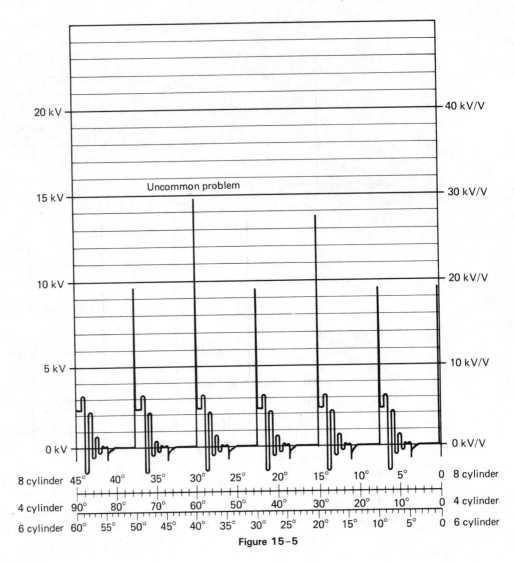

Figure 15-5

the resistance in the coil wire; (2) the amount of voltage necessary to move from the coil terminal of the distributor cap, through the rotor, and jump the air gap to a spark plug wire terminal on the cap; (3) the voltage needed to go from the distributor cap through the spark plug wire to the spark plug; and (4) the amount necessary to go through the spark plug and to bridge the gap to the ground electrode. This section is normally the largest of the sections of the firing line. By bypassing each of these sections one at a time, the cause of an abnormal firing line can be determined.

Using the firing order, identify the cylinder that is abnormal and record its required voltage. By removing the spark plug wire from the spark plug, and by grounding the end of the wire to a known good ground with a good jumper wire, the spark plug is bypassed. The scope can now display the amount of voltage necessary to overcome the resistance in all of the secondary circuit, minus the spark plug. The firing line of that cylinder will drop down to 5 kV if the problem is caused by the spark plug or something in the combustion chamber affecting the plug. If it does not drop below 5 kV, the problem is elsewhere in the secondary circuit.

If the problem is not in the spark plug circuit, the next section must be bypassed. By removing the plug wire from the distributor cap, and by putting the jumper wire to ground on the plug terminal at the distributor cap, the spark plug

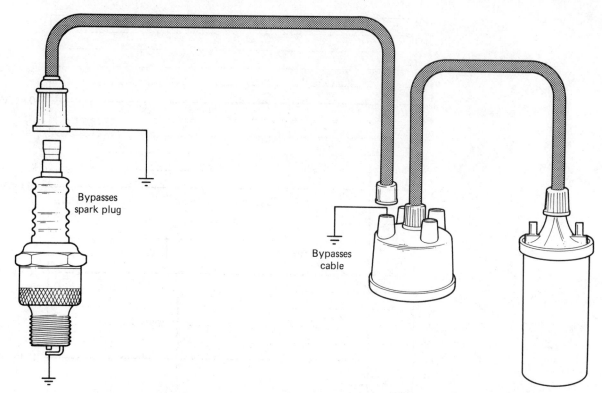

Figure 15-6 Systematically bypassing components in the secondary circuit.

wire is bypassed. If the required voltage decreases to about 5 kV, the source of resistance is the plug wire or its terminals. If the voltage does not fall to 5 kV, the problem is either the coil wire or the rotor and distributor cap. In either case these are common problems which are best determined by visual inspection and/or the use of an ohmmeter.

Care should be taken when bypassing secondary components on engines equipped with catalytic converters and electronic ignition. If equipped with a catalyst, the engine should be run for only short periods with a cylinder bypassed. When bypassing circuit components of an engine equipped with electronic ignition, use a test plug to bypass the circuit.

SPARK LINE

Once the spark plug gap is bridged by the electrical arc and there is current flow in the secondary circuit, the waveform in the scope moves to the spark line. The spark line represents the amount of voltage necessary to maintain the arc across the spark plug gap. Because there is now current flow, there are also voltage drops within the secondary. The trace drops between the 2- and 3-kV markings on the scope. Changes in resistance that take place during the time the arc continues will cause changes in the spark line. If the resistance increases halfway through the time allotted for the spark, the trace will show an increase in voltage as the coil puts out more voltage to maintain the spark. All conditions and components that affect the firing line may also affect the spark line (Fig. 15-7).

The arc is ended by the action of the rotor moving away from the castle pin inside the distributor and by the coil not having enough energy remaining to jump across the increased gap between the rotor and the distributor cap. Therefore, if

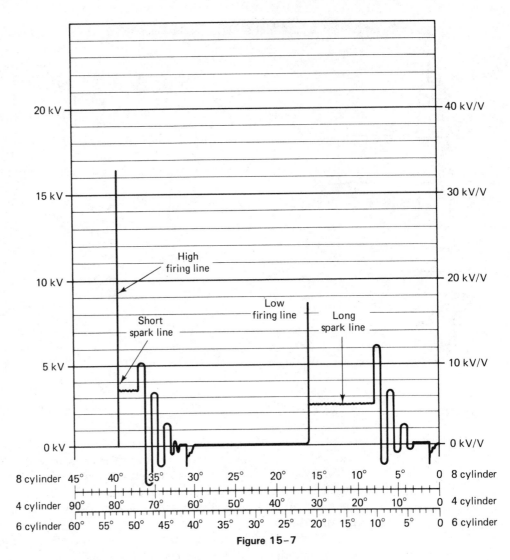

Figure 15-7

time is lost trying to overcome the resistances in the secondary, there will be less time remaining to continue the arc. Because of this, increased resistance in the secondary will decrease the time allowed for the continuance of the spark. The higher the firing line, the shorter the spark line. This relationship is another reason for the firing line to fall within the normal range; it allows for enough spark duration. Too short a spark line can be evidence for poor fuel economy, loss of power, or overall poor performance. If the firing line is kept at the correct height, the spark line will be of the correct length.

The shape and length of the spark line can best be observed in the secondary raster pattern. All the spark lines should be compared to each other. Ideally, each should be of an equal length and shape: a flat horizontal line. Like the firing line, to diagnose the conditions that are causing abnormal spark lines, the possible problem must be defined as a common or uncommon problem. If all the spark lines of an engine are affected, the problem is a common problem. If the problem is an uncommon one, it can be found by systematically bypassing the major components for the cylinders affected. If bypassing a component causes the spark line to flatten out or return to a horizontal line, that component is the source of the problem.

To adjust the scope properly to observe and measure the spark line accurately, the beginning of the firing line, which is the points-open signal, should be placed on the zero-voltage line. A normal spark line should appear as a flat horizontal line on or about the 2-kV line. Resistor spark plugs cause the spark line to have a gradual slope downward toward the coil section. Since most cars are equipped with this type of spark plug, a typical normal spark line will actually begin at a point higher than 2 kV and slope down gradually to or below the 2-kV line (Fig. 15–8).

As the resistance in the secondary is increased, the slope of the spark line will increase. Therefore, it is fair to assume that the greater the slope, the more the resistance. As with the firing line, the more resistance, the more likely this condition will affect the running of the engine. Problems such as a cracked center tower of the distributor cap or a poor connection of the terminal end and the coil wire are common problems and would therefore affect the spark lines of all cylinders (Fig. 15–9).

Opens in the spark plug cables or corroded plug towers on the distributor cap would be uncommon problems and would affect only the spark line of the cylinders with the problem. The presence of a spark line in the pattern indicates that the secondary circuit has been completed and there is current flow. If a plug circuit is open, the circuit is not complete and no spark line will appear (Fig. 15–10).

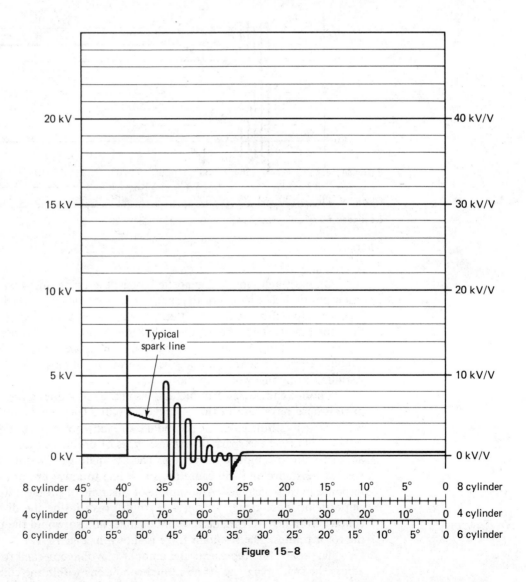

Figure 15–8

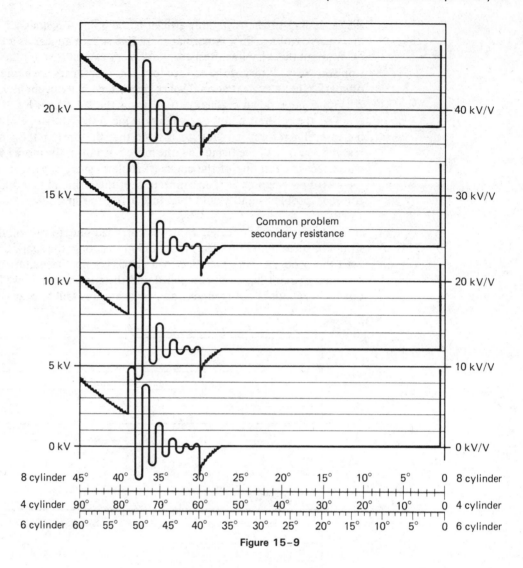

20 kV	40 kV/V
15 kV	30 kV/V
10 kV	20 kV/V
5 kV	10 kV/V
0 kV	0 kV/V

Common problem
secondary resistance

8 cylinder	45°	40°	35°	30°	25°	20°	15°	10°	5°	0	8 cylinder
4 cylinder	90°	80°	70°	60°	50°	40°	30°	20°	10°	0	4 cylinder
6 cylinder	60°	55°	50°	45°	40°	35°	30°	25°	20°	15°	10°	5°	0	6 cylinder

Figure 15–9

On most automotive diagnostic scopes, the firing line does not appear in the raster position. The length of the spark line is determined by the height of the firing line. Therefore if in the raster pattern, one or more spark lines are shorter or longer than the others, it can be assumed that the firing lines of those cylinders will be higher or shorter. An extremely long spark line is an indication of a shorted plug circuit. A bridged spark plug gap is an example of this (Fig. 15–11). The corresponding firing line would be short.

Upward and downward fluctuations in the spark line can be caused by changes in resistance anywhere in the secondary circuit. Changes in resistance can occur while current is flowing in the secondary and may not influence the firing line, but rather, will affect the spark line. These types of problems cause the voltage to take another path to complete the circuit. The new path can be one of increased or decreased resistance and will cause the spark line to move upward and downward in response to these changes. An example of this type of problem would be a broken, chipped, or loose center button or carbon brush inside the distributor cap. The contact from here to the rotor would change, and so would the path of voltage as the rotor rotates. The resultant trace is shown in Fig. 15–12.

The spark line represents the amount of voltage needed to maintain the arc across the spark plug gap. If abnormal conditions within the combustion chamber

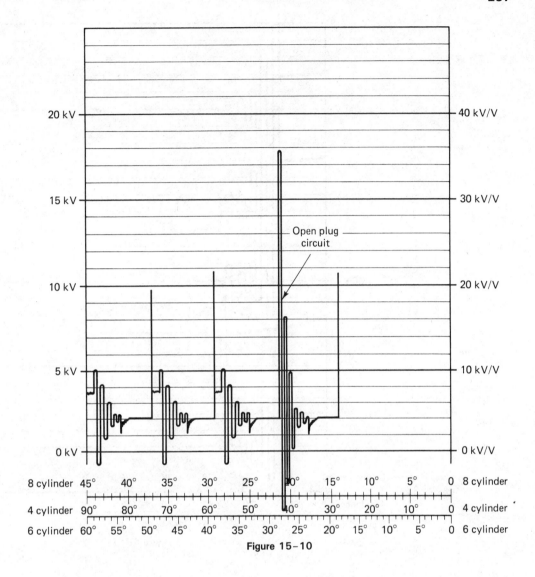

Figure 15-10

affect the quality of air within the air gap of a spark plug, the spark line trace will respond to these conditions. Combustion chamber problems will cause the spark line to slope upward into the coil section (Fig. 15-13). The trace may start at the normal 2-kV level but will slope above that level as it moves to the right of the screen. This pattern can be caused by low cylinder compression or cylinder pressure losses. All upward-sloping spark lines are caused by combustion chamber problems, and the cause can be found by conducting mechanical tests on the engine.

COIL SECTION

The coil section follows the spark line and represents the dissipation of the remaining voltage in the coil. The more voltage that is needed to start and maintain the arc across the spark plug's gap, the less voltage there is remaining in the coil. A normal coil oscillation has five distinguishable waves above the baseline that appear to be gradually diminishing (Fig. 15-14). High required voltages for the secondary will affect the amount of voltage remaining in the coil, so a high firing line can cause an abnormal coil section. When observing the coil section, the firing section must also be considered.

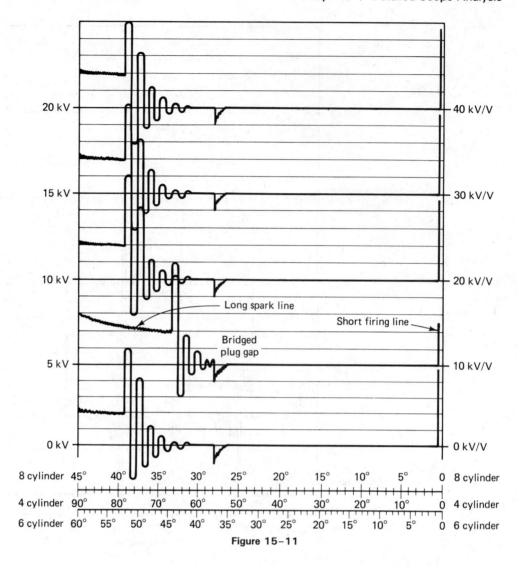

Figure 15–11

If there is a problem in the primary ignition circuit, the coil's output will be affected; therefore, the coil oscillations may not appear normal. Consider all the components in the primary that can affect coil output. Any resistance in the primary can prevent the coil from reaching the maximum value of current flow required for saturation or maximum coil output.

COIL OUTPUT TEST

The coil circuit, which includes the primary ignition system as well as the coil itself, can be tested as a unit for maximum output. This is done by removing a spark plug wire at the plug and not allowing the plug wire to contact a ground. With the engine running at about 1000 rpm, the open-circuited cylinder should display no firing or spark line and a large coil section, with the highest oscillation above the 20-kV line (Fig. 15–15). This coil section should also be composed of five good oscillations. If the trace indicates low coil voltage output or has fewer than five oscillations, a problem with the primary or the coil is indicated. It is not unusual for an electronic ignition coil to have an output of nearly 50 kV. When testing the coil, check the specs for maximum coil output.

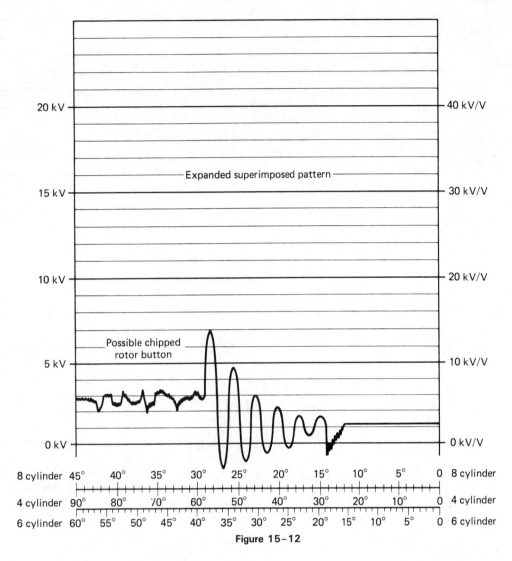

Figure 15-12

If the coil section displayed a weak coil prior to opening the plug circuit, and the results from testing the coil circuit showed good output, it can be assumed that excessive voltage is being used in the secondary, leaving less than normal ignition reserve. If when open-circuited the coil section remains weak, there is a problem in the coil or in the primary. The primary circuit is best observed by looking at the dwell section of the pattern.

SECONDARY INSULATION TEST

While testing the output of the coil, a test for secondary insulation can be made. For the output test, the upward spikes above the zero line are observed. To determine the effectiveness of the secondary insulation, the oscillations below the zero line are observed. The height of these downward spikes should be at least half as large as those above the zero line and should form an upward diagonal pattern from left to right across the screen (Fig. 15-16). If these downward spikes are less than half of the size of the upward spikes or the upward diagonal pattern is interrupted, there is voltage leaking through the insulation of the secondary. When this occurs, the voltage is finding an easier path to ground somewhere in the secondary. The

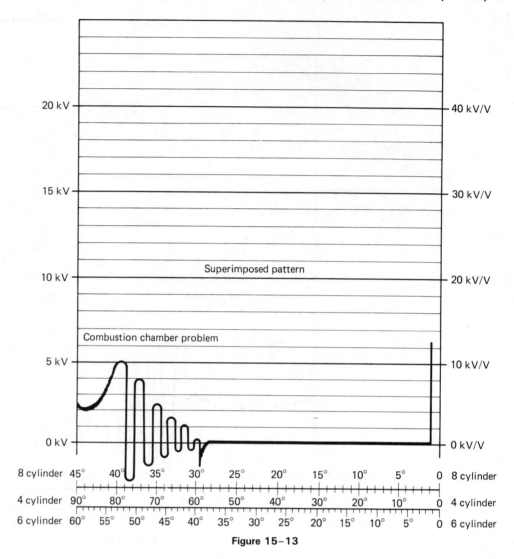

Figure 15–13

problem area can be identified by systematically bypassing the components in the secondary, as is done to define a short or high firing line.

If a weak coil section is displayed and the ignition coil is suspect, the coil should be tested with a coil tester or be tested with an ohmmeter. Measurements of resistances in the primary and secondary windings, when compared to manufacturer's specifications, can identify the cause of low coil output. However, the coil should be suspect only after the primary circuit has been carefully observed.

DWELL SECTION

The dwell section begins with the closing of the points or the start of current flow through the primary. As the trace drops downward in a straight line, this signals the actual mechanical closing of the points (Fig. 15–17). Following this signal are oscillations moving upward to a horizontal line. The point where the trace begins to move horizontally is the time where the points, acting as a switch, close electrically. The oscillations between these two events represent the current starting to flow across the points.

The dwell section ends with the opening of the points or the start of the next

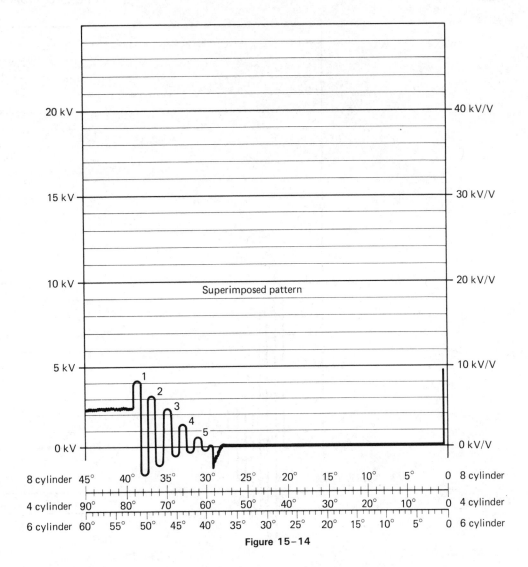

Figure 15–14

firing line. The distance, measured in degrees, between the mechanical closing and opening of the points represents the dwell period (Fig. 15–18). To check the length of the dwell period, it is best to set the scope in the superimposed pattern. This allows you to view all the cylinders at once and to compare the timing of these events on all cylinders. In the superimposed pattern, the point-close and point-open signals should appear as one pattern, showing little variation. If variations are noted, the cylinders that vary can be identified by moving the pattern into the raster position. Most manufacturers allow a variance of 2½ degrees maximum. It is important to note that this variance not only indicates different-length dwell periods, but also different spark timing for the cylinders because the opening of the points begins the firing of the plug.

If a variation was evident in the superimposed pattern and was identified as being caused by one cylinder, it is most likely that this is caused by an improperly machined or worn distributor cam lobe (Fig. 15–19). Worn cam lobes can also affect more than one cylinder, as can other faults within the distributor. But it is not difficult to determine, as not only will the point-close and point-open signals be out of vertical alignment while the pattern is in raster, but the length of dwell in the affected cylinders will be different from the rest. To verify this problem, measure the opening of the points as the rubbing block rests on the different cam lobes. If

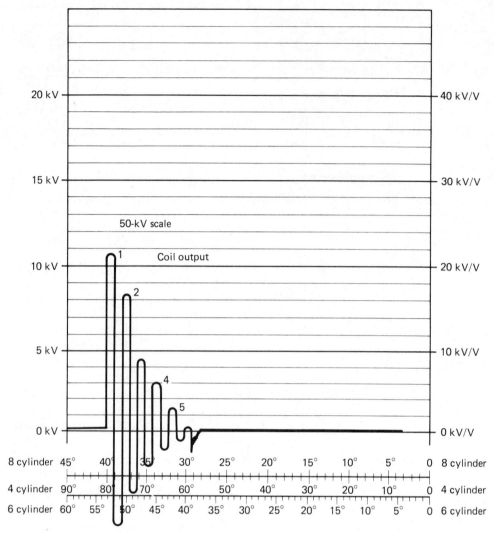

Figure 15-15

there is a difference in point gap, the distributor cam lobes were poorly machined or are worn.

If the time of the point-close and point-open signal varies between cylinders but the length of dwell appears to be equal (Fig. 15-20), it can be assumed that the problem is caused by something changing the placement of the points in relationship to the cam. A worn centrifugal advance plate can shift the placement of the points as the plate vibrates with the motion of a running engine.

Worn distributor shaft bushings will allow the shaft to wobble as it spins. This action not only causes the scope to display variances in the point-close and point-open signals, but also causes variations in the length of dwell. As the shaft wobbles, the actual amount of mechanical opening of the points will vary; thus dwell time varies. Normally, this condition causes the point-close signals to form a symmetrical curve while the pattern is in raster (Fig. 15-21).

Much information about the primary ignition circuit can be derived while observing the point-close and point-open signals. When the points close, they should close quickly and precisely, to allow the current to flow instantly and to be capable of saturating the primary windings of the coil. Any undesirable resistance on the

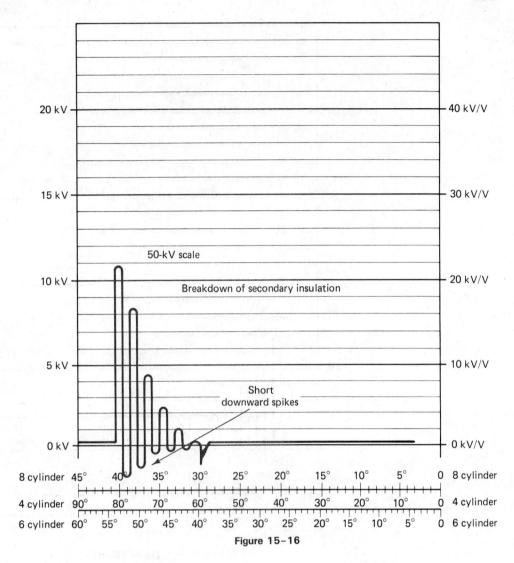

Figure 15–16

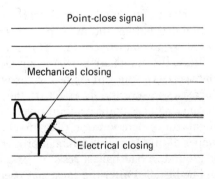

Figure 15–17

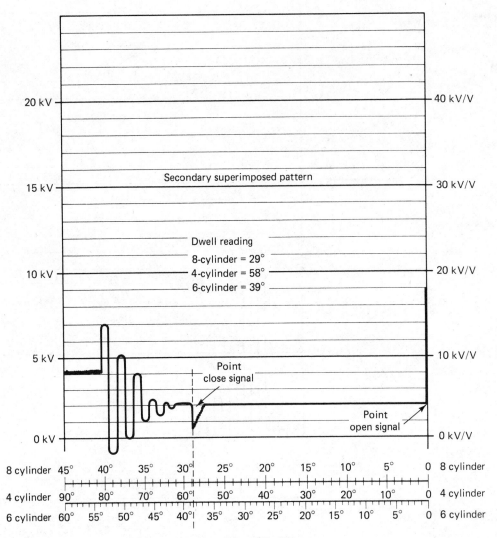

Figure 15–18

surface of the contact points, whether it be surface deterioration, partial contact of the two surfaces, or foreign matter on the surfaces, will either delay the start of current flow or prevent primary current flow from reaching maximum strength. Some faults will cause both.

If the point-close signal for a cylinder appears to have two downward spikes prior to the horizontal oscillations and movement of the trace, a partial closing of the points is followed by a complete closing. This is usually caused by misaligned or burned points. While the points are partially closed, the surface area that is not in full contact will experience some arcing as the current attempts to start flowing. As the problem increases in severity, its effect on the point-open signal increases. Traces indicating point arcing at both the point-open and point-close signals signify severe point misalignment or deterioration (Fig. 15–22). The burning of points can be caused by a number of problems, including improper dwell, a high charging rate, a poor distributor ground, a bad ballast resistor, or a defective condenser. A new set of points should be installed after the cause has been determined and corrected.

Foreign substances on the surfaces of the points decrease the chance of coil saturation by offering a resistance to current flow. The point-close signal on the scope will display this problem by showing the downward spike caused by the me-chanical closing, but followed by few, if any, oscillations moving toward the hor-

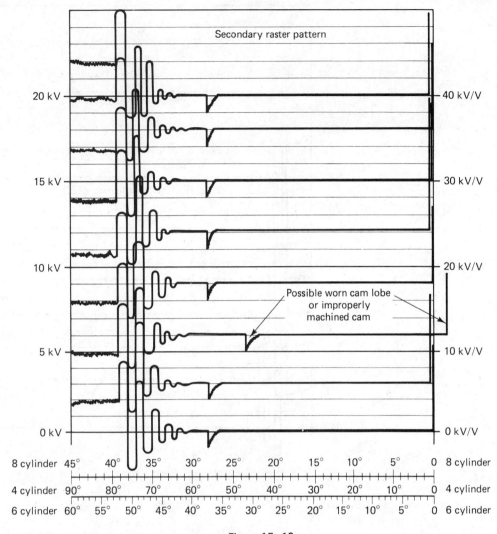

Figure 15-19

izontal dwell line (Fig. 15–23). Much time will be displayed between the mechanical and electrical closing of the points. The slope upward and the oscillations between these two events will appear to be quite lazy. Points that have this condition are referred to as being *frosted*. This term has been given to points that offer some insulation to current flow caused by oil, wax, fumes, or dirt. The surface of frosted points appears white and the points should be replaced. Frosted points will have little or no effect on the point-open signal.

If the point spring tension is incorrect or the distributor cam lobes are dry, point bounce can occur. When the points close quickly and the harshness of this closing causes them to open again briefly, the beginning of the dwell period has been interrupted. The resultant scope trace of this occurrence would display an upward spike, following the mechanical closing signal (Fig. 15–24).

Point problems affect primary current flow and therefore affect coil saturation. It is likely that the coil section will be affected by point problems (Fig. 15–25).

The main purpose of an ignition condenser is to prevent an arc at the points when they first open. This is accomplished by providing a place where current can flow until the points are fully open. Loosened or corroded connectors will increase the series resistance of the condenser, which will cause it to be slow in taking a

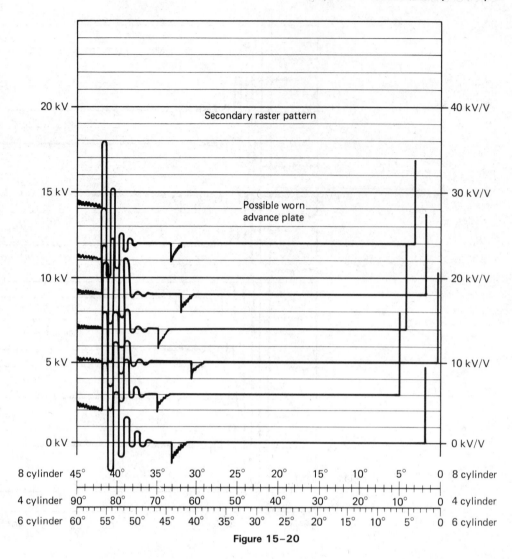

Figure 15-20

charge. The inability of the condenser to quickly absorb the high voltage from the coil will allow some arcing to take place as the points separate. This arcing will be displayed at the end of the dwell period, at the point-open signal. The scope will display some hashing or oscillations before the trace continues into the firing line. Since an open condenser will not attract the current flow from the coil, heavy arcing will occur when the points open. If there are many oscillations at the point-open signal, an open condenser could be indicated. These conditions will affect the time the next cylinder fires, as the arcing delays the establishment of the firing line and tends to retard the firing of the plug.

During the firing of the spark plug, the high voltage from the primary windings of the coil oscillates between the condenser and the coil. If the condenser leaks or discharges the voltage too quickly, the total voltage available from the coil will be affected. A leaking condenser will not be displayed at the point-open signal but would cause the coil oscillations to be few and weak (Fig. 15-26). The resultant pattern would be much like that of a poor coil. To identify the cause of this pattern, a coil output test should be conducted. If the coil tests all right, the condenser is undoubtedly bad.

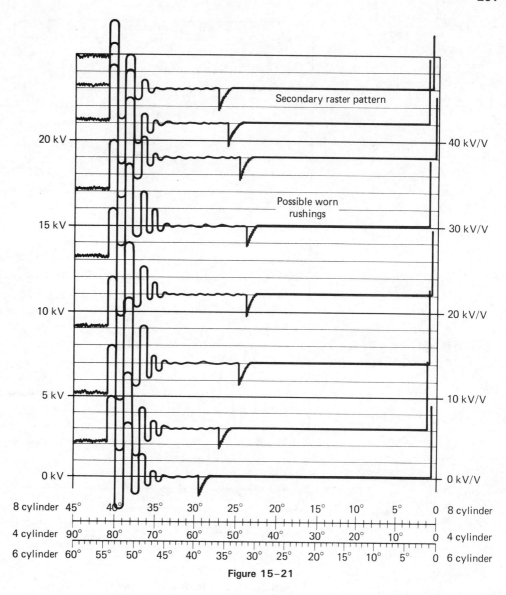

Figure 15-21

ELECTRONIC IGNITIONS

With constant reference to what is taking place in the ignition system and that event's corresponding trace on the scope, ignition problems can be diagnosed. This statement also holds true for electronic ignition systems. By correctly substituting the electronic components that replace the breaker points and condenser, primary electronics can also be checked. Currently, automobile manufacturers are not using electronics in the secondary; therefore, all techniques and logic for troubleshooting secondary circuits can be used for electronic ignition systems.

Most engine analyzers are equipped with a special adapter (Fig. 15-27) used to pick up the coil signal on GM's HEI system. This adapter replaces the regular coil pickup and fits onto the top of the distributor cap. The adapter must be used to pick up the coil's activity on HEI systems, which have no coil wire. When using the adapter, you should be aware that voltages will appear to be lower than normal. Because of the design of the adapter, it cannot accurately measure the voltages from the coil. When testing the secondary circuit of a HEI system, lower firing voltages

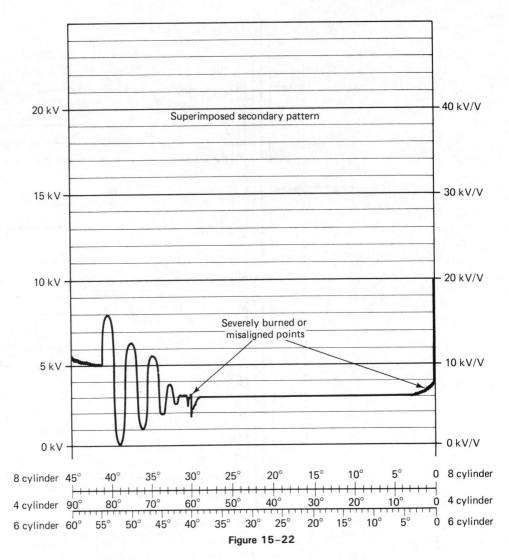

Figure 15-22

should be expected. To minimize the difference between the actual voltages and the voltages displayed, care should be taken to properly install the adapter to the distributor cap.

Other characteristics of the GM HEI system should also be discussed, not just their effect on scope patterns. The control module allows for an increase in dwell time, with an increase in engine speed. Dwell in a point-type ignition system should remain constant throughout all engine speeds. If the control module of the HEI system is functioning properly, the length of dwell should increase with an increase of speed (Fig. 15-28). Also, with an increase of speed, the coil oscillations will become closer and tend to be fewer. This is normal.

It is also normal for the dwell time to vary from cylinder to cylinder, even at a constant engine speed. These differences are the result of the control unit making the necessary changes to allow for proper ignition. Voltage ripples are usually seen on the dwell line. These ripples may or may not be present. This is also normal for the HEI system. Increases in the length of dwell or the voltage ripples with an increase in engine speed do not indicate that there is a faulty pickup or control module (Fig. 15-29).

Testing on the scope the electronic ignition systems of other automobile manufacturers does not require the use of a special adapter for the coil. In addition,

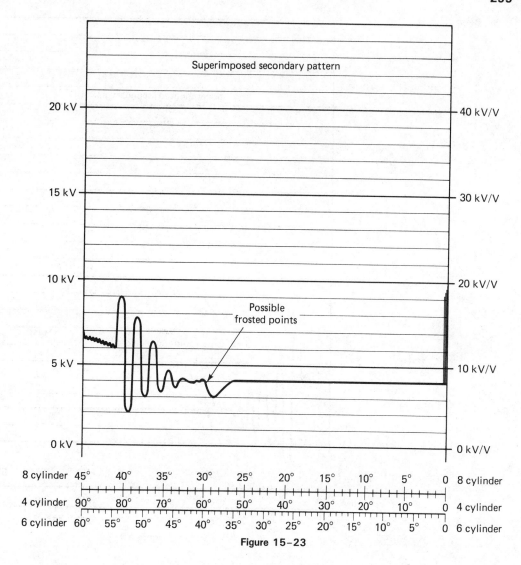

Figure 15–23

these systems do not display the variances that are normal for the HEI system. Secondary circuit diagnosis should be conducted in the same manner as for a point-type ignition system, but instead of opening a plug circuit for testing, use a test spark plug. Figure 15–30 shows the normal secondary scope patterns for the most common types of electronic ignitions.

MODULE TEST

Earlier it was pointed out that it was normal for the GM HEI to display an increase in dwell with an increase of engine speed. Other electronic ignition systems will also cause a change in the dwell line with changes of engine speed. The length of dwell is controlled by the system's control module, and by observing the length of dwell time in relationship to engine speed, the operation of the control module can be observed. Some electronic ignition systems allow for no change in dwell during changes of engine speeds, while others allow for a decrease or increase in dwell with an increase of engine speed. Table 15–1 groups the different types of electronic ignitions by the reactance of dwell to engine speeds.

Other components in the primary circuits are tested more accurately through

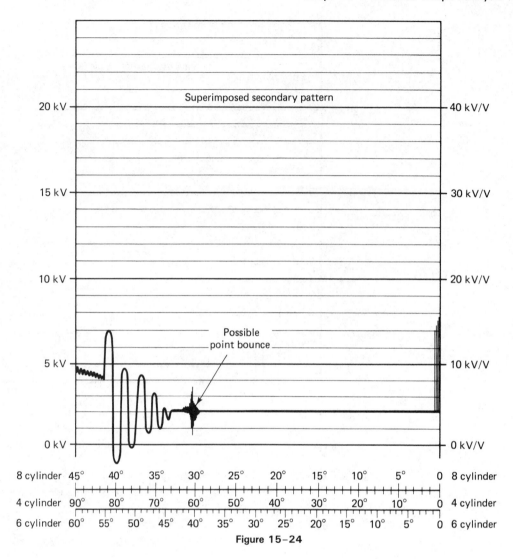

Figure 15-24

the use of an ohmmeter and voltmeter, as instructed in the shop manual. This is also true for the HEI system. The scope can be used to identify which components should be tested. In place of the points, electronic systems use a pickup coil. This coil sends signals to the control unit at the appropriate time for the start and end of current flow, through the primary windings of the coil. If the scope pattern for an engine has an abnormal point-open or point-close signal, the pickup coil should be suspected and tested.

TABLE 15-1 DWELL FOR ELECTRONIC IGNITIONS

Remains constant with speed changes	Dwell increases with speed increases	Dwell decreases with speed increases
Chrysler Hall effect with EIS	Chrysler Hall effect with ESC	Other Chrysler types
Other Ford types	Ford Thick Film Ignition (TFI)	
	Ford EEC III and IV	
	Ford Dura-Spark I	
	GM HEI systems	
	Prestolite BID systems	
	Most imported types	

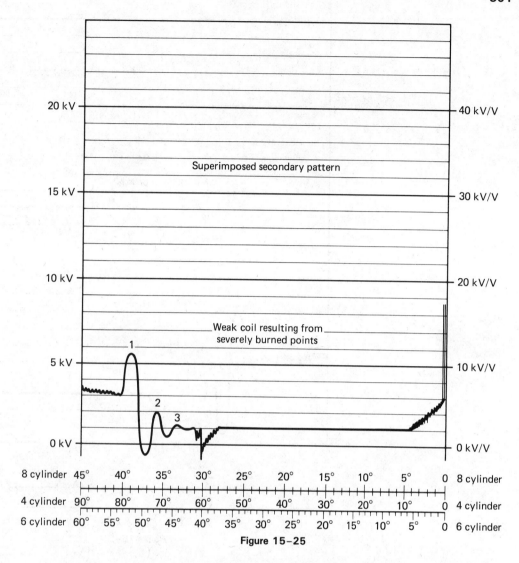

Figure 15-25

Some engine analyzers have the capability to test the pickup coil and control modules of electronic ignition systems. These components are tested when the engine is not running. When the engine is cranked by the starter and the ignition disabled, the output of the pickup coil is displayed on the scope's screen. The module screen is the result of the engine analyzer sending pulsing signals to the module when the ignition is on. The screen will display a pattern similar to the primary or secondary pattern of that type of ignition. While conducting this test, you can select to display the module pattern in the primary or secondary mode. This test is valuable for diagnosis of "won't start" problems: the analyzer acts like the pickup coil and sends signals to the control module. If the resultant pattern is fine, the control module is not the cause for the lack of spark, which is preventing the engine from starting. The most likely cause would be the pickup coil. While observing these patterns, shake the primary wiring harness. If the wires' movement causes a change in the pattern, the harness is defective and needs to be replaced.

PICKUP TEST

The pattern displayed by the pickup coil will depend on the type of triggering system it is. The three most commonly used triggering systems are the Hall effect, magnetic pulse, and proximity switch types. Proximity-switch-triggered systems are the least

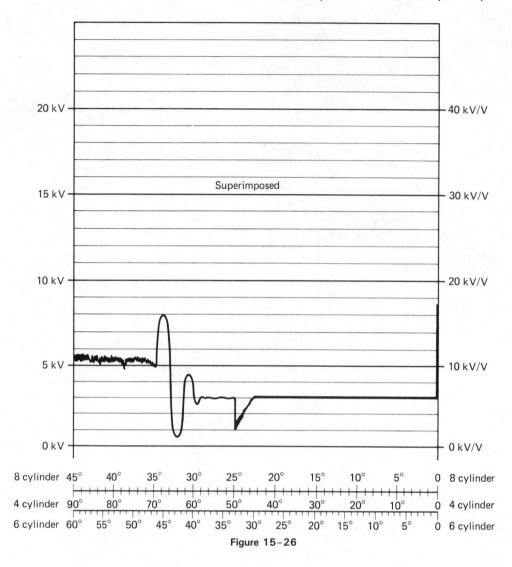

Superimposed

8 cylinder	45°	40°	35°	30°	25°	20°	15°	10°	5°	0	8 cylinder
4 cylinder	90°	80°	70°	60°	50°	40°	30°	20°	10°	0	4 cylinder
6 cylinder	60°	55°	50°	45°	40°	35°	30°	25°	20°	15°	10°	5°	0	6 cylinder

Figure 15-26

Figure 15-27 HEI adapter mounted on distributor cap.

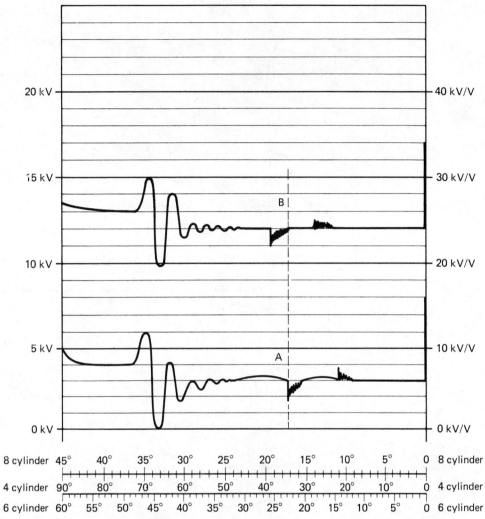

Figure 15-28 Dwell variation with variation in engine speed. A is a lower engine speed than B.

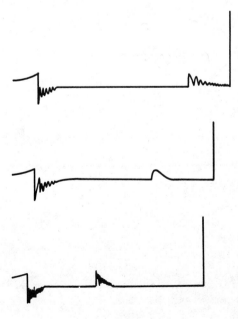

Figure 15-29 Variations in the current-limiting hump on GM HEI patterns.

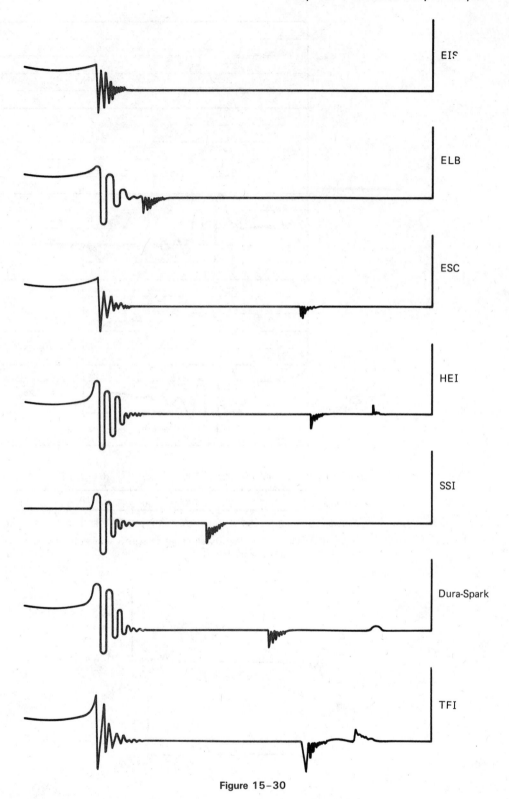

Figure 15-30

commonly used. Most often used is the Prestolite unit by American Motors. The typical trace displayed by this type of pickup is very similar to a normal alternating-current pattern. The pulsing signal from the pickup unit to the control module appears as a group of sine waves, as shown in Fig. 15–31. After a group of these waves is displayed, a small empty space appears and then another group is displayed. The groupings of the waves is the signal to the control unit to start and end the flow of current in the primary.

The Hall-effect system is gaining wide use in the automotive industry. Nearly all manufacturers have a system that incorporates a Hall-effect switch. The pattern displayed on the scope from this type of pickup is a very choppy pattern. The basic appearance of the trace is that of straight lines and squares. This pattern is the result of the sensor quickly changing the voltage signals to the control unit (Fig. 15–32).

The most commonly used triggering device is the magnetic pulse type. As the reluctor passes by the pickup, the voltage signals change from a high negative voltage to a high positive voltage. The trace displayed will reflect this change. The pattern will include a peak negative voltage and positive voltage. Connecting the two peaks is a line with a gentle curve in it, which displays the gradual change from negative to positive voltage (Fig. 15–33).

If the pattern displayed on the pickup test is not normal, the pickup unit is not functioning properly. The cause of this could be the pickup itself, the air gap between it and the reluctor, a damaged reluctor, or faulty wiring to the pickup or control module. Detailed testing of these components was covered in an earlier chapter and is given in the shop manuals for that system.

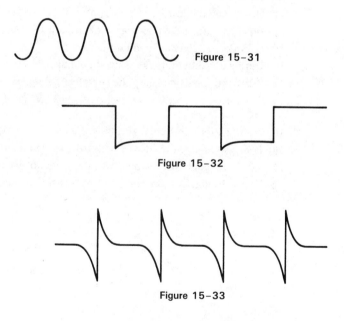

Figure 15–31

Figure 15–32

Figure 15–33

SUMMARY

The operation of a system must be known as well as the scope pattern that it normally produces. By understanding the results of malfunctioning components, you can accurately and quickly determine the cause of the problem. It is important that you use the examples of scope patterns given in this text as examples only. The intended purpose of the patterns was to aid in the understanding of the scope, not as a reference for troubleshooting. Many technicians have failed to master the scope because they attempted to memorize scope patterns given in texts or manuals. Mastery of the scope is possible only through an understanding of the waveforms and of the components they represent, together with hours of thought and experience with a scope.

REVIEW QUESTIONS

1. What causes an increase in the height of a firing line?
2. What are the two main reasons for keeping the firing line between 8 and 11 kV?
3. To begin diagnosis with a scope, what is the first pattern to be observed?
4. What is the main difference between a common and an uncommon problem?
5. Name the possible areas of resistance in a secondary circuit which may influence the height of the firing line.
6. What has a major effect on the resistance of the spark plug's gap?
7. What is the maximum allowable variance between the heights of an engine's firing lines?
8. What is the proper procedure for defining the firing line by using the systematic bypassing technique?
9. What are the four major sections that a firing line can be broken into?
10. What is taking place in the secondary during the spark line?
11. State the relationship between the height of the firing line and the length of the spark line.
12. What does the slope of the spark line tell you about the engine and its systems?
13. How many oscillations should appear in the coil section of the secondary pattern?
14. What can affect the appearance of the coil section?
15. How do you conduct a coil output test?
16. During a coil output test, what should also be observed?
17. What two events make up the complete point-close signal?
18. In a raster pattern, the point-close signals are not in vertical alignment. What does this indicate?
19. How will incorrect point spring tension affect the point-close and point-open signals?
20. How will an open condenser affect the scope pattern and the way the engine runs?
21. When using an HEI adapter with the scope, of what do you need to be aware?
22. When testing an electronic ignition system, a faulty point-open or close signal indicates _____.
23. What can be observed in the secondary pattern of an electronic ignition system?
24. If the pickup coil displays an abnormal pattern, what are the possible causes?
25. Why does a faulty diode affect the alternator's output pattern?

16

ELECTRONIC ENGINE CONTROLS TESTING

Many factors have led to the widespread use of electronics in automobiles. Perhaps the major factor is the attempt to make the engines run more efficiently. Through increased efficiency, the amount of pollutants emitted from the exhaust is decreased, fuel economy is increased, and driveability is improved. The use of electronics to achieve these results was made possible through the advances made in electronics. The development of electronic computation equipment and small reliable components has led to the development of *microprocessors*. To understand what a microprocessor is, a breakdown of the term will be helpful. The word "micro" is used to represent something small. The term "processor" refers to something that follows certain steps to accomplish something. The process is the method used to accomplish something. Microprocessors are computers. Computers process information. Computers used in conjunction with automobile engines receive information about the engine and the conditions under which it is running, process it, and send out information to the engine's systems, which cause adjustments to be made. The computer constantly monitors the engine and makes adjustments in the fuel, ignition, and emission control systems. Computerized engine controls have increased engine fuel economy and performance and reduced exhaust emissions.

Monitoring of the engine by the microprocessor allows for a constant tuning of the engine. As parts wear and the operating conditions vary, the computer causes changes in the air-and-fuel mixture, ignition timing, and emission controls. These changes are made to maintain peak engine efficiency. The purpose of a maintenance tune-up is also to maintain peak engine performance. Not all automotive computer systems have the ability to adjust the same systems. Because the computer technology on automobiles is relatively new, manufacturers have applied it to engines in different ways. Some control only the air-and-fuel mixture; others control the mixture, ignition timing and dwell, and the timing and metering of emission control devices.

INPUTS

To provide information about the engine and its operating environment and conditions, the engine and its systems are equipped with a number of *sensors* (Fig. 16-2). These sensors are attached to the engine in a variety of locations, depending on

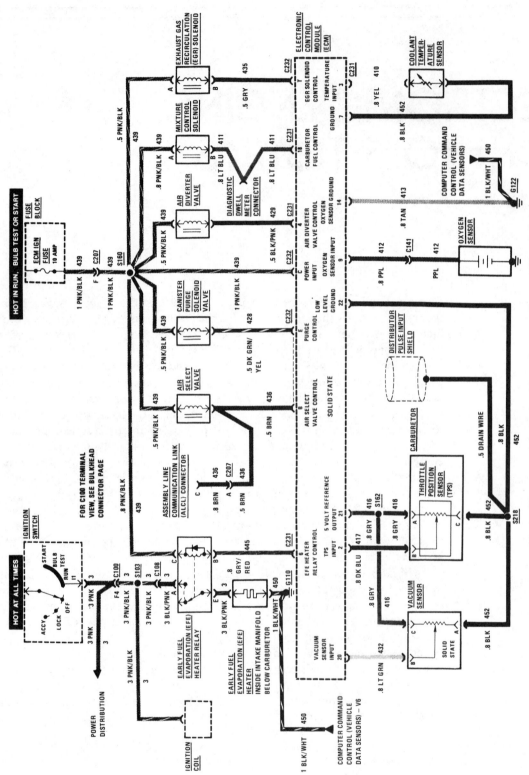

Figure 16-1 (Courtesy of Chevrolet Motor Division, General Motors Corporation)

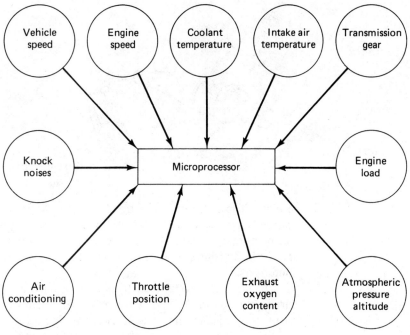

Figure 16–2

what is being monitored. The sensors are simply electrical switches and variable resistors which are connected to the computer by wires. These sensors change the rate of current flow in their circuits as certain conditions change. The computer notices the changes and interprets them in terms of information about the engine. This information is used by the computer to decide on the adjustments that need to be made. From the computer comes a certain amount of electricity to a particular engine control (Fig. 16–3), which adjusts according to the electrical information it

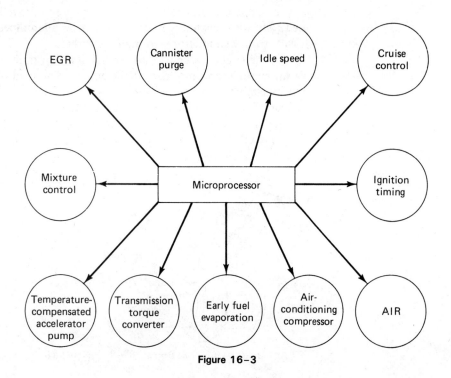

Figure 16–3

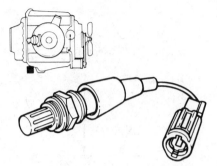

THREADED INTO REAR OF RIGHT EXHAUST MANIFOLD   **Figure 16-4**   (Courtesy of Ford Motor Company)

receives. The decisions made by the computer are based on preprogrammed information. Given certain conditions, the computer knows what adjustments to make. The decision processes of the computers are built in by the manufacturers. The high speed of electronics permits the computer to respond instantaneously to changes in conditions.

Much like a technician troubleshooting a problem, the computer needs accurate information before it can make an accurate decision. This information is received by the sensors. The sensors provide an *input* to the computer about the conditions of the engine. Different manufacturers use different sensors to control the engine. The most commonly used sensor is the oxygen sensor (Fig. 16-4), which measures the amount of oxygen in the exhaust. This is an excellent way to determine the air/fuel ratio of the mixture and the efficiency of combustion. If the computer receives information that the mixture is lean, the computer will send out information to the fuel system to supply more fuel. This allows the engine to receive the ideal air-and-fuel mixture at all times. For an engine to run efficiently, it must receive the correct amount of fuel mixed with the correct amount of air.

Another commonly used sensor is an engine temperature sensor (Fig. 16-5), which sends information to the computer about the engine's temperature. The information received by the computer from this sensor can be used to cause many things to happen. Cold engine operation can be improved, overheating can be prevented, and ideal engine temperatures can be maintained. The use of this sensor will vary from manufacturer to manufacturer.

For the engine to be able to control ignition timing and dwell, the computer must know the speed and load of the engine. The load of the engine can be monitored by engine vacuum. A sensor that receives engine vacuum and converts it to electrical information can describe the engine's load to the computer. Engine speed can be measured at the distributor by counting the times the number 1 cylinder fires, or at the crankshaft by counting the times the crankshaft turns.

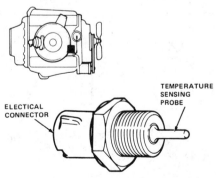

ELECTICAL
CONNECTOR

TEMPERATURE
SENSING
PROBE

THREADED INTO COOLING SYSTEM   **Figure 16-5**   (Courtesy of Ford Motor Company)

If the engine load sensor is mounted in the intake manifold, the computer determines if the engine has high vacuum because it is at idle or it is at a cruising speed. It achieves this by the engine speed sensor or by a sensor on the throttle which tells the computer how wide the throttle is open. Throttle sensor inputs are also used to tell the computer when the engine is decelerating, so it can make adjustments to the systems which prevent backfiring and excessive emissions during deceleration.

These sensors are the most commonly used. Other sensors are also used in some computer systems. To provide additional information about the engine's conditions, inputs are received by the computer from sensors which monitor the temperature of the air entering the engine, the amount of air intake, the barometric pressure of the atmosphere, the pressure and temperature of the fuel, the speed of the vehicle, and the transmission gear in which the car is operating. Some computer systems also have a detonation sensor which can hear the beginnings of detonation in the cylinders. The computer can then make the necessary adjustments to prevent the engine from "pinging" or detonating.

MICROPROCESSOR

The computer or microprocessor (Fig. 16-6) is the brain box of these electronic engine control systems. It is also the control box or module of the system. Much like the control units for electronic ignition systems, the control units of the engine control systems only cause things to happen when told to do so. The pickup unit of the electronic ignition circuit signals the control module the proper time for the firing of a plug. The control unit then stops the current flow in the primary, which collapses and causes the spark plug to fire. The control module receives information from the pickup and sends information to the coil. The control unit of a computerized system works in much the same way. Sensors send information into the control unit and the control unit sends commands or information to engine controls.

The computers or control units are located either in the engine compartment or in the passenger compartment. The more complex and sophisticated units are located within the passenger compartment to protect them from vibration and temperature. The units are usually housed in a sealed steel box. Some units have a removable programming assembly. These assemblies plug into the main microprocessor and tell it about the car. Information about the type of car, engine, and accessories is permanently stored in the assemblies. This information gives the computer additional facts on which to base its decision making. These assemblies are called the *calibration assembly* or *PROM* (programmable read-only memory). The

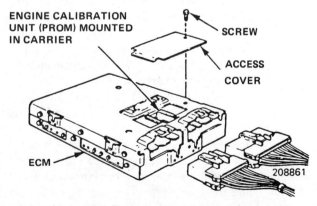

ENGINE CALIBRATION
UNIT (PROM) MOUNTED
IN CARRIER

SCREW

ACCESS

COVER

ECM

208861

Figure 16-6 (Courtesy of Chevrolet Motor Division, General Motors Corporation)

information stored in a PROM comprises data that are never changed and are used as a basis for the decisions made.

During the running of the engine, the computer is in either the open loop or the closed loop. When the computer is in the *closed loop,* it is monitoring and adjusting the engine and its control. In the *open loop,* the computer is not making decisions and the engine is operating on preprogrammed information stored in the computer. Until the engine and its related sensors are warmed up enough to provide for accurate information, the computer is in the open loop. During this time, each engine control is operating at a compromise setting. Once the engine reaches the desired temperature, the computer goes into the closed loop. The controls are adjusted constantly by the computer. The open and closed loops are the two operation modes of a computer (Fig. 16–7).

Open-loop operation simply means that the engine and the oxygen sensor are not warm enough and the computer is not controlling the air-and-fuel mixture. In closed-loop operation, the computer is controlling the mixture and is receiving accurate information from the engine and the sensors. The heating of the oxygen sensor signals the switch of the computer from the open to the closed loop. Once the engine is in the closed loop, the computer and its systems are in control of the engine. If a running problem occurs with the computer in the closed loop, the computer will attempt to compensate for it. If the engine has a problem but seems to run correctly when the computer is in the closed loop, it indicates that the computer and its systems are functioning properly and that the cause of the problem is somewhere other than in the computer.

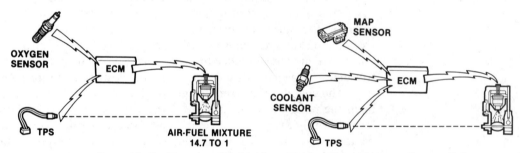

Figure 16–7   (Courtesy of Buick Motor Division, General Motors Corporation)

OUTPUTS

The computer sends its commands to the engine controls. These engine controls react instantaneously to the commands. The engine controls are activated by solenoids, motors, or other devices. The computer can also send a command directly to the ignition module to tell it to advance or retard the timing. The typical engine controls of a computerized system include controls for fuel, air, idle speed, and ignition. The operation of these controls will vary from manufacturer to manufacturer, but the method of activating them is the same—the computer.

The technician's job is to identify defective parts and replace them. The computer and its related sensors and controls cannot be repaired. Faulty components are replaced. Most computers have a self-diagnostic program built in them to aid the technician in diagnostics. This program will identify the area of the problem. It is the technician's job to pinpoint it. Too often diagnostics begins at the computer, and the other systems of the engine are overlooked. Before starting the computer into its self-diagnostics, the technician should have a reason to think that the problem is computer related. The technician should approach the diagnostics of a computerized engine in the same way as for a regular engine. There must be a reason to test a particular system. An understanding of what individual computer systems control will aid in the diagnostics of engine problems. Although the computer can control the firing of the plugs, it only controls the timing of the firing. If a cylinder has too high a required firing voltage, the problem is not computer oriented and the computer should not be checked.

It is recommended by some manufacturers that special testers (Fig. 16–8) be used to diagnose the computer circuits. These special testers make diagnostics easier but are not absolutely necessary. Most computer systems can be checked with normal tune-up and diagnostic equipment. Components such as sensors and controls can be tested with a digital volt-ohmmeter. This meter will display the exact amount of ohms and volts. Rather than use a needle moving across a scale to indicate the reading, the digital meter will show the reading digitally. For most computer systems, a digital volt-ohmmeter with an impedance of at least 10 MΩ is required. The impedance value of meters is given on the back of most meters or stated in the papers included with the meter when it is purchased.

Solenoid-activated controls can be checked with an ohmmeter. There should be some resistance through the windings of a solenoid. If the resistance of the windings is zero, the windings are shorted. If the reading of infinite is displayed, the windings are open. This is also true of most electrical loads. Solenoids and motors can also be checked by applying power to them with a jumper wire and seeing if they operate. (Care should be taken when testing a component with a jumper wire. Some computer outputs are controlled by only 5 V, not full battery voltage. Before using a jumper wire, check your service manual.) The resistance of sensors can be compared to the specs for that sensor and can be watched to see if the sensors open and close at the correct time. If all the components of the computer circuit test out all right, the computer should be replaced with a known good one to see if the problem still exists.

The basic test procedures for the most common computer systems follow. Available in most shop manuals are the more detailed procedures. Prior to testing any of these systems, you must be sure that the problem is computer related, and all the wires and connectors of the circuit should be inspected visually. Because the computer depends on changes in electrical voltage to receive its information, any change in resistance due to loose or corroded wires will affect the input to it. These

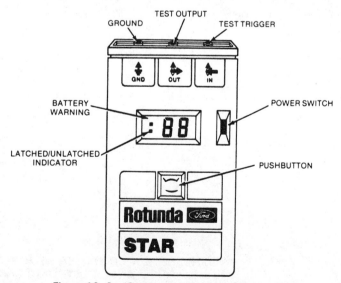

Figure 16-8 (Courtesy of Ford Motor Company)

false inputs may cause the computer to think that another problem exists, and it will send commands to the controls to correct them. A leaner mixture will not remove corrosion to the end of a wire, but the computer may think that the engine needs less fuel if that is what the input signal tells it.

Most automobiles manufactured today have some sort of computerized system. The systems used in this text are the most commonly used, and an understanding of these systems will lead to an easier understanding of other systems. All the systems have a control unit which receives information from sensors, and in turn causes some changes in the operation of the engine through the controls that it governs.

CHRYSLER ELECTRONIC FUEL CONTROL SYSTEM

Chrysler Corporation uses the Electronic Fuel Control (EFC) system on all its front-wheel-drive cars except those equipped with the electronically fuel injected 2.2-liter and the 2.6-liter engines. This system uses a Spark Control Computer (SCC) to control the air/fuel ratios and the ignition timing of the engine. To do this, the SCC receives input signals from engine speed, coolant, oxygen, engine load, and throttle position sensors. The ignition timing is directly controlled by the SCC, which signals to the ignition coil the proper time for the firing of the spark plugs. The air-and-fuel mixture is controlled by the operation of a solenoid, which meters the amount of fuel in the mixture (Fig. 16-9).

The carburetors on engines equipped with an EFC system have a solenoid which controls the amount of fuel in the air-and-fuel mixture. When the solenoid is in its off position, it provides a very rich mixture. Lean mixtures are the result of the solenoid being turned on. The controlling of the mixture occurs with the turning off and on of the solenoid. By controlling the length of time that the solenoid is activated, the computer is able to provide nearly ideal air/fuel ratios under all running conditions. In response to engine running conditions and input from the oxygen sensor, the computer controls the amount of time the solenoid is activated, which controls the air/fuel ratio.

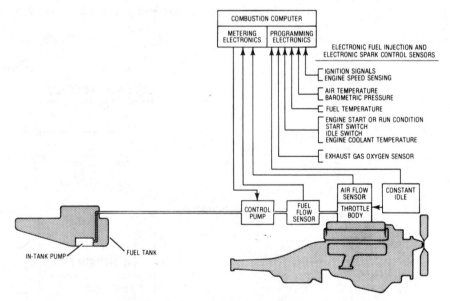

COMBUSTION COMPUTER WITH ELECTRONIC FUEL INJECTION AND ELECTRONIC SPARK CONTROL

Figure 16-9   (Courtesy of Chrysler Corporation)

The oxygen sensor is located in the exhaust manifold and signals to the computer the amount of oxygen present in the exhaust. The amount of oxygen present determines the amount of voltage that is sent back to the computer. The more oxygen present, the more voltage the computer receives from the sensor. The computer then sends commands to the solenoid in the carburetor which controls the mixture ratio. The computer will monitor and adjust the mixture's ratio in response to the oxygen sensor during all operating conditions except idle.

During idle conditions the computer does not control the mixture. Because of the constant speed there is no need for changes in ignition timing; therefore, the computer cancels its controls on the engine's systems. Located on the idle stop solenoid is a carburetor switch that tells the computer when the throttle (Fig. 16–10) is closed and the engine is at idle. When the throttle is closed, the switch grounds and current flows through the circuit. The signal to the computer that the engine is at idle is the flowing of current through the carburetor switch.

The idle speed is increased by a throttle control system (Fig. 16–11). This system increases engine speed when the air conditioning or rear window defogger are operating. This system also controls the throttle opening during deceleration and engine warm-up after starting. Two timers are used in this system to control or delay throttle closing. One timer is used during the starting of the engine to raise the idle speed, and the other is used during deceleration to delay the closing of the throttle plates by 2 seconds.

The speed sensor is located inside the distributor and supplies a basic timing signal to the computer. This speed sensor is also the pickup unit for the Hall-effect ignition system. The signals from the pickup tell the computer when the cylinders are on their compression stroke. This information is used to determine the speed of the engine.

The signals from the Hall-effect pickup unit are received by the computer and the timing of the ignition is based on them. The amount of advance allowed by the computer based on speed is preprogrammed in the computer. As the engine speed is increased, the amount of ignition advance is slowly increased. Engine vacuum also influences the amount of advance the ignition timing will have. The computer will advance the ignition timing in response to the engine's load, measured by engine vacuum and the engine's speed. This allows the computer to determine the exact moment that ignition is needed. When the throttle is closed, the carburetor switch prevents the change in ignition timing.

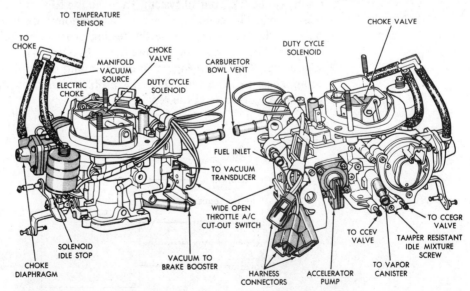

Figure 16–10 (Courtesy of Chrysler Corporation)

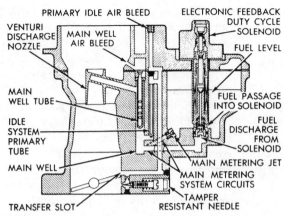

PRIMARY IDLE AIR BLEED

ELECTRONIC FEEDBACK
DUTY CYCLE
SOLENOID

VENTURI
DISCHARGE
NOZZLE

MAIN WELL
AIR BLEED

FUEL LEVEL

MAIN
WELL TUBE

FUEL PASSAGE
INTO SOLENOID

IDLE
SYSTEM
PRIMARY
TUBE

FUEL
DISCHARGE
FROM
SOLENOID

MAIN METERING JET

MAIN WELL

MAIN METERING
SYSTEM CIRCUITS

TRANSFER SLOT

TAMPER
RESISTANT NEEDLE

Figure 16–11   (Courtesy of Chrysler Corporation)

Engine vacuum is monitored by a vacuum transducer (Fig. 16–12), which converts the amount of vacuum into electrical information for the computer. The vacuum transducer is mounted on the computer in the engine compartment near the battery. The transducer receives vacuum signals from the engine through a tube connected to the intake manifold. The vacuum signals provide information to the computer for the determination of ignition timing and the mixture required by the engine.

A coolant sensor or switch is located on the thermostat housing. The purpose of this switch is to tell the computer when the engine has reached operating temperature. When this temperature is reached, the computer controls the ignition timing and the mixture of the engine. When the engine is cold, the information from the switch's circuit will prevent the computer from controlling the engine.

The EFC system does not have self-diagnostic ability. Diagnostics should be conducted in a very logical manner. Because the system controls only ignition timing and the air/fuel ratio of the engine, the computer should be of concern only when there is a mixture or timing problem. A visual inspection of all wires, connectors, and vacuum lines should be followed by a thorough scoping of the engine. The use of a scope will help identify the area of the problem. If the problem is found to be in the computer and its systems, the spark control system should be tested prior to testing of the fuel control system. Faults in the EFC system can cause starting problems, rough idle conditions, poor fuel economy, or rough running at particular speeds. Shop manuals include the testing procedures for each of the sensors and controls. These should be followed step by step. All the tests do not have to be conducted. Your knowledge of the system and the problem will dictate to you the components to be tested. Once the problem has been identified, the component should be replaced and the repair verified.

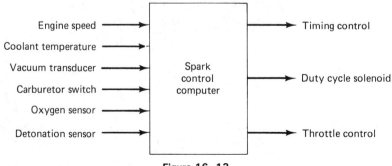

Figure 16–12

Chrysler uses a system similar to EFC on their rear-wheel vehicles. The major differences between the two systems is the location of the SCC and the type of ignition used. The rear-wheel-drive vehicles have the computer mounted on the air cleaner assembly. The functions of the computer are the same as those of the front-wheel-drive vehicles. Additional sensors are used on the rear-drive systems to monitor detonation and intake air temperature. Instead of using the Hall-effect type of ignition, these vehicles use dual magnetic pickups. One pickup is used during engine starting only; the other is used at all times while the engine is running. The operation of this EFC system is quite similar to the system discussed previously. The diagnostics and testing of the two EFC systems are also similar.

Chrysler also uses different computer systems on certain vehicles. Engines equipped with electronic fuel injection use a more complex system. This type is similar to other computer systems that are used more commonly by other manufacturers. These systems will also be covered in this chapter.

FORD MOTOR COMPANY MCU SYSTEM

Ford Motor Company uses two basic types of computerized engine control systems. One type, the Electronic Engine Control (EEC), has appeared in four varieties. Each variety developed from the system EEC I, which was the first model. The number of inputs and outputs have increased the complexities of the system and its efficiency as well. The most commonly used type today is the EEC IV. The other variety of computerized control systems is the MCU or Microprocessor Control Unit (Fig. 16–13). The MCU system is commonly used on four- and eight-cylinder engines that are not equipped with electronic fuel injection.

The microprocessor (Fig. 16–14) or computer of the MCU system, located in the engine compartment, is capable of controlling the ignition timing, air/fuel ratios, air injection systems, and idle speeds. The MCU has three modes of operation: initialization, open loop, and closed loop. The initialization mode works only when the engine is being started. During this mode, the engine receives an extremely rich mixture for easier starting. After the engine starts, the computer goes into the open loop until the oxygen sensor and engine are at operating temperatures. At this time, the computer switches to the closed loop and begins to control the engine.

To provide the information needed by the computer, the system includes four main sensors (Fig. 16–15), which monitor and measure the speed of the engine, the temperature of the coolant, the load on the engine, and the oxygen in the exhaust. To sense engine speed, the MCU receives electrical pulses directly from the negative (Tach Test) side of the ignition coil. The computer receives these pulses and is able to calculate engine speed. Knowing the engine's speed, the computer can make more accurate decisions for air-and-fuel mixture and ignition timing control. A coolant temperature switch is used to signal to the MCU the temperature of the engine. When the engine is cold, the computer adjusts the mixture to allow for improved driveability. The coolant switch is mounted on the Ported Vacuum Switch (PVS) near the thermostatic housing.

The position of the throttle and the amount of vacuum in the intake are the sources of information to the computer regarding engine load. An idle tracking switch is used to send signals to the computer when the throttle plates are closed. A vacuum switch is used to send a signal to the MCU when the engine's vacuum is low. Under wide-open throttle conditions, the engine's vacuum will be low and the computer will receive this information. The MCU receives signals when the throttle plates are fully opened or closed. Because the computer receives no signal regarding engine load during part-throttle operation, the engine speed sensor input governs the mixture's ratio at these times.

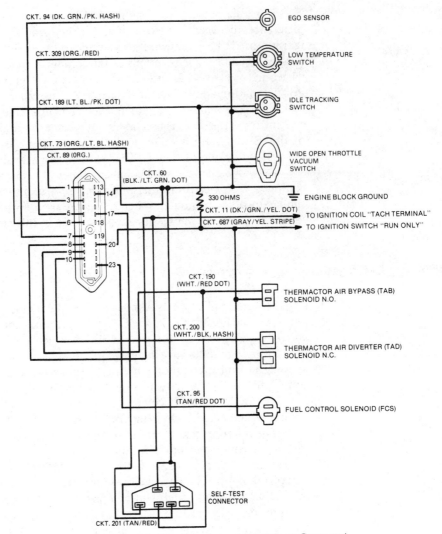

Figure 16-13   (Courtesy of Ford Motor Company)

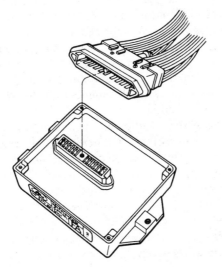

Figure 16-14   (Courtesy of Ford Motor Company)

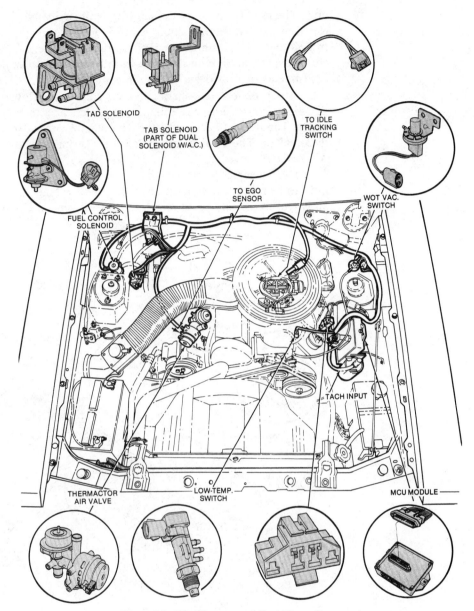

TAD SOLENOID

TAB SOLENOID
(PART OF DUAL
SOLENOID W/A.C.)

TO EGO
SENSOR

TO IDLE
TRACKING
SWITCH

WOT VAC.
SWITCH

FUEL CONTROL
SOLENOID

TACH INPUT

THERMACTOR
AIR VALVE

LOW-TEMP.
SWITCH

MCU MODULE

Figure 16–15   (Courtesy of Ford Motor Company)

The oxygen sensor of the MCU system uses low voltage signals to inform the computer about the mixture that has been burned. If the mixture was lean, the sensor would send a signal of less than 0.2 V to the computer. The computer would react by adding more fuel to the mixture for the next burning. If the mixture was rich, the signal would be about 0.6 V. Voltage should never be applied to the oxygen sensor because it could ruin the sensor's calibration. Before connecting and using a voltmeter, make sure that the meter has a high impedance.

Some engines equipped with the MCU system have a knock sensor. The purpose of this sensor is to reduce detonation. When detonation is heard by the sensor, it passes a voltage signal to the computer which causes the timing to retard. This is the only control the MCU has on ignition timing. The MCU works with the normal electronic ignition module to cause this retarded spark on some engines.

Based on the information received by the computer from the sensors, the MCU controls and operates devices that improve overall engine performance and economy. The use of these devices may vary with engine types, but they are all operated electronically. On engines equipped with a knock sensor, a signal is sent to the control module which retards the ignition timing to prevent further detonation. Other commands from the computer are sent directly to a solenoid or motor which makes adjustments or provides a change in conditions (Fig. 16–16).

A pair of solenoid-operated valves control the flow of air from the air pump. The air is sent to either the exhaust manifold, the atmosphere, or the catalytic converter. The flow of vacuum to the Thermactor Air Valve assembly is controlled by these valves. These valves are called the Thermactor Air By-pass and Thermactor Air Diverter valves.

During engine warm-up, the air from the air pump is sent to the exhaust manifold to help heat the exhaust gases before they reach the converter. Normally, the air is sent to the catalyst to improve its effectiveness. Upon deceleration and during

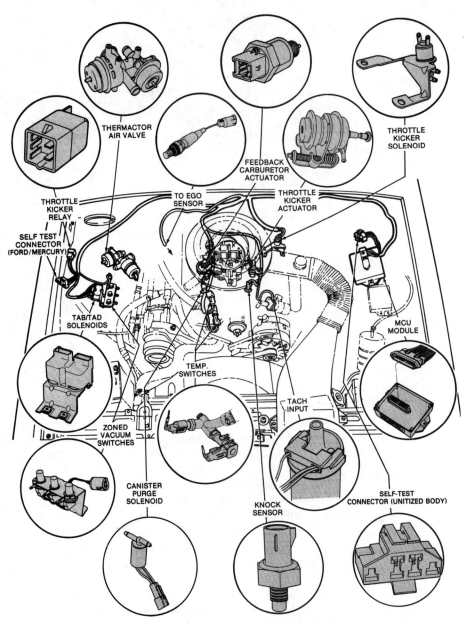

Figure 16–16 (Courtesy of Ford Motor Company)

long periods of idle speed operation, however, the air is sent back to the atmosphere. This prevents the objectionable backfiring that can take place when there is a high concentration of HC in the exhaust and a large amount of air present to cause secondary combustion processes.

Some engines equipped with the MCU system have a fuel vapor canister purge mode. During the purge of the canister, fuel vapors are delivered to the intake system to be burned by the engine. Mounted at the fuel vapor canister is a solenoid which opens to allow the vapors to enter the intake. This purging occurs only when the computer senses that the conditions are right for the introduction of the fuel vapors.

The air/fuel ratio is controlled in one of two ways: by a motor or a solenoid. On eight-cylinder engines equipped with the MCU system, the carburetor is equipped with a motor that moves a shaft in the air-bleed orifice of the carburetor, changing the size of the orifice and thereby changing the mixture. Six-cylinder engines have a fuel control solenoid which controls the amount of fuel in the mixture. This solenoid functions much the same as the solenoid used in Chrysler's EFC system. Four-cylinder engines use a fuel control solenoid/vacuum regulator. The movement of the solenoid opens or closes the path of vacuum to a mixture control diaphragm in the carburetor (Fig. 16–17).

The MCU has a self-diagnosis mode. Most of the common operating problems can be diagnosed by the computer. The computer will send out voltage pulses which can be monitored with a special tester, as recommended by Ford, or with a volt-ohmmeter. If a problem does exist in the system, the computer will send out a pattern of voltage pulses which will indicate a particular service code. Most shop manuals will contain a list of the codes and their meaning. Counting the number of pulses from the computer will be your way of "listening" to what the computer has to tell you about the system.

To set the MCU into its self-test mode (Fig. 16–18), turn the ignition off and locate the self-test connector in the engine compartment. Insert a jumper wire between the "trigger" and ground sockets of the connector. Connect the positive lead of the voltmeter to the positive post of the battery and the negative lead to the self-test output socket. Turn the ignition switch on. Do not start the engine. Keep your eyes on the needle of the voltmeter. Within 5 to 30 seconds, the service codes or voltage pulses should begin. The first digit of the service code is given by a series of voltage pulses. Count the pulses for the first digit. When this series is complete,

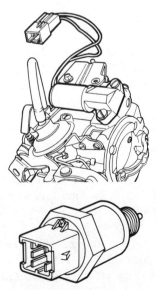

Figure 16–17 (Courtesy of Ford Motor Company)

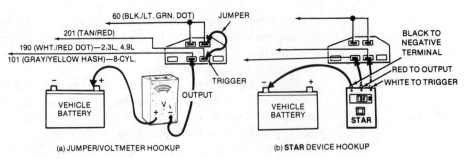

(a) JUMPER/VOLTMETER HOOKUP   (b) STAR DEVICE HOOKUP

- Engine OFF.

- Connect jumper from self-test trigger input to ground at self-test terminal (see Figure 22), or connect STAR device white lead to self-test trigger. Voltmeter may show battery voltage when connected, until key is turned ON.

- Connect analog voltmeter, set for 15- or 20-volt range, between self-test output terminal and positive terminal of battery, or connect STAR device red lead to output terminal. Connect STAR device black lead to vehicle battery negative terminal. Turn tester ON and latch pushbutton (: displayed).

- On 2.3L engines, disable the canister purge system by disconnecting and plugging the hose from the engine side of the canister purge valve.

- On 8-cylinder engines, remove the PCV valve from the valve cover.

Figure 16-18   (Courtesy of Ford Motor Company)

the needle of the meter will drop to zero for about 2 seconds. The second digit will now appear. If the system has more than one problem, the first service code may be followed by others. Record all codes displayed by the MCU. When all the codes are displayed, there will be a 4-second pause and the service codes will be repeated (Fig. 16-19).

Together with a list of the meanings of the service codes, most shop manuals will include the test procedures for further testing of the system. The self-diagnostic capability of the MCU is a great aid for timely and effective diagnostics of the system.

FORD MOTOR COMPANY EEC IV SYSTEM

The EEC IV (Fig. 16-20) system is becoming widely used on many engines from Ford Motor Company. This system controls three major areas of the operation of the engine: the ignition, the air/fuel ratio, and the emission controls. To control these systems, the EEC IV receives inputs from many sensors and electrical devices. The controlling unit of this system is the Electronic Control Assembly (ECA). The ECA has a calibration assembly mounted inside it to provide the information about the engine and the car needed by the computer. The ECA is located in the passenger compartment to protect it from temperature and vibrations. The EEC IV system is a far more complex system than the MCU system. The ignition system is controlled by the ECA through a Thick Film Ignition module and the air/fuel ratios are controlled by a solenoid on carbureted engines and by the firing of the individual fuel injection nozzles.

Together with a coolant and oxygen sensor, the EEC IV system has a Barometric Pressure sensor (Fig. 16-21). This sensor measures the barometric pressure of the atmosphere. This information is used by the ECA to determine the proper

Code	2.3L	4.9L	8-Cyl.
--	Self-Test Not Functional (P.32)	Self-Test Not Functional (P.44)	Self-Test Not Functional (P.60)
11	System OK	System OK	System OK
12	N.A.	N.A.	Idle Speed Incorrect (P.56)
25	N.A.	N.A.	Knock Detection Inoperative (P.59)
33	RUN Test Not Initiated (P.31)	RUN Test Not Initiated (P.31)	N.A.
41	EGO Always Lean (P.24)	EGO Always Lean (P.38)	EGO Always Lean (P.52)
42	EGO Always Rich (P.26)	EGO Always Rich (P.39)	EGO Always Rich (P.54)
44	Thermactor Air System Problem (P.28)	Thermactor Air System Problem (P.40)	Thermactor Air System Problem (P.57)
45	Thermactor Air Always Upstream (P.30)	Thermactor Air Always Upstream (P.42)	Thermactor Air Always Upstream (P.58)
46	Thermactor Air Not Bypassing (P.31)	Thermactor Air Not Bypassing (P.43)	Thermactor Air Not Bypassing (P.58)
51	LOW-Temp. Switch Open (P.21)	LOW-Temp. Vacuum Switch Open (P.36)	HI/LO Vacuum Switch(es) Open (P.50)
52	Idle Tracking Switch Open (P.22)	Wide-Open-Throttle Vacuum Switch Open (P.37)	N.A.
53	Wide-Open-Throttle Vacuum Switch Open (P.23)	Crowd Vacuum Switch Open (P.37)	DUAL-Temp. Switch Open (P.50)
54	N.A.	N.A.	MID-Temp. Switch Open (P.50)
55	N.A.	N.A.	MID-Vacuum Switch Open (P.50)
56	N.A.	Closed-Throttle Vacuum Switch Open (P.37)	N.A.
61	N.A.	N.A.	HI/LO Vacuum Switch(es) Closed (P.51)
62	Idle Tracking Switch Closed (P.22)	Wide-Open-Throttle Vacuum Switch Closed (P.37)	N.A.
63	Wide-Open-Throttle Vacuum Switch Closed (P.23)	Crowd Vacuum Switch Closed (P.37)	N.A.
65	N.A.	N.A.	MID-Vacuum Switch Closed (P.51)
66	N.A.	Closed-Throttle Vacuum Switch Closed (P.37)	N.A.

Figure 16–19   (Courtesy of Ford Motor Company)

amounts of fuel needed to maintain the ideal air/fuel ratio. Barometric pressures vary with weather changes. These changes cause slight changes in atmospheric pressure.

To measure the pressure of the mixture in the intake manifold, the system uses a Manifold Absolute Pressure sensor. The pressure in the intake manifold is usually lower than atmospheric pressure. This sensor is actually measuring the vacuum of the engine. Through a comparison of the barometric pressure and the manifold pressure, the ECA can precisely determine the amount of vacuum being produced by the engine. The sensor for manifold pressure is located on the inner fender in the engine compartment, together with the barometric pressure sensor.

Engines equipped with fuel injection have an Air Flow Meter. The assembly is mounted between the air cleaner and the throttle body. The purpose of this sensor is to give the ECA the information it needs to determine the total mass of air that is entering the engine. To do this, the sensor measures the rate of airflow and the temperature of the air that is entering the intake. By sensing the exact amount of air that is entering the intake, the computer can determine the exact amount of fuel that is needed (Fig. 16–22).

If the car is equipped with air conditioning, the EEC IV system is equipped

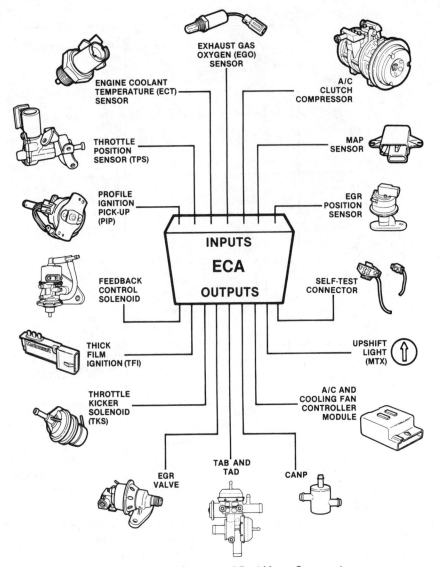

Figure 16-20 (Courtesy of Ford Motor Company)

Figure 16-21 (Courtesy of Ford Motor Company)

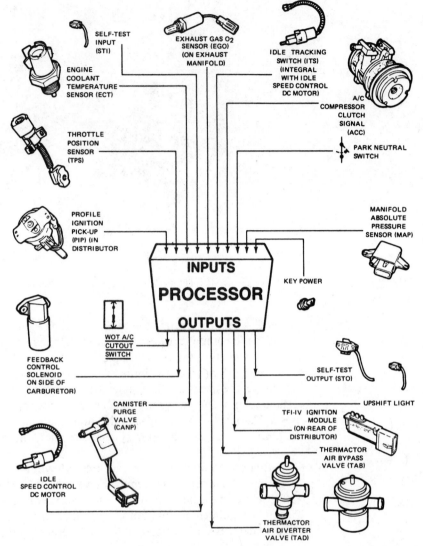

Figure 16-22 (Courtesy of Ford Motor Company)

with an air-conditioning Compressor Clutch signal switch. This switch sends a signal to the ECA when the compressor clutch is activated, so that the ECA can maintain a constant idle speed. The activation of the air conditioning is an extra load on the engine and the ECA adjusts the ignition and fuel system to compensate for the load.

The distributor used in the EEC IV system has no vacuum or mechanical advance built in. The computer controls all ignition timing changes. The distributor is adjustable to reset the initial timing, if this is ever necessary. To give the computer inputs about the engine's speed and crankshaft location, the system uses a Hall-effect switch. This switch and the assembly it is contained in are called the Profile Ignition Pick-up (PIP) (Fig. 16-23). Engine speed and crankshaft location inputs are also used to determine the necessary length of dwell for the ignition.

As an aid to the engine speed sensor or the PIP, engines equipped with fuel injection have a throttle position sensor. The TPS is mounted on the side of the throttle body and is connected to the throttle shaft. The signals from the TPS inform the computer of normal cruising, closed-throttle, and open-throttle conditions.

Some engines are equipped with an EGR Valve position sensor. This sensor informs the computer about the position of the EGR valve. The ECA can shut the

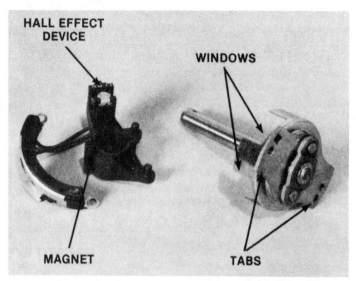

Figure 16-23   (Courtesy of Ford Motor Company)

valve off when it is open if this will allow the engine to perform better. This sensor also tells the computer when the valve is open, so that the computer can judge the amount of mixture that should enter into the cylinders.

The ECA receives the inputs from these sensors and from other switches and determines the needs of the engine. The computer then sends out signals that cause certain components to react and cause the necessary changes. The EEC IV system uses a Thermactor Air By-pass valve and Thermactor Air Diverter valve just as in the MCU system, but the EEC system uses many more outputs than the MCU (Fig. 16-24).

Engines equipped with a carburetor have a solenoid built in to control idle speeds and the mixture, as commanded by the ECA (Fig. 16-25). Engines equipped with fuel injection have the injection nozzles controlled in pairs by the computer. The ECA regulates the amount of time that the injector is open to spray fuel into the cylinders. Each injector is solenoid operated and the computer sends electrical

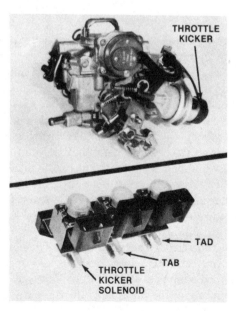

Figure 16-24   (Courtesy of Ford Motor Company)

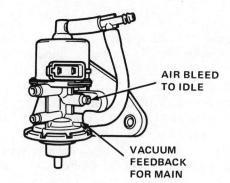

AIR BLEED
TO IDLE

VACUUM
FEEDBACK
FOR MAIN

Figure 16-25   (Courtesy of Ford Motor Company)

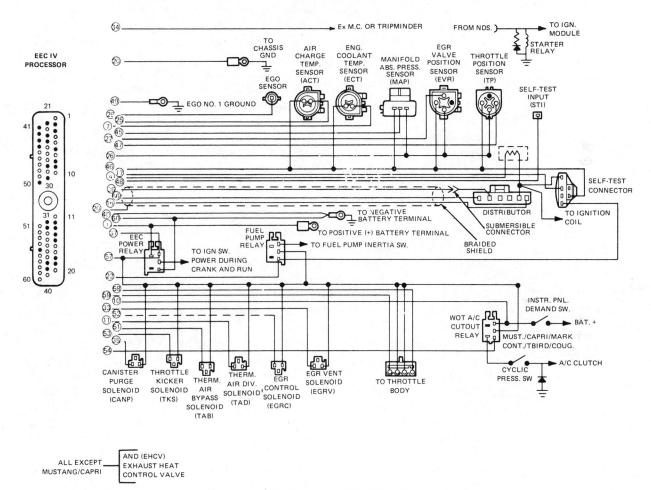

Figure 16-26   (Courtesy of Ford Motor Company)

signals to the injectors at the time it is determined that they should open. Each injector is wired and activated through the ignition circuit, and the ECA controls the amount of time that the injectors will be open. Each injector opens two times for each intake stroke. Each opening of the injector delivers one-half of the amount of fuel needed for that cycle (Fig. 16-26).

Engines equipped with a carburetor have a solenoid built in to control the air-and-fuel mixture. This solenoid is called the Feedback Carburetor (FBC) Solenoid. The carburetor is also equipped with a Temperature Compensated Accelerator Pump (TCP), which is responsible for increased fuel from the accelerator pump when the engine is cold. The ECA controls the TCP solenoid, which sends a vacuum signal to the carburetor, causing the increased shot from the accelerator pump. The ECA also controls the amount of time that the choke stays closed during engine warm-up. The computer controls the time voltage is supplied to the choke. The opening and closing of the choke is aided by the choke relay, which is used to switch the voltage from the computer on and off.

The computer, from a signal from the PIP, determines the crankshaft's position. The PIP has an armature with four windows and four metal tabs that rotate past the Hall-effect switch. When the tabs pass by the Hall-effect switch, a signal is sent to the computer indicating that the crankshaft is at 10 degrees BTDC. The ECA then calculates the engine's timing and dwell needs and sends a signal at the appropriate time to the TFI (Thick Film Ignition) module. When the module receives this signal, it interrupts the flow of primary current in the ignition coil and the spark plug fires.

The ECA also controls the operation of some emission control equipment. The EGR Control Solenoid switches engine vacuum to the EGR valve when it receives the signal from the computer. When the solenoid is activated by the signal, vacuum is applied to the EGR valve and opens it. The EGR valve can also be closed by the ECA. An EGR Shut-off solenoid is an electrically operated vacuum switch used to bleed off the vacuum going to the EGR valve, causing it to close. This solenoid is located between the EGR control valve and the EGR valve. The solenoid controls a vacuum bleed which depletes the vacuum in the line going to the EGR valve. The use of these two control devices allows the ECA to operate the EGR valve only when it will not hinder overall performance of the engine. A Canister Purge Solenoid is also used and controlled by the computer to allow for the purging of fuel vapors from the cannister to occur at the best time.

The ECA has the ability to diagnose itself. The self-test is an internal function of the computer in which the condition of every circuit and component is checked. Self-test procedures are built into the ECA to allow the system to perform a continuous check and to display service codes indicating problem areas. Prior to setting the system in the self-test, there must be some certainty that the problem exists in the computer system. The best approach in determining this is to proceed as if the system did not have the computer controls. The computer does not have the ability to test systems that are not included in the computer system. The engine operates in the same basic way as an engine that does not have computer controls. The use of the computer merely improves the overall efficiency of the engine.

The recommended test equipment for the self-test on EEC IV systems are a STAR tester (this is a special piece of test equipment produced by Rotunda) or a volt-ohmmeter with the range 0 to 20 V, and a timing light. The STAR tester comes with a harness that connects directly to the self-test connector. If the voltmeter is used in place of the special tester, two jumper leads must be made. These jumper leads should be about 5 in. long and have male blade terminals on each end. To connect the voltmeter to the connector for the self-test, insert one end of a jumper wire into the number 4 terminal of the self-test connector. The negative lead of the meter should be connected to the other end of the jumper wire. The positive lead of the meter should be connected to the positive post of the battery. The other jumper wire should be connected to the number 2 terminal of the connector, and inserted into the self-test input connector. The self-test and input connectors are located under the hood in the engine compartment, usually near the right shock tower. The shop manual will indicate the exact location of the connectors.

After the ignition is turned on and the engine off, service codes will be displayed on the STAR tester or by the counting of the sweeps of the needle of the voltmeter. All the service codes are two-digit codes. The number of pulses indicate the digit. Each digit is separated by a 2-second display. To determine the service code, count the number of sweeps until there is a pause and then count the next set of sweeps. The service code 34 would be represented by three sweeps followed by a 2-second pause and then four sweeps of the needle. If more than one service code is signaled by the computer, there is a 4-second pause between the pulsing. Each service code will refer to a particular service test that should be performed (Fig. 16–27). These tests will determine the exact location of the problem. The procedures for conducting the self-test should be followed as presented in the service manual.

The self-test is a functional test of the EEC system and consists of 10 basic test steps. These steps must be followed carefully to correctly diagnose the engine:

1. *Visual inspection.*
2. *Equipment hook-up.*
3. *Service code descriptions:*
   a. Fast codes: serve no purpose in field.
   b. Engine identification codes: serve no purpose in field.

1984 SELF-TEST SERVICE CODES

11 System "pass"	61 ECT input too low
12 Rpm out of spec (extended idle)	63 TPS input too low
13 Rpm out of spec (normal idle)	64 ACT (VAT) input too low
14 PIP was erratic (continuous test)	65 Electrical charging over voltage
15 ROM test failed	66 MAF (VAF) input too low
16 Rpm too low (fuel lean test)	67 Neutral drive switch — drive or accelerator on (engine off)
17 Rpm too low (upstream/lean test)	68 ITS open or AC on (engine-off test)
18 No tach	72 No MAP change in "goose test"
21 ECT out of range	73 No TPS change in "goose test"
22 MAP out of range	76 No MAF (VAF) change in "goose test"
23 TPS out of range	77 Operator did not do "goose test"
24 ACT out of range	81 Thermactor air bypass (TAB) circuit fault
25 Knock not sensed in test	82 Thermactor air diverter (TAD) circuit fault
26 MAF (VAF) out of range	83 EGR control (EGRC) circuit fault
31 EVP out of limits	84 EGR vent (EGRV) circuit fault
32 EGR not controlling	85 Canister purge (CANP) circuit fault
33 EVP not closing properly	86 WOT A/C cut-off circuit fault (all 3.8L and 5.0L Continental)
34 No EGR flow	87 Fuel pump circuit fault
35 Rpm too low (EGR test)	88 Throttle kicker circuit fault (5.0L)
36 Fuel always lean (at idle)	89 Exhaust heat control valve circuit fault
37 Fuel always rich (at idle)	91 Right EGO always lean
41 System always lean	92 Right EGO always rich
42 System always rich	93 Right EGO cooldown occurred
43 EGO cooldown occurred	94 Right secondary air inoperative
44 Air management system inoperative	95 Right air always upstream
45 Air always upstream	96 Right air always not bypassed
46 Air not always bypassed	97 Rpm drop (with fuel lean) but right EGO rich
47 Up air/lean test always rich	98 Rpm drop (with fuel rich) but right EGO lean
48 Injectors imbalanced	
51 ECT input too high	
53 TPS input too high	
54 ACT (VAT) input too high	
55 Electrical charging under voltage	
56 MAF (VAF) input too high	
58 Idle tracking switch input too high (engine running test)	

NOTE: *Many of the codes are programmed into all engines equipped with EEC-IV. Other codes, such as codes 86 and 88, are only programmed into certain engines. The important thing to remember is that each code has only one interpretation, wherever it appears. The same code will NEVER mean two different things on two different engines.*

Figure 16–27 (Courtesy of Ford Motor Company)

c. Engine service codes: results from self-test.

d. Continuous codes: memory.

4. *Key ON/engine OFF self-test:* checks ECA inputs against calibrated sensor values for Key ON/Engine OFF.

5. *Continuous self-test:* checks the sensor inputs for shorts and opens.

6. *Output cycling test:* ECA activates all outputs when the throttle is cycled for additional diagnostics.

7. *Computed timing check:* verifies the system's ability to compute and maintain 30-degree spark timing in self-test.

8. *Engine running test:* checks sensors and actuators under actual operating conditions.

9. *Wiggle test:* checks wiring harness.

10. *Pinpoint routines:* corrective action to be entered only when instructed by the functional test.

The Key ON/Engine OFF and engine running tests are functional tests which detect faults only at the time of the self-test. Continuous testing is an ongoing test that stores information in the memory for retrieval at a later time.

GENERAL MOTORS COMPUTER COMMAND CONTROL

Since 1981, all General Motors gasoline engines have been equipped with the Computer Command Control (CCC) system. The CCC system is used to monitor as many as 10 engine functions, including the fuel system, ignition system, emission control systems, and automatic transmission torque converter operation.

The systems controlled by the computer will vary depending on the model of the car, the size of the engine, and the accessories on the vehicle. The computer in the system is called the Electronic Control Module (ECM). Like other computer systems, the ECM receives inputs from a variety of sensors, and responds to these inputs by controlling and adjusting certain engine controls (Fig. 16–28). Although the use of sensors varies a lot throughout the line of GM cars, all CCC systems have the sensors necessary to monitor and control the air/fuel ratio of the mixture. This mixture control is accomplished either with an electronically controlled carburetor or with electronic fuel injection.

The ECM is equipped with a PROM which is removable, and gives the computer its limits of control based on the description of the vehicle in which it is installed (Fig. 16–29). The PROM contains information about engine size, type of transmission, final drive gear ratio, and tire size. This information allows the computer to make exact decisions for the vehicle in which it is installed. The inputs from the sensors and the information stored in the PROM allow the ECM to improve engine efficiency during all operating conditions.

The CCC system uses many of the same sensors as the other manufacturers. Vehicles equipped with a minimum-functioning CCC system use only a coolant temperature sensor, an oxygen sensor, and a throttle position sensor. As the computer is used to control more engine functions, the number of sensors used increases. Many of the sensors that may be used in the more complex systems are similar to those of the Ford EEC IV system; however, some of the sensors are located in different points on the engine, and the information is used to control different components.

Regardless of the location and function of the sensors, all sensors send inputs to the computer to provide the information needed by the computer. GM vehicles can be equipped with any or all of the following sensors: an oxygen sensor, a throttle

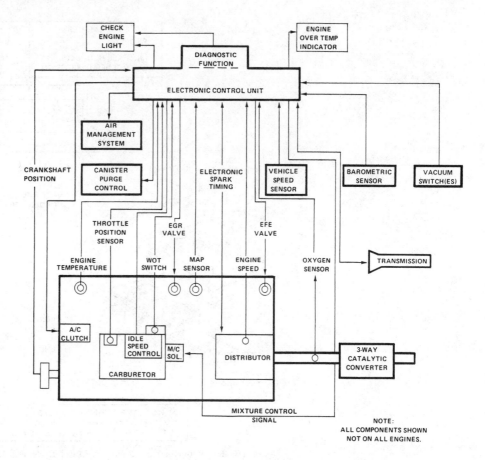

Figure 16-28 (Courtesy of Chevrolet Motor Division, General Motors Corporation)

position sensor, a Manifold Absolute Pressure Sensor (MAP), a Barometric Pressure Sensor (BARO), a Coolant Temperature Sensor (CTS), a Vehicle Speed Sensor (VSS), and a High Gear switch. These sensors supply signals to the computer, which in turn adjusts the air/fuel ratio and ignition timing (Fig. 16-30).

Ignition timing control is possible only on vehicles that are equipped with a full-function system. The timing is controlled by one of two systems: the Electronic Spark Timing (EST) or the Electronic Spark Timing with Electronic Spark Control (EST/ESC). The major difference between the two systems is that the EST/ESC system uses a Detonation Sensor which signals to the computer that detonation is taking place. The EST/ESC can then retard the timing to prevent further detonation. The EST system does not have this ability. Both systems are governed by the ECM and control the ignition module of an HEI distributor. The module used in these systems is very much like that used in a noncomputerized HEI with the exception of the hook-up to the ECM. To provide this input from the computer, the ignition module has two additional terminals. The noncomputerized systems use a five-terminal module; the computerized systems use a seven-terminal module. These modules cannot be interchanged.

All carbureted engines are equipped with a feedback carburetor which contains an electrically operated mixture control, an M/C solenoid. This solenoid controls the metering rods which control the mixture. When the solenoid is activated, the fuel flow through the carburetor is reduced and the mixture becomes leaner. A rich mixture results from the deactivation of the solenoid. The ECM cycles the so-

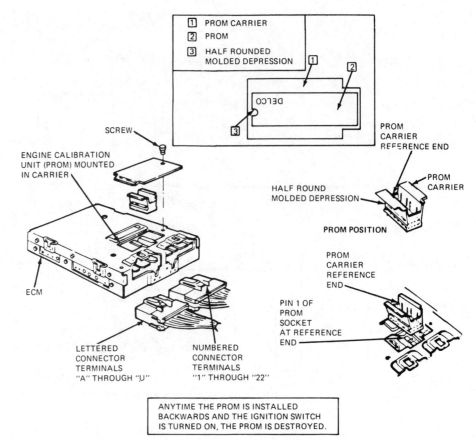

Figure 16-29 (Courtesy of Chevrolet Motor Division, General Motors Corporation)

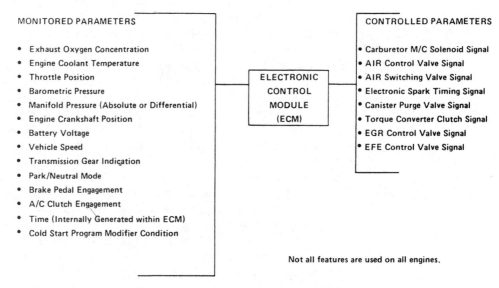

Figure 16-30 (Courtesy of Chevrolet Motor Division, General Motors Corporation)

lenoid 10 times per second to maintain an ideal mixture during the operation of the engine.

Engines equipped with fuel injection have electrically pulsed injectors located in the throttle body. The ECM controls the amount of time the injectors are on, and this controls the mixture. The pulse width of the injector is controlled by the ECM, in response to the inputs from the sensors, to provide the engine's air/fuel requirements. The longer the pulse, the more fuel that enters the intake manifold, and therefore the richer the mixture. During the open loop, the ECM controls the injector according to a preprogrammed sequence. This sequence is determined by the engine's temperature and vacuum (Fig. 16–31).

The ECM also controls the following emission control devices: Air Management system (AIR), EGR, Early Fuel Evaporation (EFE), and Evaporative Emission Control system (EECS). The AIR system is used to heat the catalytic converter and oxygen sensor quickly during cold engine operation. This helps reduce the amounts of HC and CO in the exhaust and allows the computer to go into its closed loop quicker. The ECM activates an air control solenoid which allows air to flow into the engine's exhaust ports. During warm operation, the air is directed to the catalytic converter (Fig. 16–32).

To increase the engine's efficiency during cold operation, some vehicles are equipped with an EFE system. This system simply heats the fuel as it enters the intake manifold. This is accomplished either by a vacuum-operated valve or by a ceramic heater mounted at the base of the carburetor. During warm operation the EFE is turned off by the computer (Fig. 16–33).

Canister purging is accomplished by the EECS. The ECM allows the purging of the carbon fuel vapor canister to occur only when it is in the closed loop and the engine is operating above a particular speed. The system uses an electrically operated vacuum valve. When the valve is activated, no purging will take place. When the proper conditions are present for purging, the valve is deactivated and allows for vacuum to be applied to the purge valve. The ECM controls the vacuum signals to the EGR valve with a solenoid valve. The solenoid is activated when the engine is cold, and this blocks the vacuum to the EGR. The EGR opens when a predetermined temperature is reached and the ECM turns the solenoid valve off.

The CCC system has the ability for self-diagnostics. All vehicles equipped with

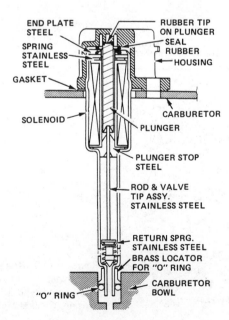

Figure 16-31 (Courtesy of Chevrolet Motor Division, General Motors Corporation)

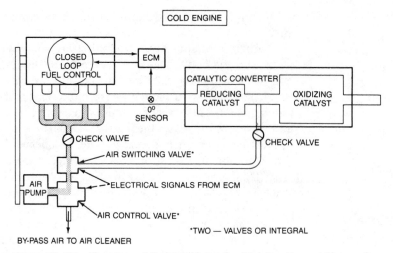

Figure 16-32 (Courtesy of Cadillac Motor Car Division, General Motors Corporation)

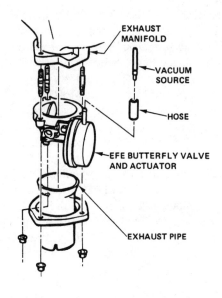

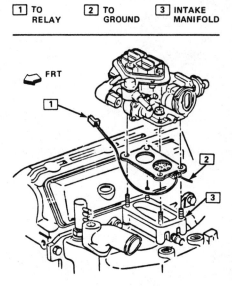

Figure 16-33 The top EFE system is a vacuum servo type, while the lower system is an electrically heated type. (Courtesy of Chevrolet Motor Division, General Motors Corporation)

the system are equipped with a "Check Engine" warning light mounted in the dash (Fig. 16–34). This light will be lit only when the computer senses that something is not working properly in the system. Because other engine systems can affect the input from the sensors, diagnostics on this system should begin in the same manner as any other engine. Before suspecting the computer and its related systems, tests should be performed on the ignition's secondary circuit and checks made on the vacuum and other systems. Faults in these systems will cause the computer to sense a problem. The problem is not in the computer system, however; rather, the signals from the sensors are out of the normal range, causing the computer to suspect its own system. If the engine has a fouled spark plug, the exhaust will contain an excessive amount of HC. The engine's vacuum will also be affected. Operating with this problem, the computer would receive signals that are not within the normal range, and may suspect that the oxygen sensor or manifold pressure sensor are not functioning properly.

If the "Check Engine" light stays on during operation of the engine, it is an indication that there is a permanent problem in the system. GM refers to these problems as "hard" problems. If the light flashes on occasionally, this indicates that there is an intermittent problem. The computer will store the intermittent and hard problems in the computer's memory. By putting the computer into its self-diagnostic mode, these problems will be displayed by the "Check Engine" light. The light will flash in a set pattern representing a particular trouble code. The trouble codes are two-digit numbers and the set of flashes separated by a pause will represent the trouble code. By counting the flashes, the trouble code can be determined (Fig. 16–35).

To put the ECM into the diagnostic mode, turn the ignition on with the engine off. The check engine light should be on. This tests the lamp in the warning light. Next locate the Assembly Line Data Link (ALDL) connector, attached to the wiring harness to the ECM, located under the instrument panel (Fig. 16–36). (On later models, this diagnostic connector is located under the steering column.) Insert a male spade connector across the "test" and ground terminal of the connector. On late-model vehicles, these terminals are marked "A" and "B." Most ALDL connectors have a slot molded into the plastic to allow for easy connection between the test and ground terminals. If no slot is provided, the test terminal needs to be grounded through the ground terminal of the connector or at another suitable ground.

This should cause the "Check Engine" light to flash code "12." This code will be repeated two more times and be followed by any code stored in the memory of the computer. Each code is repeated three times, with the lowest number displayed first. As the codes come up, they should be recorded. To determine whether or not the codes are for permanent or intermittent failures, the memory of the com-

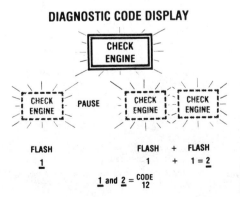

DIAGNOSTIC CODE DISPLAY

Figure 16–34 (Courtesy of General Motors Corporation)

TROUBLE CODE IDENTIFICATION

The "CHECK ENGINE" light will only be "ON" the malfunction exists under the conditions listed below. It takes up to five seconds minimum for the light to come on when a problem occurs. If the malfunction clears, the light will go out and a trouble code will be set in the ECM. Code 12 does not store in memory. If the light comes "ON" intermittently, but no code is stored, go to the "Driver Comments" section. Any codes stored will be erased if no problem reoccurs within 50 engine starts.

The trouble codes indicate problems as follows:

TROUBLE CODE 12 No distributor reference pulses to the ECM. This code is not stored in memory and will only flash while the fault is present.

TROUBLE CODE 13 Oxygen Sensor Circuit — The engine must run up to five minutes at part throttle, under road load, before this code will set.

TROUBLE CODE 14 Shorted coolant sensor circuit — The engine must run up to five minutes before this code will set.

TROUBLE CODE 15 Open coolant sensor circuit — The engine must run up to five minutes before this code will set.

TROUBLE CODE 21 Throttle position sensor circuit — The engine must run up to 25 seconds, at specified curb idle speed, before this code will set.

TROUBLE CODE 23 Open or grounded M/C solenoid circuit.

TROUBLE CODE 24 Vehicle speed sensor (VSS) circuit — The car must operate up to five minutes at road speed before this code will set.

TROUBLE CODE 32 Barometric pressure sensor (BARO) circuit low, or altitude compensator low on J-Car.

TROUBLE CODE 34 Manifold absolute pressure (MAP) or vacuum sensor circuit — The engine must run up to five minutes, at specified curb idle speed, before this code will set.

TROUBLE CODE 35 Idle speed control (ISC) switch circuit shorted. (Over 50% throttle for over 2 sec.)

TROUBLE CODE 41 No distributor reference pulses to the ECM at specified engine vacuum. This code will store in memory.

TROUBLE CODE 42 Electronic spark timing (EST) bypass circuit or EST circuit grounded or open.

TROUBLE CODE 43 ESC retard signal for too long; causes a retard in EST signal.

TROUBLE CODE 44 Lean exhaust indication — The engine must run up to five minutes, in closed loop, at part throttle and road load before this code will set.

TROUBLE CODE 44 & 55 (At same time) — Faulty oxygen sensor circuit.

TROUBLE CODE 45 Rich exhaust indication — The engine must run up to five minutes, in closed loop, at part throttle and road load before this code will set.

TROUBLE CODE 51 Faulty calibration unit (PROM) or installation. It takes up to 30 seconds before this code will set.

TROUBLE CODE 54 Shorted M/C solenoid circuit and/or faulty ECM.

TROUBLE CODE 55 Grounded V ref (terminal "21"), faulty oxygen sensor or ECM.

Figure 16-35 (Courtesy of General Motors Corporation)

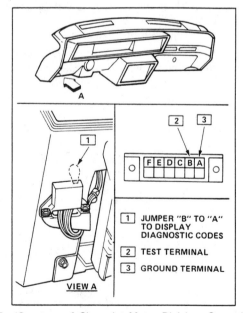

Figure 16-36 (Courtesy of Chevrolet Motor Division, General Motors Corporation)

puter must be cleared. To do this, turn the ignition off, remove the test ground from the ALDL, and remove the ECM fuse from the fuse block for at least 10 seconds. Reinstall the fuse and start the engine. The car should be run for at least 10 minutes. If the problems were "hard," the "Check Engine" light will be on and the computer will repeat the same trouble codes. If the problems were intermittent, the light will not be on and the codes will not reappear. By using the list of trouble codes in the shop manual, the faults identified by the computer may be defined. The shop manual will also give the proper procedures for testing the individual components of the system. These procedures should always be followed.

If more than one trouble code is flashed, you should diagnose the area indicated by the lowest code number first, then work up to the higher code numbers. The only exception to this is code 51, which indicates that there is a problem in the PROM unit. Code 51 takes precedence over all other codes, and if displaced, you should check to see if the PROM is installed in the ECM properly or if it has malfunctioned.

A dwell meter can be used to test the performance of the CCC system. It is used to analyze the operation of the M/C solenoid circuit. The operation of that circuit is controlled by the ECM, which uses information from the systems' sensors. The dwell meter, after it is set on the six-cylinder scale, is connected to the dwell pigtail connector in the M/C Solenoid wiring harness. You must use the six-cylinder scale on all engines equipped with CCC, whether the engine you are working on is a four-, six-, or eight-cylinder engine. The dwell meter will read the time, in degrees, that the ECM closes the M/C Solenoid circuit, allowing voltage to operate the M/C Solenoid.

After connecting the dwell meter to a warm-running engine, the dwell will read between 5 and 55 degrees and will be varying. Needle movement up and down the scale indicates that the engine is in the closed-loop (Fig. 16–37). The M/C Solenoid moves the carburetor's metering rod up and down 10 times per second. The duration of the solenoid on-time determines the air/fuel mixture being delivered to the cylinders. When the solenoid is "on," the metering rod is down and is restricting fuel flow. When a lean mixture is desired, the ECM will have the M/C Solenoid turn "on" and restrict fuel flow 90% of the time. The dwell reading of this lean command will be about 54 degrees (90% of 60 degrees). A rich command will have a dwell reading of about 6 degrees, as the M/C Solenoid is "on" only 10% of the time. The ideal mixture will be displayed by the dwell reading varying continuously up and down the scale. This indicates that the mixture is being monitored and controlled by the ECM. If there is no variation in the dwell reading, either the engine is operating in an open loop or there is a fault in the ECM circuit or the oxygen sensor, the ECM, and the M/C Solenoid.

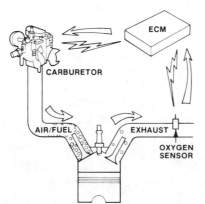

Figure 16–37 (Courtesy of Chevrolet Motor Division, General Motors Corporation)

Figure 16-38 Aftermarket computer system diagnostic tester.

As the CCC system became more widely used, tool manufacturers produced diagnostic meters to read the ECM's self-diagnostic codes and to monitor individual parts of the system. These testers typically plug into ALDL (Fig. 16-38) and have a selection switch which enables the technician to select the area of the system to be monitored. These testers offer quick and precise diagnostics on the CCC system.

GENERAL MOTORS DISTRIBUTORLESS IGNITION

One current computer system introduced by General Motors does not use a distributor for the ignition. The ECM controls the sequence of the firing of the cylinders. The system consists of three ignition coils which are used to fire the six cylinders of the engine (Fig. 16-39). Each coil is responsible for the firing of two cylinders. These cylinders are paired so that as one cylinder is on the compression stroke, the other is on the exhaust. The spark plugs of the two cylinders fire at the same time, but since one plug is firing into the exhaust, it is doing nothing. The system uses two Hall-effect switches to monitor engine speed and crankshaft position. A Hall-effect switch is used to monitor the movement of the engine's camshaft. This sensor "watches" the position of the number 1 piston and sends this information to the coil module assembly. The module must know where piston 1 is to know where the other pistons are. The cam sensor sends a pulse to the module when the number 1 piston is at 25 degrees ATDC on the power stroke. If the cam sensor is defective, the engine will not start. The module will not know where number 1 is and cannot start the ignition system.

The timing of the spark from each coil is controlled by the ECM, which is receiving information about engine speed and crankshaft location from a Hall-effect sensor on the crankshaft pulley. You should be aware that any air that enters the engine after the Mass Air Flow sensor will not be used to control the sequential firing of the fuel injection nozzles. Therefore, any air leaks in the system can cause the engine to run poorly or not at all. The testing procedure for most of this system is the same as the conventional CCC system, except that additional codes have been

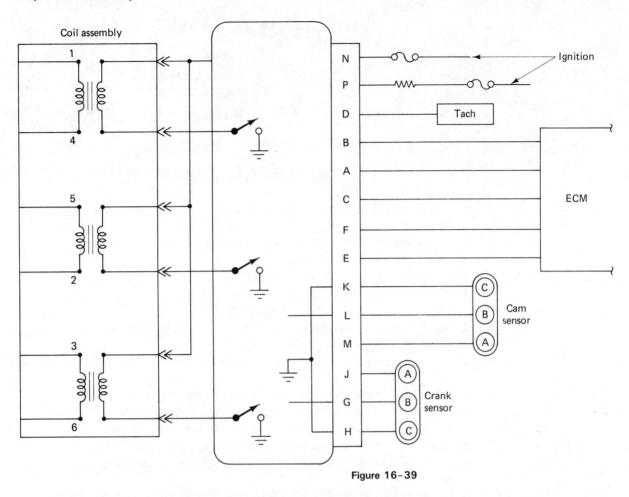

Figure 16-39

added to the diagnostic codes. These additions allow for diagnostics of this more complex computer-controlled system.

SUMMARY

There are many other computer systems used on automobiles today. The systems mentioned in this chapter are to serve as representatives of the various systems. In all cases the computer received information from a group of sensors and used that information to calculate the needed adjustments to the engine for it to run efficiently.

Tune-up procedures for the computer systems are much like that of the electronic ignition systems. The only time you should be concerned with the fact that the system has a computer is when you have reason to believe that the computer or its system has a problem. The only operation that is unique to some computer systems is the adjustment of initial ignition timing. For those systems that control ignition timing, the initial timing must be set with the computer in its open loop. To prevent the system from going into the closed loop while checking or adjusting the timing, a particular procedure must be followed. Once the system is in the open loop, the timing can be checked and changed in the same manner as in a noncomputer system. The procedure for doing this is also given in the shop manual.

TERMS TO KNOW

Microprocessor	PROM
Sensor	Closed loop
Input	Open loop
Output	Impedance

REVIEW QUESTIONS

1. Why has the use of electronics in engine controls allowed for increased engine efficiencies?
2. What is meant by the computer's ability to constantly tune the engine?
3. What is the purpose of an input?
4. What is the purpose of an oxygen sensor?
5. What sensors are used to determine the engine's load?
6. What is the purpose of a detonation sensor?
7. Where is the microprocessor or computer located in most automobiles?
8. What is the purpose of a PROM?
9. Automotive computers have usually two modes of operation. What are these called, and when is the computer in them?
10. Give a simple explanation of how a computer works.
11. What should be done before the computer and its inputs and outputs are tested?
12. Why should only meters with high impedance be used with computer systems?
13. How is ignition timing controlled by most computer systems?
14. How is the air-and-fuel mixture controlled in the Chrysler Electronic Fuel Control system?
15. During engine idle speeds, what is being controlled by Chrysler's EFC system?
16. What inputs are used to determine the proper air/fuel ratio for Chrysler's EFC system?
17. Chrysler engines that are equipped with Electronic Spark Control use what type of pickup for ignition timing and engine speed?
18. What are the three modes of operation for Ford's MCU system, and how do they differ?
19. What are the four major sensors used as inputs to the computer in Ford's MCU system?
20. What is the proper procedure for bringing the computer of the MCU system into self-diagnostics?
21. How are the problems identified by the MCU system signaled to the technician?
22. What three major engine systems are controlled by Ford's EEC IV system?
23. In the EEC IV system, what type of sensor is used to monitor engine speed and crankshaft position?
24. How is fuel controlled on engines equipped with the EEC IV system?
25. What is the purpose of the cannister purge solenoid?
26. What does Ford call the PROM that is installed in the EEC computer?
27. Both carbureted and fuel-injected systems of GM's CCC have the air-and-fuel mixture controlled by the computer. How is this done?
28. What is the purpose of the early fuel evaporation system used in GM's CCC system?
29. How is GM's CCC system brought into the self-diagnostic mode?
30. What is the purpose of GM's "Check Engine" light?
31. What is indicated by the "Check Engine" light staying on?
32. State briefly the operation basics for GM's distributorless ignition.

17

ELECTRONIC FUEL INJECTION DIAGNOSIS

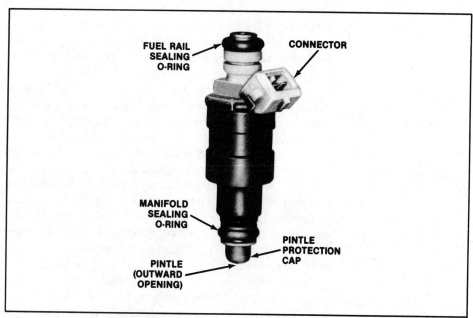

Figure 17-1 (Courtesy of Ford Motor Company)

The use of computerized engine controls has led to a widespread use of electronic fuel injection. Electronic fuel injection allows for more precise control of the air-and-fuel mixture, especially when the injection system is controlled by a computer. The amount of fuel sprayed into the combustion chamber can be regulated to provide an ideal mixture. The inputs from the various sensors used in a computerized engine control system inform the computer of the conditions under which the engine is operating. With this information, the computer controls the amount of time that the injectors are activated, which controls the amount of fuel sprayed from them. Each time an injector is activated, fuel is sprayed under a high pressure. The length of time that the injector is activated is its *pulse width*. The longer the pulse width, the richer the air/fuel ratio will be.

Electronic fuel injection controlled by a microprocessor has many advantages over the typical carburetor. These advantages include an increase in engine efficiency and power, improved fuel economy, improved driveability under all operating conditions, and a decrease in the amount of pollutants in the exhaust. These advantages are the result of precise metering of the amount of fuel that is delivered to a given amount of air. From the input of its sensors, the computer knows the amount of air that is entering the engine and the load of the engine, and calculates the fuel requirement for those conditions. Within a very short time, the computer is controlling the pulse width of the injectors, and thereby controlling the air/fuel ratio.

When the injectors are located right before the intake valve, the system is referred to as a *port injection system* (Fig. 17-2). Fuel is injected into the intake valve port just outside the combustion chamber. Each cylinder has its own injection nozzle, and through the use of the computer, the mixture for each cylinder has the potential of being precisely controlled.

Another type of electronic fuel injection is the *single-point injection system* (Fig. 17-3). This system does not have individual injectors for each cylinder, but rather, has one or two injectors mounted in a throttle plate assembly. The injection nozzles spray fuel into the engine past the throttle plates, and the vacuum of the engine distributes the fuel and air into the individual cylinders.

Many imported engines use an electronically controlled mechanical injection system. Fuel is continuously being delivered by the injectors; therefore, the mixture is not controlled by altering the pulse width of the injectors. To control the mixture, inputs to the computer are used to calculate the fuel volume necessary to provide for the correct ratio. Fuel volume is regulated by the fuel pump pressure, which controls the rate of fuel flow to the injectors.

The *continuous injection system* (CIS) (Fig. 17-4) is a mechanical fuel injection system used on many engines. Control of the fuel volume by a microprocessor employs an oxygen sensor referred to as a *Lambda sensor*. The CIS/Lambda system is a variance of the basic CIS system and has been commonly used on imported engines since 1977. This is the electronically controlled version of the CIS.

The variety of fuel injection systems include the *Digital Fuel Injection* (DFI) and the *Sequential Fuel Injection systems* of General Motors. The DFI is similiar to a normal GM TBI system, except in the display of the diagnostic codes from the computer. This system has the feature of displaying the service codes in a digital readout rather than the normal flashing of the "Check Engine" light. The Sequential system is a port-type injection system that activates the injectors one at a time in the sequence of the firing order of the cylinders. General Motors introduced this system in 1984. The system is quite similar to the Multi-Point Injection system introduced by Ford Motor Company in 1983.

Two other types of electronic fuel injection systems are used today. The names of these systems refer to the system that is used to provide for the inputs needed by the microprocessor to determine fuel needs. The Air Flow Control (AFC) system measures the amount of air passing by an airflow sensor to control fuel flow. The Manifold Pressure Controlled (MPC) system uses manifold vacuum to determine

Figure 17-2   Port injection system.

Figure 17-3 Throttle body injection unit.

the engine's load. In this system the load of the engine dictates the fuel requirements.

There are many types of fuel injection systems used in engines today. Most are derivatives of the basic systems mentioned above. The diagnostics and service for each of these systems are similar, regardless of the manufacturer of the engine. To work on these systems, the type of fuel injection system must first be identified, and then the proper procedures for testing the appropriate type of injection system

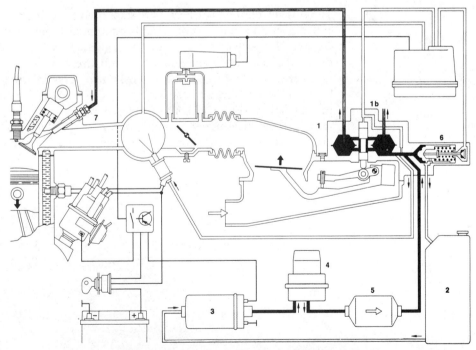

Figure 17-4 Schematic diagram of the CIS: 1, mixture control unit; 1b, fuel distributor; 2, fuel tank; 3, electric fuel pump; 4, fuel accumulator; 5, fuel filter; 6, pressure regulator; 7, fuel injection valve. (Courtesy of Robert Bosch Company)

should be followed. Representative examples of the variety of systems will be covered in this chapter. Volkswagen's version of the MPC system will be discussed. This system is similar to the systems used on early models of Renaults, Jaguars, Mercedes-Benzs, Porsches, and Saabs. Volkswagen's version of the CIS will also be discussed. Audis, Porsches, BMWs, Mercedes-Benzs, and Saabs have been equipped with this system as well. In 1980, Volkswagen began the use of a lambda sensor with the CIS. This system will also be discussed. Similar systems are used on Porsches, Audis, Saabs, Volvos, Peugots, Mercedes-Benzs, and BMWs. Since 1979, Toyota has used an AFC fuel injection system. This type of system is also used on Nissans (Datsuns), Fiats, and some AMC Alliances, Renaults, Porsches, Jaguars, BMWs, and Volkswagens. The Toyota version of this system will be included in the discussion of fuel injection. Also discussed in this chapter will be GM's version of the TBI. Chrysler, AMC, and Ford Motor Companies use their own versions of this type of injection. The discussion of TBI will include the DFI, which is unique to GM. Ford's Multi-Point Injection system will be used as a basis for the ported-type injection systems. General Motors and Chrysler use systems that are similar to Ford's.

This discussion of electronic fuel injection systems is actually a continuation of the study of computerized engine controls. Many of the systems' inputs and outputs that are used to control ignition timing, dwell, and mixture control are used to govern the fuel injection systems. The normal engine checks and visual inspections should be completed before testing any of the fuel injection systems. The self-diagnostic features of some of the computer systems will also display faults in the fuel injection systems. Before you test the components of the fuel injection system, you should be sure that the problem is the air/fuel mixture.

BOSCH CIS SYSTEM

The Bosch Continuous Injection System used by Volkswagen is a mechanically operated fuel injection system. Fuel is continuously metered to the cylinders of the engine. This metering of the fuel is regulated by the amount of air drawn into the intake. This air is measured by a mixture control unit, which also distributes the correct amount of fuel to the cylinders. This mixture control unit is located between the air filter and the throttle body. It consists of an airflow sensor and a fuel distributor (Fig. 17–5). The fuel distributor is mechanically linked to the airflow sensor, which measures the volume of air flowing into the engine. This linkage allows for the fuel distributor to supply the correct amount of fuel to the injectors.

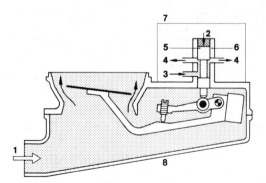

Figure 17–5 Fuel distributor with barrel and metering slots: 1, intake air; 2, control pressure; 3, fuel intake; 4, fuel metered to cylinders; 5, control plunger; 6, barrel with metering slits; 7, fuel distributor; 8, airflow sensor. (Courtesy of Robert Bosch Company)

The fuel distributor consists of a fuel control unit, a system pressure regulator, and pressure regulating valves for each cylinder. The fuel control unit consists of a slotted metering cylinder. This cylinder has a slot for each cylinder of the engine, and fuel flows through each slot to the cylinders. The amount of opening, provided by the slots, determines the amount of fuel that will be delivered. The size of these slots is determined by a fuel control plunger. This plunger moves, in relation to the movement of the plate of the airflow sensor. As more air passes by the sensor, more fuel is allowed to flow through the control valve.

To maintain a constant pressure of fuel, regardless of the opening size of the slots in the fuel control valve, a pressure regulating valve is used for each cylinder. This constant pressure allows for the delivery of a volume fuel that is proportional to the opening of the slots. Excess fuel delivered to the fuel distributor is returned to the fuel tank by the pressure regulator.

The fuel injectors open at a predetermined pressure and are located so that they spray fuel into each cylinder at the intake valve (Fig. 17–6). Once the engine is started, the injectors stay open and spray the amount of fuel determined by the fuel control valve and the fuel pressure. To help atomize and break up the fuel, each injector has a vibrating pin. The movement of this pin is dependent on the fuel passing through the injector. Therefore, when the engine is off, the pin closes off the fuel circuit, due to the decrease in fuel pressure at the injector. The closing of the injector by the pin tends to keep some pressure in the system when the engine is turned off, which allows for quicker restarting.

The system (Fig. 17–7) is equipped with a cold start valve, which is a electro-magnetic fuel injector valve. This valve receives an electrical signal from the starter circuit. The circuit that connects the starting system with the cold start valve is interrupted by a thermo-time switch. This thermo switch allows for the delivery of the current from the starter to the cold start valve for a short period of time while the engine is being started. The time that the cold start valve is in operation depends on the starter's current draw and the temperature of the engine. The thermo-time switch is a bimetalic switch, and when it reaches a particular temperature, it will cause an open circuit between the starter and the cold start valve. This open circuit stops the current flowing to the valve, which stops it from operating.

Before testing this system for faults, the system must be checked for loose fuel fittings and leaky lines. If there is a fuel leak, the engine must be run under a load

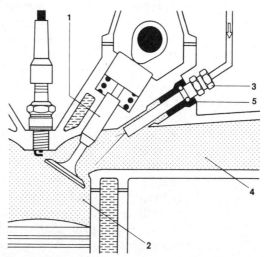

Figure 17–6 1, Intake valve; 2, combustion chamber; 3, fuel injection valve; 4, intake manifold; 5, heat-isolating mount. (Courtesy of Robert Bosch Company)

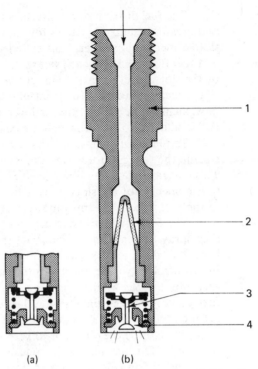

Figure 17-7 Fuel injection valve: 1, valve housing; 2, filter; 3, valve needle; 4, valve seat (a, inoperative; b, during injection). (Courtesy of Robert Bosch Company)

after the repair has been made to normalize the system. This system is extremely sensitive to vacuum leaks, and therefore all troubleshooting of the system should begin with an inspection of all vacuum lines and fittings. Because the injection of fuel by the injectors is continuous, the pressure of the fuel during certain operating conditions is very important. A faulty fuel control valve or fuel pump may supply the incorrect amount of fuel to the cylinders. This may cause an engine to run rough, have poor performance or efficiency, or be hard to start. The initial test on this system is the fuel pump pressure test. After the fuel pressure tester is connected to the system according to the procedures outlined in the shop manual, usually between the warm-up compensator and the fuel distributor, the fuel system must be purged. The purging of the system allows any air that may have entered the system to escape. The presence of air in the system may allow for inaccurate readings on the fuel pressure gauge. The procedure for purging the fuel system is also given in the manual. This procedure is much like the common service procedure of bleeding the brakes of a vehicle.

The idle speed screw on the throttle-body housing bleeds air into the manifold as you increase idle speed. When you change idle speed, therefore, you change idle mixture. To adjust the idle mixture, you need a 3-mm Allen wrench and the proper CO idle specs. The mixture is brought within specs and the idle speed recorrected. Care should be taken not to adjust the idle screws until the oil temperature is within its normal range, about 140 to 176°F.

Other individual components of the system can be checked. While testing these, follow the procedures given in the manual. Because each cylinder has its own injector, a malfunctioning injector may cause the engine to have a poor power balance. Before testing and comparing the performance of individual injectors, the ignition system of the engine should be tested. Poor balance can be caused by factors other than the injectors.

The use of a Lambda sensor with the CIS allows for more precise fuel metering. The Lambda sensor is an oxygen sensor located in the engine's exhaust (Fig. 17-8). The addition of this sensor not only improves the engine's performance, but also decreases the pollutants in the exhaust. This system also uses a three-way catalytic converter to reduce emissions further. The CIS/Lambda system was introduced by Volkswagen in 1980 and is currently used on many models of VW and other imported cars.

The amount of oxygen in the exhaust varies according to the air/fuel ratio of the mixture and the efficiency of the engine's combustion. The Lambda sensor measures the amount of oxygen in the exhaust and sends a signal to the control unit. The control unit then sends a series of electrical signals to "frequency valves." The electrical pulses cause the frequency valve to open and close, which causes the fuel pressure in the fuel distributor to vary. The frequency valves are solenoid-type valves which are connected between the lower chambers of the pressure regulator valves of the fuel distributor and the return lines to the fuel tank. When the frequency valves are open, fuel is allowed to return through the fuel return line. This causes the pressure in the lower chamber to decrease, allowing the spring pressure in the upper chamber to press the steel diaphragm of the fuel distributor down, which causes an increase in the amount of fuel sprayed by the injectors. When the frequency valve is closed, lower chamber pressure is increased. The pressure pushes the diaphragm upward, which reduces the amount of fuel to the injectors. By controlling the amount of time the frequency valves are open and closed, the control unit regulates the air/fuel mixture.

The system includes many enrichment switches and relays. These operate the cold start valve to allow for additional fuel when the engine is warming up, accelerating, and operating under full throttle. The use of these switches and relays will vary from model to model. The location of these will be noted in your shop manual. Since this system is basically a CIS with the addition of the Lambda sensor, the test procedures for the basic injection system are the same as that of the CIS without the sensor. If the sensor malfunctions, problems in starting when the engine is warm, poor acceleration and overall performance, or poor fuel economy can result.

To test the Lambda sensor, an exhaust analyzer and a dwell meter are used. The dwell meter should be set on the four-cylinder or 90-degree scale. It is connected to the test point connector, which is located in the wiring harness of the control unit. The exhaust gas analyzer is used to measure the amount of CO in the exhaust. After the test equipment is set up, disconnect the wires going to the Lambda sensor. Start the engine and read the dwell meter. The reading should be steady and between 40 and 50 degrees. (This spec is for VWs; different applications have different specs. The specs are listed in your shop manual under "Duty Cycle Specifications.") If the reading is correct, set the engine to the recommended idle speed and the mixture to the proper CO level. Now reconnect the oxygen sensor. The CO reading should

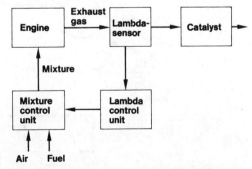

Figure 17-8 (Courtesy of Robert Bosch Company)

drop and the dwell meter needle should fluctuate slightly but still be within the specs. If the readings on either meter do not meet the specs, there is a problem in the Lambda sensor circuit. The cause could be in the wiring and connectors or the sensor itself. The circuit should be completely tested before the sensor is condemned and replaced.

MPC ELECTRONIC FUEL INJECTION

Volkswagen has used the Bosch Manifold Pressure Controlled fuel injection system on many models since 1971 and calls it the D-Jetronic system (Fig. 17–9). This system regulates only the injection pulse width to control the air/fuel ratio. The MPC system measures the amount of incoming airflow by monitoring the vacuum in the intake manifold. Engine speed and temperature are also used to determine the proper pulse width of the injectors. For cold engine operation, the system also uses an auxiliary air valve, a thermo-time switch, and a cold start injector. The main system consists of one injector per cylinder and a fuel pressure regulator, which maintains fuel pressure at 28 psi. The electronic control unit regulates the activation of the injectors, based on information it receives about the engine's speed, temperature, and manifold vacuum.

Manifold vacuum is sensed by a pressure sensor located in the engine compartment. This sensor is connected to the intake manifold by a vacuum tube. The sensor controls the basic amount of fuel to be delivered to the cylinders according to the load and vacuum of the engine. An air-temperature sensor sends information about air temperature to the control unit. The control unit will supply extra amounts of fuel when the air temperature is low or cold. An engine temperature sensor provides the control unit with information about the coolant's or engine's temperature. This information is used to determine the fuel requirements of a cold engine.

To determine when the cylinders are on the intake stroke, the system has triggering contacts located in the distributor (Fig. 17–10). These contacts provide signals that determine when a particular cylinder is ready for the injection of fuel. These contacts are also used to determine engine speed.

A throttle valve switch is mounted to the throttle assembly. It informs the control unit of the throttle position and movement. This switch informs the control unit when the engine is accelerating and decelerating, so that the control unit can

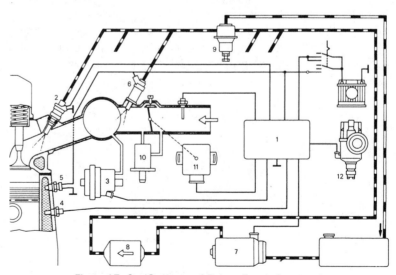

Figure 17–9 (Courtesy of Robert Bosch Company)

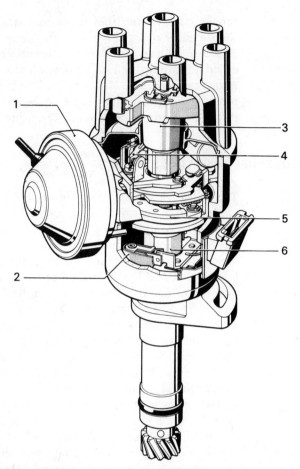

Figure 17-10 1, Vacuum unit; 2, trigger contacts; 3, distributor rotor; 4, distributor contact points; 5, mechanical ignition timing adjustment mechanism; 6, cam. (Courtesy of Robert Bosch Company)

supply more or less fuel to the injectors. An auxiliary air valve is used to allow more air to enter into the intake. This valve is used during cold starts and is closed after the engine reaches about 140°F.

Diagnosis of this system is rather straightforward. If a fault is found in only one cylinder of the engine, the problem must pertain only to that cylinder. The only items in the fuel system that are not common to all cylinders are the individual fuel injectors, including the fuel line and wires to each. If there is a problem with an injector, the lines and wires to it should be inspected and repaired/replaced if necessary or that injector replaced.

If none of the injectors are working, the engine will probably not start. The cause of this problem must be something common to all cylinders. The triggering contacts could be burned or closed and not provide the control unit with the information it needs to activate the injectors. The fuel pump is also a likely cause of this sort of problem.

It is important that you define the problem before any tests are conducted on this system. An understanding of the system and the problem will enable you to use the testing procedures given in the shop manual to identify the exact causes of the malfunction. If the engine has excessive fuel usage, the problem could be in the cold operation system, or an injector can be leaking fuel.

The cold-operation system supplies extra fuel to the engine and if it is malfunctioning will allow extra fuel when it is not desirable. The switches that allow for the change from cold to warm operation may not be working properly. The shop manual will give you details for testing the concerned components for the engine you are testing. A power balance test will not only identify the injector and cylinder that may be leaking but will also tell you whether the problem is related to one cylinder or to all cylinders. The basics of diagnosis will lead you to the appropiate component that needs to be tested, and the shop manual will lead you through the proper procedures for testing that component.

AFC ELECTRONIC FUEL INJECTION SYSTEMS

The Bosch Air Flow Controlled fuel injection system is an electronically controlled system that maintains an ideal air/fuel ratio by monitoring the intake airflow and the oxygen content in the exhaust. The air intake system cleans the incoming air, measures the amount and temperature of the air, and regulates the flow of air to the engine. An electric fuel pump supplies the fuel at a constant pressure to the fuel injectors located in the intake ports of the engine. The injectors spray a metered quantity of fuel into the intake according to the signals supplied by the EFI computer. The EFI computer receives input signals from sensors that monitor intake air, coolant temperature, engine vacuum, and exhaust oxygen content (Fig. 17-11). From this information the computer calculates the injector pulse width needed to maintain an ideal mixture ratio. A signal is sent to the solenoid-controlled injectors to allow for the delivery of one-half the amount of fuel needed. The injectors are pulsed two times during one intake stroke. The EFI computer is located in the passenger compartment behind the left kick panel on most Toyotas.

The main sensor for the regulation of the mixture is the airflow meter. This sensor is located between the air filter and the throttle plates. It consists of a tunnel with a measuring flap and a compensation flap, which is mounted at a right angle to the measuring flap. The measuring flap swings with the flow of incoming air. The movement of the flap works against the pressure of a spiral spring connected to a potentiometer (a variable resistor). The movement of the flap causes the potentiometer to change its voltage signal to the computer. This signal informs the

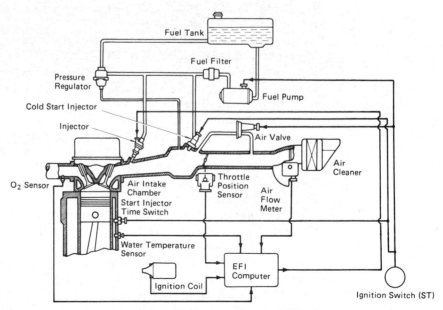

Figure 17-11   Toyota's AFC system.

computer of the incoming airflow and the engine's load. Located in the airflow meter is the air-temperature sensor and the idle mixture screw. This screw regulates the amount of air that can enter the intake without moving the measuring flap.

An ohmmeter can be used to test the airflow meter (Fig. 17–12). To do so, disconnect the wiring harness to the meter. Measurements will be taken across each terminal of the harness connector on the meter. The shop manual will give the specs for this test. Measurements will also be taken while you move the measuring plate. If the resistance of the meter is not within specs, the voltage signals to the computer will not be indicative of the actual flow of air and will cause the system to be unable to maintain the ideal mixture ratio.

Air leaks in the intake system can cause various driveability problems. To check for leaks, disconnect the air line from the auxiliary air valve to the intake manifold. Apply a soapy water solution to any connections or joints where leaks are likely to occur. Then plug the exhaust pipe with a shop towel or rag. Apply no more than 15 psi of compressed air at the air line. Open the throttle plate. The presence of bubbles or foam at any point in the system indicates the location of air leaks.

Other components of this system can be checked with an ohmmeter. Toyota recommends the use of a special tester to diagnose the system properly. However, the most common cause of problems is poor electrical connections. Diagnosis of this system should begin with a visual inspection of all connections. The least probable cause of a problem is the computer; therefore, all other components of the system should be tested before replacing the computer.

The Toyota EFI Checker (SST No. 09991–00100) (Fig. 17–13) tests the input and output signals of the computer. The following signals are checked by the EFI checker: battery voltage, airflow meter signals, water temperature signals, intake air temperature signals, throttle position sensor signals, oxygen sensor signals, injector signals, and a signal from the starter circuit. Most of these systems can be tested through the conventional means of a volt-ohmmeter. The procedures for doing so are given in the shop manual for the model and engine you are troubleshooting.

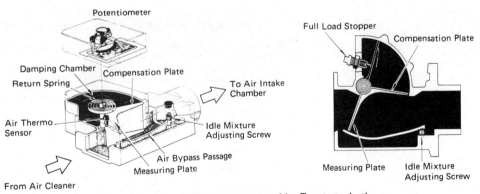

Figure 17–12 Airflow meter as used by Toyota and others.

AFC AND MPC DIAGNOSIS

Except for the monitoring systems necessary for fuel control, the layout and circuitry of the AFC and MPC injection systems are rather similar. Comparable also are the guidelines for basic troubleshooting of the systems. What follows are some common causes of problems in the AFC and MPC injection systems. Included with each cause is a simple explanation for the testing of these components.

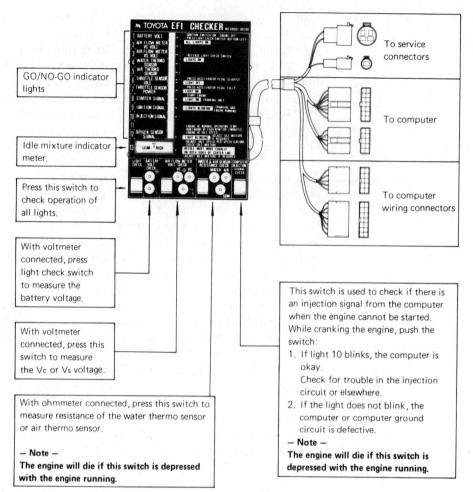

GO/NO-GO indicator lights

Idle mixture indicator meter.

Press this switch to check operation of all lights.

With voltmeter connected, press light check switch to measure the battery voltage.

With voltmeter connected, press this switch to measure the Vc or Vs voltage.

With ohmmeter connected, press this switch to measure resistance of the water thermo sensor or air thermo sensor.

— Note —
The engine will die if this switch is depressed with the engine running.

To service connectors

To computer

To computer wiring connectors

This switch is used to check if there is an injection signal from the computer when the engine cannot be started. While cranking the engine, push the switch:

1. If light 10 blinks, the computer is okay.
   Check for trouble in the injection circuit or elsewhere.
2. If the light does not blink, the computer or computer ground circuit is defective.

— Note —
The engine will die if this switch is depressed with the engine running.

Figure 17-13 Toyota's EFI checker.

A. *Control unit defective:* Use known good unit to confirm defect.

B. *Mechanical defect in engine:* Check compression, valve adjustment, and oil pressure.

C. *Defect in ignition system:* Check battery, distributor, plugs, coil, and timing.

D. *Loose connection in wiring harness or system ground:* Check and clean all connections.

E. *Trigger contacts in distributor defective:* Replace trigger contacts.

F. *Fuel pump not operating:* Check pump fuse, pump relay, and pump.

G. *Relay defective; wire to injector open:* Test relay; check wiring harness.

H. *Blockage in fuel system:* Check fuel tank, filter, and lines for free flow.

I. *Fuel system pressure incorrect:* Test and adjust at pressure regulator.

J. *Defective injection valve:* Check valves individually for spray.

K. *Idle speed incorrectly adjusted:* Adjust idle speed with bypass screw.

L. *Cold start valve defect:* Check for spray or leakage.

M. *Thermo-time switch defective:* Test thermo-time switch for correct function.

N. *Auxiliary air valve not operating correctly:* Must be open with cold engine, closed with warm.

O. *Leaks in air intake system:* Check all hoses and connections: eliminate leaks.

P. *Airflow sensor defective:* Check pump contacts: test flap for free movement.

Q. *Throttle butterfly does not completely close or open:* Readjust throttle stops.

R. *Throttle valve switch incorrectly adjusted or defective:* Adjust as necessary or replace.

S. *Pressure sensor defective:* Test with ohmmeter.

T. *Temperature sensor II defective:* Test for 2 to 3 kΩ at 68°F.

U. *CO concentration incorrectly set:* Readjust CO with screw on airflow sensor.

Use the letter that precedes the items above to identify the possible causes of the following problems.

Problem: The engine cranks but will not start.
Possible causes: A, B, C, D, F, G, H, I, L, M, N, O (AFC only), P (AFC only), S (MPC only), T

Problem: Engine starts but then dies
Possible causes: A, B, C, D, E (MPC only), G, H, I, K, L, N, O, P (AFC only), S (MPC only), T

Problem: Engine has rough or unstable idle
Possible causes: A, B, C, D, H, I, J, K, L, N, O, P (AFC only), Q, R, S (MPC only), T (AFC only), U (AFC only)

Problem: Incorrect idle speed
Possible causes: B, C, J, K, L, N, O, P (AFC only), Q, R

Problem: Incorrect CO valve
Possible causes: I, J, K, L, O (AFC only), P (AFC only), S (MPC only), T

Problem: Engine runs erratically
Possible causes: A, B, C, D, I, J, L, O, P (AFC only), R, S (MPC only), U (AFC only)

Problem: Engine misfires while driving
Possible causes: A, B, C, D, E (MPC only), I, J, O

Problem: Engine has poor power
Possible causes: A, B, C, H, I, J, O, P (AFC only), Q, R, S (MPC only)

Problem: Excessive fuel consumption
Possible causes: B, C, I, L, P (AFC only), S (MPC only), T

SINGLE-POINT FUEL INJECTION SYSTEMS

Single-point injection systems are used by Chrysler, Ford, and General Motors. GM uses this system on many models of engines. Particular applications dictate the detail of the system, but all applications are quite similar. The most notable difference in single-point injection applications is the use of one or two throttle body assemblies. Most engines are equipped with a single throttle body. Dual throttle bodies are used on some engines installed in Corvettes, Camaros, and Firebirds. This system is called the cross-fire injection system (Fig. 17–14). The control and operation of this system is similar to the single units and therefore the cross-fire system will not be discussed in great detail. The single-unit TBI (Throttle Body Injection) types are not only the systems used most commonly by GM, but similar single units are used by the other automobile manufacturers.

The fuel from this system is supplied to the engine from an injector located in the throttle body mounted on top of the intake manifold. The injector is controlled by electrical pulses from the Electronic Control Module (ECM). The ECM regulates the amount of fuel sprayed from the injector according to the information

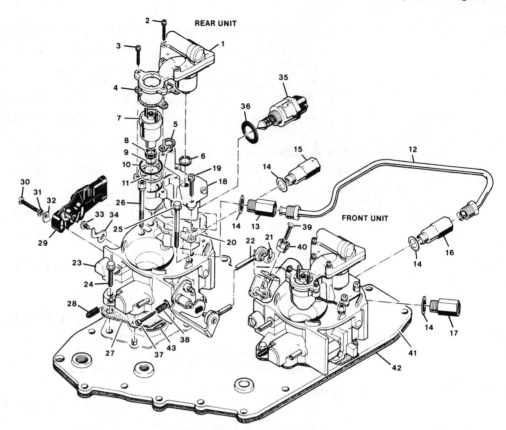

Figure 17-14 (Courtesy of Chevrolet Motor Division, General Motors Corporation)

it receives from a variety of engine sensors (Fig. 17-15). These sensors are the same ones as those used for the Computer Command Control system. The ECM not only controls the mixture's ratio, but also the ignition dwell and timing. The details of this computerized system were discussed in Chapter 16 and will not be discussed here. The action of the injection system will be discussed, as well as the basic tests of the fuel injection system.

TBI FUEL INJECTION SYSTEMS

The TBI assembly (Fig. 17-16) is composed of two castings: a throttle body with a plate to control the airflow to the engine, and a fuel body, consisting of an injector and fuel pressure regulator. The fuel pressure regulator is operated by the pressure of the injector acting on one side of a diaphragm and the incoming air pressure acting on the other side. The regulator maintains a pressure differential between these two pressures. The injector is a solenoid-controlled device that is activated by the ECM. An electric fuel pump delivers fuel to the lower part of the injector. The solenoid of the injector is activated by a voltage signal from the ECM, and allows pressurized fuel to be sprayed above the throttle plates. Engine vacuum draws in the intake air and the injected fuel, and distributes the mixture to the appropriate cylinders.

During engine starting, the ECM causes the injector to be activated once, for each reference pulse it receives from the ignition distributor. The spraying of fuel is synchronized with the ignition system. The injector will operate in a nonsynchronized mode when the engine is running and the computer is in the closed loop.

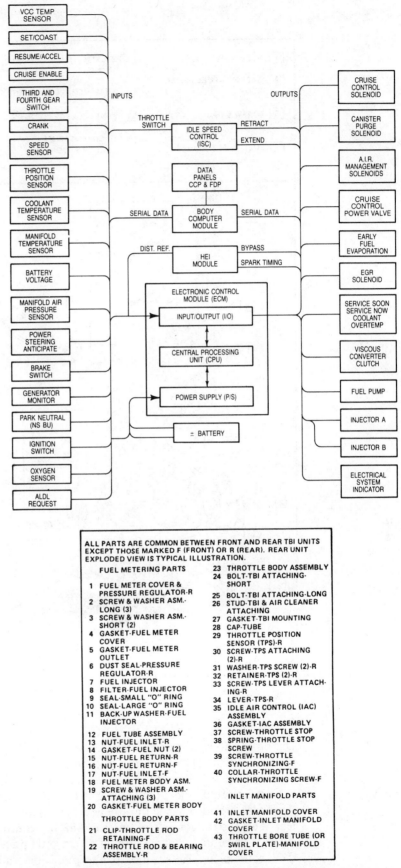

VCC TEMP SENSOR
SET/COAST
RESUME/ACCEL
CRUISE ENABLE
THIRD AND FOURTH GEAR SWITCH
CRANK
SPEED SENSOR
THROTTLE POSITION SENSOR
COOLANT TEMPERATURE SENSOR
MANIFOLD TEMPERATURE SENSOR
BATTERY VOLTAGE
MANIFOLD AIR PRESSURE SENSOR
POWER STEERING ANTICIPATE
BRAKE SWITCH
GENERATOR MONITOR
PARK NEUTRAL (NS BU)
IGNITION SWITCH
OXYGEN SENSOR
ALDL REQUEST

INPUTS

THROTTLE SWITCH
IDLE SPEED CONTROL (ISC) — RETRACT / EXTEND

DATA PANELS CCP & FDP

SERIAL DATA — BODY COMPUTER MODULE — SERIAL DATA

DIST. REF. — HEI MODULE — BYPASS / SPARK TIMING

ELECTRONIC CONTROL MODULE (ECM)
INPUT/OUTPUT (I/O)
CENTRAL PROCESSING UNIT (CPU)
POWER SUPPLY (P/S)

± BATTERY

OUTPUTS

CRUISE CONTROL SOLENOID
CANISTER PURGE SOLENOID
A.I.R. MANAGEMENT SOLENOIDS
CRUISE CONTROL POWER VALVE
EARLY FUEL EVAPORATION
EGR SOLENOID
SERVICE SOON SERVICE NOW COOLANT OVERTEMP
VISCOUS CONVERTER CLUTCH
FUEL PUMP
INJECTOR A
INJECTOR B
ELECTRICAL SYSTEM INDICATOR

ALL PARTS ARE COMMON BETWEEN FRONT AND REAR TBI UNITS EXCEPT THOSE MARKED F (FRONT) OR R (REAR). REAR UNIT EXPLODED VIEW IS TYPICAL ILLUSTRATION.

FUEL METERING PARTS

1 FUEL METER COVER & PRESSURE REGULATOR-R
2 SCREW & WASHER ASM.-LONG (3)
3 SCREW & WASHER ASM.-SHORT (2)
4 GASKET-FUEL METER COVER
5 GASKET-FUEL METER OUTLET
6 DUST SEAL-PRESSURE REGULATOR-R
7 FUEL INJECTOR
8 FILTER-FUEL INJECTOR
9 SEAL-SMALL "O" RING
10 SEAL-LARGE "O" RING
11 BACK-UP WASHER-FUEL INJECTOR
12 FUEL TUBE ASSEMBLY
13 NUT-FUEL INLET-R
14 GASKET-FUEL NUT (2)
15 NUT-FUEL RETURN-R
16 NUT-FUEL RETURN-F
17 NUT-FUEL INLET-F
18 FUEL METER BODY ASM.
19 SCREW & WASHER ASM.-ATTACHING (3)
20 GASKET-FUEL METER BODY

THROTTLE BODY PARTS

21 CLIP-THROTTLE ROD RETAINING-F
22 THROTTLE ROD & BEARING ASSEMBLY-R

23 THROTTLE BODY ASSEMBLY
24 BOLT-TBI ATTACHING-SHORT
25 BOLT-TBI ATTACHING-LONG
26 STUD-TBI & AIR CLEANER ATTACHING
27 GASKET-TBI MOUNTING
28 CAP-TUBE
29 THROTTLE POSITION SENSOR (TPS)-R
30 SCREW-TPS ATTACHING (2)-R
31 WASHER-TPS SCREW (2)-R
32 RETAINER-TPS (2)-R
33 SCREW-TPS LEVER ATTACHING-R
34 LEVER-TPS-R
35 IDLE AIR CONTROL (IAC) ASSEMBLY
36 GASKET-IAC ASSEMBLY
37 SCREW-THROTTLE STOP
38 SPRING-THROTTLE STOP SCREW
39 SCREW-THROTTLE SYNCHRONIZING-F
40 COLLAR-THROTTLE SYNCHRONIZING SCREW-F

INLET MANIFOLD PARTS

41 INLET MANIFOLD COVER
42 GASKET-INLET MANIFOLD COVER
43 THROTTLE BORE TUBE (OR SWIRL PLATE)-MANIFOLD COVER

Figure 17–15 (Courtesy of Cadillac Motor Car Division, General Motors Corporation)

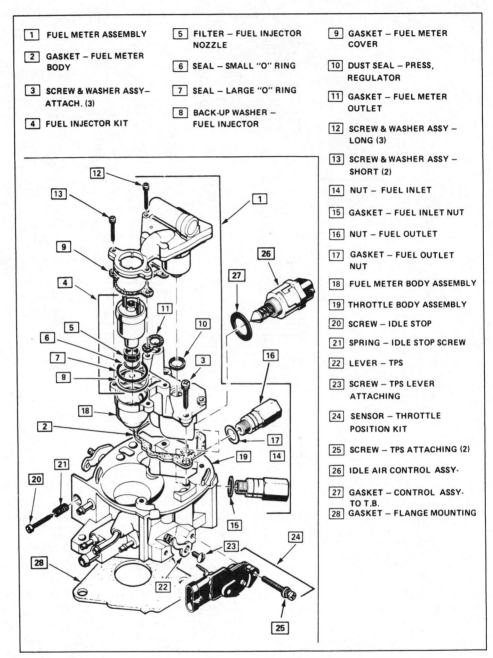

1 FUEL METER ASSEMBLY	**5** FILTER – FUEL INJECTOR NOZZLE	**9** GASKET – FUEL METER COVER
2 GASKET – FUEL METER BODY	**6** SEAL – SMALL "O" RING	**10** DUST SEAL – PRESS, REGULATOR
3 SCREW & WASHER ASSY– ATTACH. (3)	**7** SEAL – LARGE "O" RING	**11** GASKET – FUEL METER OUTLET
4 FUEL INJECTOR KIT	**8** BACK-UP WASHER – FUEL INJECTOR	**12** SCREW & WASHER ASSY – LONG (3)

13 SCREW & WASHER ASSY – SHORT (2)
14 NUT – FUEL INLET
15 GASKET – FUEL INLET NUT
16 NUT – FUEL OUTLET
17 GASKET – FUEL OUTLET NUT
18 FUEL METER BODY ASSEMBLY
19 THROTTLE BODY ASSEMBLY
20 SCREW – IDLE STOP
21 SPRING – IDLE STOP SCREW
22 LEVER – TPS
23 SCREW – TPS LEVER ATTACHING
24 SENSOR – THROTTLE POSITION KIT
25 SCREW – TPS ATTACHING (2)
26 IDLE AIR CONTROL ASSY·
27 GASKET – CONTROL ASSY· TO T.B.
28 GASKET – FLANGE MOUNTING

Figure 17-16 (Courtesy of Chevrolet Motor Division, General Motors Corporation)

During the nonsynchronized mode, the injector is pulsed once every 6.25 to 12.5 milliseconds (0.00625 to 0.0125 second). The pulse width is calculated and controlled by the ECM and the inputs it receives from the engine's sensors. During the nonsynchronized mode, the injector is operating without reference to the ignition.

Also included in the throttle body assembly is an Idle Air Control (IAC) system. This system consists of an electronically controlled actuator which positions the IAC valve in a channel around the throttle plates, which allows air to bypass the throttle plate. The position of the IAC is determined by the ECM to maintain a desired idle speed during all operating conditions.

Normal engine checks should be completed before tests are performed on the

TBI system. These checks should include the battery, ignition system, engine vacuum, and fuel supply and pressure. Also, visual inspection should be performed on all vacuum lines and electrical wires and connections of the system. As with any CCC system, the computer should be put into its self-diagnostic mode, and attention should be initially on any service codes displayed by the ECM and the "Check Engine" light.

Diagnosis on the fuel system should begin with a fuel pressure test. Before connecting any test on the fuel system, the existing pressure must be released from the system. To do this on GM's TBI, you must remove the "Fuel Pump" fuse from the fuse block and start the engine. When the engine quits, the fuel that was remaining in the system has been released. With the engine's ignition off, replace the fuse and remove the air cleaner assembly. Plug the thermal vacuum port on the throttle body assembly and remove the steel fuel line between the throttle body and the fuel filter. Install a high-pressure (0 to 100 psi) fuel pressure gauge between the fuel filter and the throttle body. Start the engine and compare the gauge reading with specs. When the test is completed, depressurize the system before removing the fuel pressure gauge. The depressurizing of the fuel system on Ford's single-point unit does not require the removal of a fuse. At the fuel inlet at the injector assembly is a Schrader valve which is used to relieve the pressure in the fuel system. This valve is also a convenient place to connect a fuel pressure gauge.

The testing of the injection system's components should be according to the problem of the engine. An understanding of the inputs and outputs of the computer will help determine what components should be tested when a particular problem occurs. What follows are some typical problems and the components that could cause them. These components are the items that should be tested to identify the cause of the particular problem. The exact procedures for testing these components for particular applications are given in service manuals.

Problem: The engine is hard to start when cold or hot.
Possible causes:

1. High resistance in the coolant sensor circuit
2. Incorrect fuel pressure
3. Injector leaking when the engine is off
4. Sticking TPS

Problem: The engine surges while operating at a constant speed.
Possible causes:

1. An intermittent open or short in the TCC circuit
2. An intermittent open or short in the EST circuit
3. An intermittent open or short in the HEI bypass circuit

Problem: The engine is sluggish, hesitates, or gets poor gas mileage.
Possible causes:

1. Incorrect fuel pressure
2. Sticking TPS
3. Vacuum leak in the MAP circuit or sensor
4. Injector leaking

Problem: The engine tends to stall.
Possible causes:

1. Incorrect fuel pressure.
2. An intermittent open or short in any of the following circuits:
   a. Fuel pump circuit

   b. IAC circuit
   c. HEI reference circuit
   d. Injector activation circuit
3. Clogged fuel filter
4. Crimped fuel line

MULTI-POINT FUEL INJECTION SYSTEM

Ford Motor Company's Multi-Point Injection system is incorporated with the EEC IV system (Fig. 17–17). This system monitors many engine functions and has the ability to control the ignition and the ratio of the mixture. The Electronic Control Unit receives inputs from many sensors located on the engine, and calculates this information to determine the requirements of the engine. The sensors that affect the fuel injection system directly are the oxygen sensor, the engine coolant temperature sensor, the throttle position sensor, and the air vane meter sensor. The Air Vane Meter Assembly is mounted between the air cleaner and the throttle body. It measures the amount and temperature of the air that enters the engine. With this

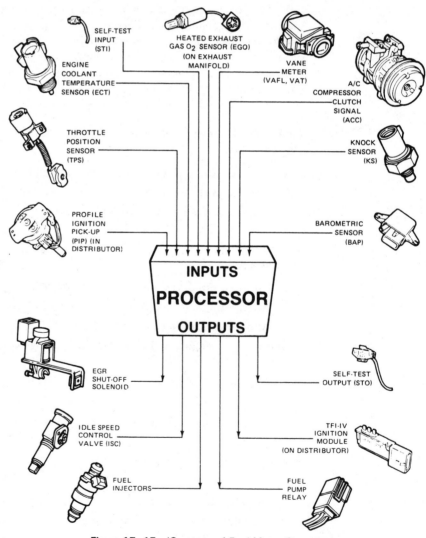

Figure 17–17 (Courtesy of Ford Motor Company)

information, the computer can determine the actual mass of air that is entering the intake system. This assembly is much like the Air Flow Meter used in AFC systems.

Fuel is delivered to the fuel injectors by a high-pressure fuel pump (Fig. 17–18). To maintain constant pressure at the injectors, a pressure regulator is used. Any excess fuel is returned to the fuel tank. All the injectors are activated once every crankshaft revolution. The pulse width of the injectors is determined and controlled by the ECM.

The rate of airflow into the engine is governed by the opening of the throttle plates. To allow for idle conditions when the plates are closed, the throttle body is manufactured with an air bypass channel. The airflow through this channel is governed by an Idle Speed Control Solenoid, which is controlled by the ECM.

Diagnosis of this system begins in the same way as for other fuel-injected and computerized systems. Before any detailed tests are conducted on the components, you should be certain that the cause of the problem is in the delivery of the fuel or air. Because of the complexity of the EEC IV with Multi-Point Fuel Injection, special testers (Rotunda T83L-50 and 07-0004) are required to perform a complete test on the system. However, some components can be tested with normal diagnostic equipment. Begin your diagnosis on the injection system by bringing the computer into its self-test mode, as explained in Chapter 16. Respond to any trouble codes that may be displayed.

Port-type injection systems do not have the spray of the injectors distributed to the cylinders according to the vacuum of the engine. Each injector is located in the intake port of an individual cylinder (Fig. 17–19). Each injector is common only to a particular cylinder. If a faulty injector is suspected, a power balance test can be conducted by the opening of each injector's circuit. As each injector harness is disconnected, the engine's rpm value should be recorded and compared to the others. The decrease in engine speed should be the same for all cylinders. If the engine's speed does not decrease when one injector is disconnected and there are no problems with the ignition system or the engine itself, the injector or its connecting wires are faulty.

If the vehicle's battery and starter motor are good but the engine will not crank with the starter, it is possible that the injectors are leaking fuel into the cylinders while the engine is off. The droppings of fuel can cause a hydrostatic lock-up, which would prevent the engine from turning over. This results from the fuel filling a cylinder. Since a liquid cannot be compressed, the fuel serves as a block, preventing

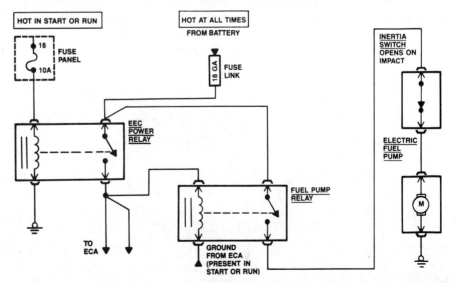

Figure 17–18 (Courtesy of Ford Motor Company)

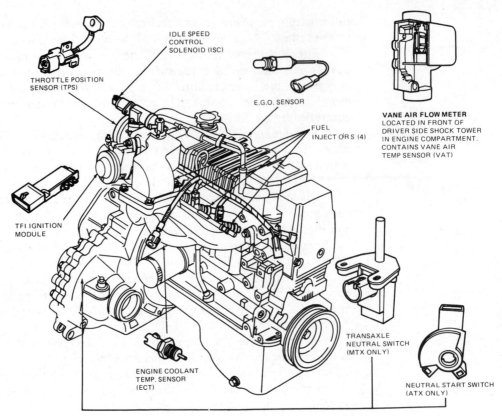

Figure 17-19 (Courtesy of Ford Motor Company)

the piston from moving upward. To identify which injector is leaking, each one should be removed and their cylinders inspected for fuel. Once identified, the leaking injector should be replaced.

If the engine cranks but does not start, the fuel pump pressure and operation should be tested. To conduct a fuel pressure test, connect the fuel pressure gauge to the pressure relief valve on the fuel rail assembly. Start the engine and let it run for a short time. Observe the fuel pressure. The gauge readings should be 35 to 45 psi. If the readings are not correct, visually check the wires from the ignition coil and check the injector hoses and lines for leaks. If no pressure was read, check the fuel pump circuit. Vehicles equipped with this injection system are equipped with an Inertia Switch, which opens the fuel pump circuit in the event of a collision (Fig. 17-20). It is possible that the engine is not able to start because this switch has been

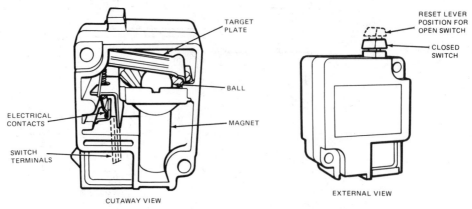

Figure 17-20 (Courtesy of Ford Motor Company)

tripped. This switch can be reset to allow for operation of the fuel pump. If the vehicle was in an accident, the fuel lines should be inspected for leaks prior to resetting the inertia switch.

If the problem has not been discovered through these tests, it is necessary to test the individual components of the injection system. The test procedures for testing these components are given in the shop manual for the vehicle on which you are working. It is important that you follow these procedures. Failure to do so may damage a component or prevent you from identifying the cause of the problem.

SUMMARY

In this chapter we have discussed the various types of fuel injection systems. Descriptions were given about the systems, and some basic tests to perform were also given. Detailed tests were not part of this discussion simply because this information is readily available in shop manuals. The purpose of this chapter is to provide an understanding of the types of fuel injection systems so that you can better follow the procedures for diagnostics and service. You should have noticed that the diagnostics for all these systems began with the basic diagnostic checks discussed in previous chapters. The ability to accurately perform and interpret the results of these tests, and an understanding of the systems you are testing, will enable you to diagnose any and all engine systems. When working on a system for the first time, take the time to learn about the system. The time spent will be *well* spent.

TERMS TO KNOW

Pulse width

Ported injection

Single-point injection

TBI

DFI

CIS

Lambda sensor

AFC

MPC

Multi-Point Injection

SFI

REVIEW QUESTIONS

1. What advantage does fuel injection have over carburetion?
2. What advantage does electronic fuel injection have over mechanical fuel injection?
3. Explain how the air/fuel ratio is controlled by controlling the injectors' pulse width.
4. How is the air/fuel ratio controlled in an MPC fuel injection system?
5. How can the results of a power balance test aid in the diagnosis of a fuel injection system?
6. How is the fuel controlled in a CIS system?
7. What aids in the atomization of fuel as it is released from an injector in the CIS fuel injection system?
8. What is monitored in an AFC fuel injection system to control the air/fuel ratio?
9. How is the fuel pressure regulated in GM's throttle body injection system?
10. GM's TBI system can operate in two modes. What are these modes, and when does the system operate in each?

11. What should be done first if a problem is suspected in the TBI system?

12. What is the purpose of the air vane meter assembly in Ford's Multi-Point Injection system?

13. What can result from a leaking injector nozzle?

14. What is the purpose of the inertia switch installed on some cars equipped with electronic fuel injection?

18

AUTOMOTIVE DIESEL DIAGNOSTICS AND SERVICE

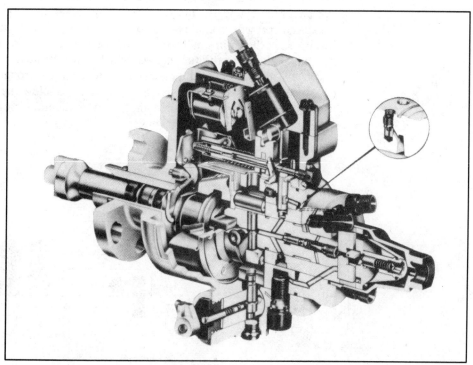

Figure 18-1 (Courtesy of Pontiac Motor Division, General Motors Corporation)

The number of automobiles that have a diesel engine as an option has grown to a substantial amount recently. This growth is a consequence of the concern for improved fuel economy. Any model of car will use less fuel per mile with a diesel engine than it would with a gasoline engine. A diesel engine is more efficient than a gasoline engine because more of the heat produced during combustion is used to perform work. There is less heat lost through the exhaust. Because a diesel engine has high compression, the air in the cylinder reaches a high temperature and pressure as the volume of the cylinder rapidly decreases during the compression stroke. Similarly, the volume rapidly increases during the power stroke, allowing the gases that result from combustion to expand quickly. This rapid expansion of the gases causes a rapid decrease in the temperature of the gases. By the time the exhaust stroke begins, the gases are quite cool. Most of the heat produced during combustion is used to push the piston down, and little is left to exit the exhaust. A cool exhaust is evidence of this fact.

ADVANTAGES OF DIESEL ENGINES

Together with better fuel economy, a diesel engine offers other advantages over a gasoline engine. Because the diesel engine uses compressed air for ignition, there are no spark plugs or other ignition parts that need periodic attention. This reduces the cost of maintenance; however, the application of electronic and computerized ignition systems on gasoline engines has lengthened the time intervals for ignition service; therefore, the advantage of the diesel is only slight. This advantage is reduced even more by the fact that diesel engines must have their oil changed more frequently. Although maintenance costs are only a small advantage of a diesel engine, this, coupled with good fuel economy, keeps the operating costs of a diesel engine below that of a gasoline engine.

Not only is the exhaust of a diesel engine cool, but it also contains less pollutants. Low amounts of CO and HC in the exhaust are the result of an abundance of air in the cylinders, and the near complete combustion that occurs. The major pollutant present in the exhaust of a diesel engine is solid matter. This pollutant leaves the exhaust in small particles of the solid called *particulate matter*. The cause of this pollutant is the fuel it uses, which is less refined than gasoline (Fig. 18–2).

The construction of a diesel engine tends to be stronger than that of a gasoline engine. This heavier construction is necessary to allow the engine to withstand the high pressures formed during compression and combustion. The construction of a diesel engine allows it to last longer than the typical gasoline engine.

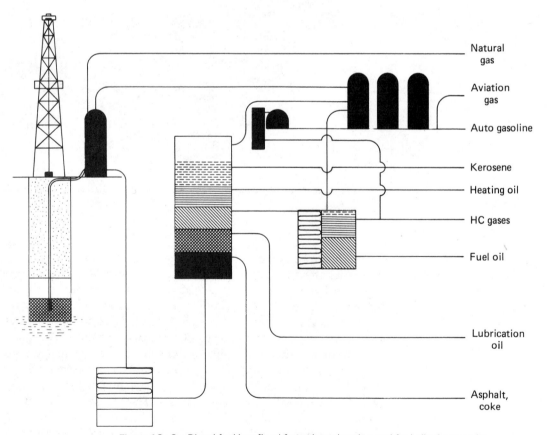

Figure 18–2   Diesel fuel is refined from those heating and fuel oils that require less refining than automotive or aviation gasoline.

DISADVANTAGES OF DIESEL ENGINES

Together with the advantages of the diesel engine come some disadvantages. Many find the noise of the engine objectionable. The noise of a diesel is the result of the normal combustion process for the engine, and is considerably louder than a gasoline engine.

Because the engines are built stronger, they weigh more. Vehicles must be equipped with stiffer springs to support this added weight of the diesel engine. The result is a harsher ride than would be experienced with a gasoline engine. Manufacturers of many diesel-powered automobiles have worked with the spring rates and achieved an excellent ride, in spite of the increased weight of the engines.

Although diesel engines are more fuel efficient than a gasoline engine, the maximum power produced by a diesel engine is much less than the amount produced by a gasoline engine of the same size. Given identical displacements, a diesel engine will be able to produce about three-fourths the power produced by a gasoline engine. The result is slower acceleration and decreased driveability compared to a gasoline engine of the same size.

The decision of whether an automobile should be purchased with a diesel engine or a gasoline engine should be based on the values of each of the advantages and disadvantages. Although only a small percentage of the automobiles on the road today are equipped with a diesel engine, they are out there in significant numbers. A technician should be able to diagnose and repair any problems that might occur with a diesel engine. For this reason, the basic operation, diagnostics, and service of diesel engines is included in this text.

DIESEL ENGINE FUNDAMENTALS

All the diagnostics and testing procedures discussed so far were on engines that use electricity at a spark plug for ignition. A diesel is a two- or four-stroke internal combustion engine that does not use electricity for ignition. Rather, the heat of compression ignites the fuel (Fig. 18–3). As air is compressed, not only does the pressure increase, but so does the temperature of the air. In a diesel engine, on the compression stroke, the air is compressed much more than in a spark-ignited gasoline engine. When the compression stroke is nearly complete, fuel is squirted under high pressure into the compressed air. The extreme heat that was produced by the high pressure causes the fuel to ignite. The heat of compression causes the air-and-fuel mixture to ignite instead of the heat generated by a spark plug. Because of this, diesel engines are not equipped with an ignition system.

Diagnostics of a diesel engine are concerned primarily with the delivery of air and fuel. The heat needed for ignition is the result of the sealed container. Compression must be good in order to cause ignition. The timing of ignition is determined by the time of fuel delivery to the cylinders. The fuel is delivered to the cylinders by a mechanical fuel injection system. Like the ignition of a gasoline engine, the timing of the injection of fuel should allow for combustion to occur when the piston is near TDC, and be completed shortly after the piston begins moving toward BDC. The timing of the fuel injection is critical to the overall efficiency of the engine. To have complete combustion in a diesel engine, the correct amount of fuel must be delivered, at the correct time, into the correct amount of heat and air in a sealed container. All diesel engine diagnostics should be based on this statement!

Because heat is necessary for combustion, diesel engines are difficult to start when the air is cold. To assist the engine in starting during cold weather, diesels are

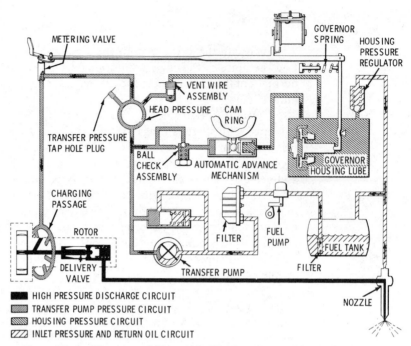

Figure 18-3   (Courtesy of Oldsmobile Division, General Motors Corporation)

equipped with small electrical heaters that are installed in the intake air port or combustion chamber. These heaters are called *glow plugs* (Fig. 18-4). Their purpose is to glow and warm the air in the cylinders, to allow for easier starting, and to plug a hole in the combustion chamber. These plugs also warm the fuel that is injected into the cylinder during starting, to help vaporize the fuel. Once the engine starts, the need for the glow plugs no longer exists, and they are turned off. These plugs do not serve the same purpose as spark plugs. The glow plugs do not cause ignition; they merely warm the fuel and air in the cylinder, to allow for easier ignition when the engine and outside air is cold.

The glow plugs, and their connecting circuits, are the electrical system of a diesel engine. Because they operate only during engine starting and warmup, malfunctions in this system affect the engine's performance only during those operating conditions. Two other systems of the diesel engine have a drastic effect on the operation of the engine: the fuel system, which delivers the fuel at the correct time, and the air intake system, which provides the correct amount of air. The engine is also equipped with a cooling and lubricating system, which operates in much the same way as that of a gasoline engine.

Figure 18-4

DIESEL FUEL INJECTION SYSTEM

The fuel for the fuel injection system is supplied by a mechanical and/or electrical fuel pump. This fuel supply is directed through fuel filters to the injection pump. The injection pump pressurizes, meters, and distributes the fuel to the individual cylinders. At each cylinder is a fuel injection nozzle or injector which has its tip extended into the combustion chamber of the cylinder. The injection pump delivers high-pressure fuel to the injector, causing it to open and spray fuel into the chamber. The injector contains a spring-loaded check valve that remains closed until the high-pressure fuel is delivered to the injector. The high pressure opens the valve and allows the fuel to be injected into the cylinder through small holes at the tip of the injector (Fig. 18–5). These holes are precisely located to allow fuel to be sprayed at the ideal points within the combustion chamber. When the pressure of the fuel decreases at the injector, the check valve closes and the fuel is no longer delivered to the cylinder. The injector pump is responsible for delivery of the right amount of fuel to the cylinders at the correct time. As engine speed increases, the injection timing is advanced with the increase of speed. This occurs for the same reason that ignition timing is advanced in gasoline engines: to allow combustion to be complete when the piston is 23 degrees after TDC.

The air/fuel ratio of a diesel engine varies greatly, according to engine speed. At idle speeds the ratio could be as lean as 100:1 or 20:1 for full throttle speeds. This extremely lean mixture results in the high efficiency of diesel engines. The burning of this lean a mixture is possible only in an engine with very high compression. The high compression of a diesel engine increases the temperature and pressure of the air present in the cylinder to a point that is greater than the fuel's ignition point. As soon as the fuel is introduced to the hot air, ignition begins. The air has been heated and pressurized by the combustion and compression processes, but it has not mixed with the fuel. The fuel does not explode when introduced into the air, but rather is highly heated and expands due to this heat (Fig. 18–6). The resultant high pressure is what pushes the piston down on the power stroke. Fuel will expand a certain amount, regardless of the space it is contained in. This expansion of the fuel takes up space in the combustion chamber, which forces the air into a smaller area and increases its pressure.

Air is delivered to the individual cylinders through an air cleaner assembly, which filters out the dirt and dust. An intake manifold then distributes the air to the cylinders. Diesel engines do not use a throttle body to govern intake air; outside air is not restricted in route to the cylinders.

Without a throttle body or a control for air intake, engine speeds are not governed by air. Rather, the injector pump (Fig. 18–7) is equipped with a governor to control engine speed. The governor meters the amount of fuel necessary to maintain a particular engine speed. The throttle or "gas pedal" of the car controls the governor, which controls engine speed. The pressure on the throttle determines the desired engine speed. To allow for an increase or decrease in engine speed, the gov-

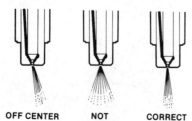

OFF CENTER   NOT UNIFORM   CORRECT

Figure 18–5 Nozzle atomizing patterns. (Courtesy of Ford Motor Company)

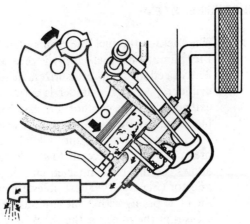

Figure 18-6 (Courtesy of Oldsmobile Division, General Motors Corporation)

ernor supplies the injectors with more or less fuel. The regulation of fuel delivery by the governor provides for the operation of the engine at a variety of speeds. If the engine speed decreases with an increase in pressure on the gas pedal, as it does under heavy-load conditions, the governor allows more fuel to be sent to the injectors. The opposite occurs when the engine speed increases and the pressure on the gas pedal does not. The amount of fuel delivered is decreased, and the engine will tend to maintain the desired speed.

The governor uses weights that rotate inside the injector pump assembly (Fig. 18-8). The injector pump is driven by the engine's camshaft and its rotor rotates at camshaft speed. The weights of the governor rotate with this rotor and pivot outwardly as the speed increases. The weights, while pivoting, work against a spring pressure and control the fuel metering assembly of the injector pump. The spring pressure is controlled by the position of the gas pedal. Therefore, engine speed and throttle position determine the amount of fuel delivered to the injectors and the

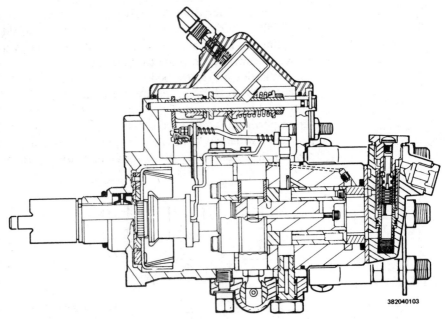

Figure 18-7 (Courtesy of Pontiac Motor Division, General Motors Corporation)

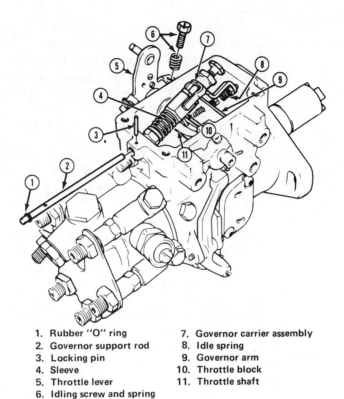

1. Rubber "O" ring
2. Governor support rod
3. Locking pin
4. Sleeve
5. Throttle lever
6. Idling screw and spring
7. Governor carrier assembly
8. Idle spring
9. Governor arm
10. Throttle block
11. Throttle shaft

Figure 18-8 (Courtesy of Pontiac Motor Division, General Motors Corporation)

engine's speed. At about 4000 engine rpm, the governor allows for a 100% fuel flow and will limit maximum engine speed to about 4500 rpm. At low speeds, the governor allows for little fuel flow.

The injection pump is normally mounted at the top of the engine. It is a rotary distributor type pump that injects a metered amount of fuel to the cylinders at the proper time. The fuel lines that connect the pump to the individual injectors are all the same length, to prevent any difference in timing between cylinders. These lines are special high-pressure lines capable of holding extremely high pressures. The injection pump delivers the fuel to the injectors at nearly 2000 psi. Through the action of the injectors, the fuel pressure is increased to as much as 4000 psi. This high delivery pressure is necessary for the fuel to enter the combustion chamber when the injector opens. If the fuel was delivered at a pressure lower than the pressure present in the combustion chamber, the fuel would not enter and combustion could take place in the injector. The fuel must be highly pressurized in order for the fuel to enter the cylinder. To keep the compressed air in the cylinder from entering through the injector when fuel is not being injected, a check valve is used. This valve seals the injector and prevents combustion leaks. The pressure of the fuel must be greater than the pressure of the spring in the check valve before the injector is allowed to open. When open, the injector sprays a fine mist of the fuel into the cylinder, and combustion begins.

To start a diesel engine, a starter motor and an ignition switch are used. Actually, the ignition switch is not for the ignition, but rather, for the fuel injector pump. In the "on" position, the switch connects the battery to a solenoid mounted in the injector pump. This solenoid activates the fuel metering system and allows fuel to be delivered to the cylinders. To cause the delivery of the fuel to the cylinders, the rotor of the pump must rotate. The rotor rotates with the camshaft of the engine; therefore, the engine must be cranked to supply fuel to the cylinders.

Because of the high compression of a diesel engine, the current needed by the starter to crank the engine is much greater than that of a gasoline engine. Most diesel-powered vehicles are equipped with one or two heavy-duty batteries. The normal current draw from the starter will drain the battery and not allow for the operation of the injector pump solenoid or the glow plugs. To increase the chances of starting the engine, diesels are equipped with batteries that have the ability to meet the demands of the starter, and have power left over for the operation of the fuel solenoid and the glow plugs.

The glow plugs heat the incoming air and fuel in the cylinders. This allows for the starting of the engine when outside temperatures are low. These plugs also serve as access holes to the cylinders (Fig. 18-9). To test individual cylinders, a technician will connect the test equipment to the holes. An example of this would be in performing a compression test.

Diagnosis of diesel engine problems should begin in much the same way as for a gasoline engine: with knowledge of the engine and of the problem. Before troubleshooting can be discussed, the similarities and differences of automotive diesel engines by various manufacturers should be discussed. To represent the diesels used in the automobile industry, the Volkswagen, Nissan, and General Motors diesel engine will be discussed. Although many other manufacturers offer a diesel engine option in their automobiles, the systems used in the three selected for this discussion have many similarities to the systems used by others.

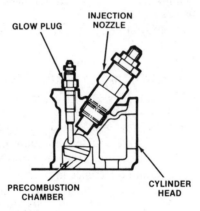

Figure 18-9   (Courtesy of Ford Motor Company)

VOLKSWAGEN DIESELS

The system used by Volkswagen consists of a distributor-type injection pump, glow plugs, a governor, and throttling pintle injection nozzles. The fuel is drawn from the fuel tank to the injector pump through a filter by a fuel pump built into the injector pump. The injector pump then supplies fuel under high pressure to the injection nozzles according to the engine's firing order. The low-pressure fuel pump delivers fuel to the injection pump assembly, which increases the pressure of the fuel to approximately 2000 psi. The high-pressure fuel is distributed to the cylinders through a line connecting the injector pump and the injection nozzles.

Volkswagen diesel engines use Bosch injection nozzles. They are pintle-type injectors (Fig. 18-10). When the high-pressure fuel is delivered to the injectors, the injector opens and sprays a pressurized mist of fuel into a round swirl chamber separated from the main combustion chamber. The fuel swirls around this chamber and mixes with hot air that is compressed to a ratio of 23:1. Combustion begins for this rich mixture in the swirl chamber, and continues through a small connecting passageway to the main combustion chamber. The purpose of the swirl chamber is

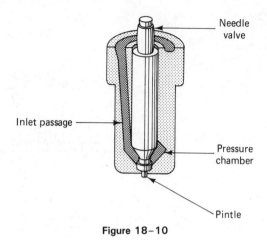

Needle valve

Inlet passage

Pressure chamber

Pintle

Figure 18-10

to reduce the force exerted on the piston top when maximum pressure is reached during combustion. This maximum pressure occurs in the swirl chamber to receive the shock resulting from the expansion of the fuel.

The glow plugs are installed with their tips extended into the swirl chambers (Fig. 18-11). During cold starts, the glow plugs are activated and preheat the swirl chamber. When the desired temperature has been reached in this chamber, the engine is ready to be started, and the glow plugs turn off shortly after the engine starts. The amount of time that the glow plugs are on is governed by a temperature sensor, connected to a timer circuit in the glow plug relay. Once the glow plugs reach a particular temperature, they will remain energized for approximately 2 minutes. This time should be enough to allow the engine to start and run while cold. After the engine is started and warmed, there is no reason for the glow plugs to remain on.

The amount of time that the injector pump (Fig. 18-12) delivers high-pressure fuel to the nozzles determines the amount of fuel injected into the cylinders. This time is controlled by the position of a metering sleeve located around the distributor plunger. Normally, the sleeve covers a relief port in the plunger, and when this port is uncovered, fuel is no longer delivered to the cylinders. The position of the sleeve is controlled by the gas pedal and the governor. The governor rotates with the in-

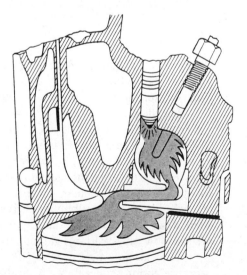

Figure 18-11   (Courtesy of General Motors Corporation)

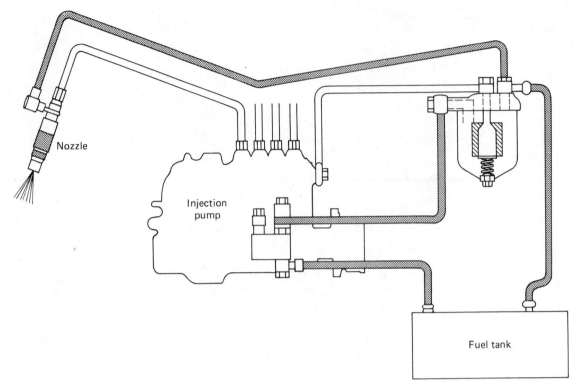

Nozzle

Injection
pump

Fuel tank

Figure 18-12 Bosch injector circuit.

jection pump rotor, and its weights, moved by centrifugal force, adjust the position of the sleeves. The engine's speed will increase only if the pressure applied to the gas pedal is increased. If the engine speed is great and there is little pressure on the gas pedal, such as during deceleration, the governor will restrict the amount of time that fuel is delivered to the injector nozzles and will allow the engine to slow down. The rotor of the injection pump is driven by the engine through a belt connected to the crankshaft. The rotor, however, rotates at camshaft speed through the use of a larger pulley. This arrangement allows for the ignition of each cylinder, one time for every two turns of the crankshaft.

NISSAN DIESELS

Nissan diesel engines also have the injector pump driven by a belt at camshaft speeds. Many of the features of the Nissan system are similar to that used by Volkswagen, with the exception of two items: the glow plug circuit and the fuel injection nozzles.

The glow plug system, as shown in Fig. 18-13, uses a glow plug control unit, two relays, and a current flow resistor to control the current draw of the glow plugs. Current is applied by the circuit to the glow plugs, for an amount of time determined by a temperature sensor, which measures engine coolant temperature. The glow plugs remain on after the engine is started, for the amount of time needed by the engine to reach a particular temperature. When the engine is sufficiently warm, the control unit stops the supply of current to the plugs. A timer circuit is not used to control the time; however, the engine's temperature usually allows for the shutting off the glow plugs within 1 minute after starting.

The fuel injection nozzles (Fig. 18-14) used by Nissan are different in appearance and in serviceability from those used by Volkswagen. Through the use of

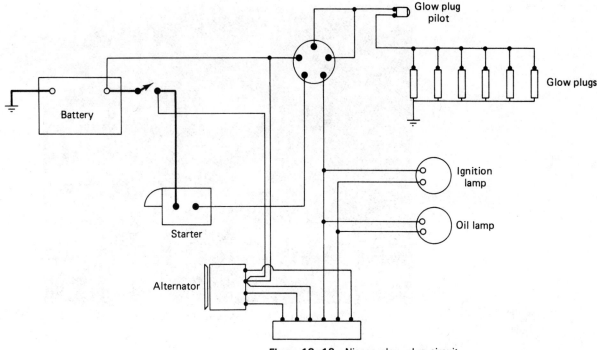

Figure 18-13 Nissan glow plug circuit.

ADJUSTING SHIM

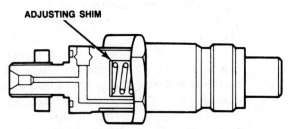

Figure 18-14 (Courtesy of Ford Motor Company)

shims, the spray pattern of the nozzles can be adjusted, as can the opening pressures. Because the adjustment and measurement of the nozzles' opening pressures require a special tester, the techniques for adjustment will not be given in this text. It is important to know that the nozzles used by Nissan are adjustable. Incorrect opening pressures can cause poor running conditions. These nozzles do not have to be replaced if they are out of specs. Only shops that do much diesel work will be equipped with the special tester. If your shop is so equipped, follow the instructions of the tester's manufacturer and of Nissan, to conduct this service.

GENERAL MOTORS DIESELS

The most common American automotive diesel engines are the products of General Motors. These engines use an injector pump that is mounted at the top of the engine and is driven directly by the camshaft (Fig. 18-15). This pump has a built-in pressure regulator and delivery (transfer) fuel pump. These supply the main injector pump with a constant supply of fuel. The fuel passes into the injector pump at a low pressure, until it exits the metering assembly. At this time, the pressure of the fuel is increased and distributed in the injector nozzles. The operation of the injector

Figure 18-15   GM 350 diesel engine.

pump and its governor are similar to the other systems. The major difference between GM diesels and the others lies in the method of regulating the glow plugs (Fig. 18-16).

Typically, a GM diesel uses 6-V glow plugs which are energized by 12 V; this causes them to heat up rapidly. This rapid heating decreases the wait that is necessary on some engines to start a cold diesel engine. Some GM diesel engines were equipped with 12-V glow plugs and required longer warm-up periods. It is important for a technician to be able to identify the voltage rating of the glow plugs prior to testing them. Shop manuals will give the details needed for the proper identification and testing of the types of glow plugs.

DIESEL DIAGNOSTICS

Although the diagnostics of a diesel engine should begin in the same way as diagnostics for a gasoline engine, many of the tests that can be conducted on a gasoline engine cannot be conducted on a diesel. Without an ignition system, there is no need for a scope. Because the intake air enters and fills the cylinders without a restriction, like a throttle plate, vacuum readings will be low and worthless for diagnostics. The key to effective diagnosis of a problem with a diesel engine is the understanding of the engine and the problem. What follows are some of the typical problems that occur with a diesel engine and the most probable causes of them.

Problem: The engine cranks normally with the starter but will not start.
Possible causes: The glow plug circuit, or the glow plugs themselves, could be inoperative; the fuel supply to the injector pump could be blocked off by a plugged fuel filter, a malfunctioning fuel delivery pump, or clogged or damaged fuel delivery lines; the fuel solenoid on the injector pump may not be receiving voltage, which prevents fuel from being delivered to the injectors; the injection pump may be out of time; or the engine may have low compression.

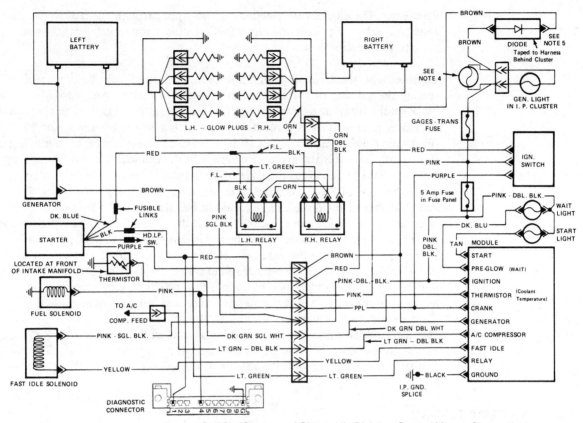

Figure 18-16 (Courtesy of Oldsmobile Division, General Motors Corporation)

Problem: The engine starts but does not continue to run at idle.

Possible causes: The idle speed may be adjusted too low, the injection pump timing may be out of specs, the glow plug control circuit may be allowing the plugs to turn off too soon, there may be a restriction in the fuel lines to the injection pump or a clogged fuel filter, the engine may have low compression, or the vehicle may have the wrong type or contaminated fuel. (Water in the fuel is a common problem; most automobiles are equipped with a means to drain off the water from the fuel filter.)

Problem: The engine starts but has a rough idle.

Possible causes: The injection timing may be retarded, there may be air in the fuel system, the pressure from the injector pump or delivery pump may be out of specs, or the injection nozzles are malfunctioning.

Problem: The engine idles fine but misfires at other speeds.

Possible causes: The fuel supply could have a restriction such as plugged fuel filter, the injector pump timing could be off, or the fuel is contaminated or is of the wrong type.

Problem: The engine runs fine but will not return to idle.

Possible causes: The injector pump is allowing too much fuel to be delivered to the injectors; this can be caused by a malfunctioning injection pump or by the throttle linkage binding or sticking.

Problem: There is a general lack of power from the engine.

Possible causes: The injector timing could be out of specs, there can be a fuel restriction, the exhaust system can be collapsed or clogged, the fuel could be contaminated, the engine could have low compression, or the intake air supply could be restricted, possibly caused by a plugged air filter.

Problem: The engine runs fine but will not shut off with the ignition key.
Possible cause: The injection pump fuel solenoid is not shutting off the fuel supply
to the injector pump.

Together with these running condition problems, diesels can be diagnosed by observing the exhaust of the engine and the noise it makes. An unusual amount of
black smoke from the exhaust typically indicates that the engine is running too rich.
This over-richness can be caused by a restriction in the air supply or by a malfunctioning fuel injection pump. If the exhaust smoke is noticeably white in color,
this usually indicates the presence of water in the cylinders. Water in the cylinders
can be caused by a bad head gasket or by a cracked cylinder head or block.

The noise of a diesel is quite objectionable to most people; however, the degree
and character of the noise can be an aid in the diagnostics of the engine. Incorrect
injection pump timing can cause the noise of the engine to be extra noticeable. This
extra loud noise can also be caused by the injection pump delivering too high pressure to the injectors. In this case the injection pump should be replaced. If the noise
of the engine consists of the normal sound with a regularly occurring "rap" or loud
knock, this usually indicates that an injection nozzle is stuck open or has too low
an opening pressure. Air in the high-pressure fuel lines will also cause this "rap."
Care should be taken when listening to the noise of a diesel engine; often, mechanical problems and the associated noises caused by those problems are hidden
by the normal noise of a diesel. To use noise as a diagnostic tool on diesel engines,
you should be familiar with the sounds of a normally operating diesel engine.

DIESEL SYSTEMS TESTING

Throughout the discussion of probable causes of problems, four components were
mentioned often. The glow plug operation affected not only engine starting but also
continued running of the engine when it was cold. Because air can be compressed,
its presence in the fuel system affects the delivery of the fuel to the individual injectors. The presence of air in the fuel system will cause rough-running conditions.
Poor compression will cause starting and running problems. Improper injection timing will affect overall engine performance. Because of the variety of methods used
by the manufacturers to control glow plug operation, the exact procedure for testing
the system will not be discussed. These procedures are available in the shop manuals.
However, certain points of the system can be tested outside the normal suggested
sequence of the manufacturers. Since all glow plugs are electrical heaters, they can
all be tested in the same way as any electrical load. A normal glow plug will have
some resistance that can be measured with an ohmmeter. Most shop manuals will
give you the exact specs for this resistance.

If the red lead of an ohmmeter is connected to the terminal at the glow plug
and the black lead connected to ground, a normal glow plug will cause the meter
to show some resistance (Fig. 18–17). If the plug is burned out, the meter will show
an infinite reading. This will indicate that the plug is open and needs to be replaced.
A reading of zero ohms resistance indicates that the glow plug is shorted and should
be replaced. This simple means of testing a glow plug does not determine if the plug
is operating within the specs; however, it can be used to identify a bad glow plug.

To test the glow plug circuit, a voltmeter should be used. With the ignition
key "on," there should be voltage present at the glow plugs. The amount of regulated voltage that should be present will depend on the application. Check your
service manual for the specs of the engine you are testing. If there is no voltage
there, you have discovered the cause for the engine not starting. To find the exact
cause, the control circuit and its individual components should be tested. If the

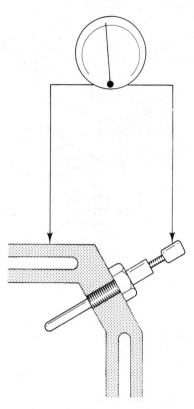

Figure 18–17 Testing a glow plug with an ohm-meter.

engine starts but does not continue to run, the problem may be the control circuit. Using the voltmeter, a reading of battery voltage should remain at the plugs for a short time after the engine is started. The length of time will vary between manufacturers, but if the voltage is not present when the engine first starts, you have identified the area that needs to be tested in detail, the control circuit.

Air in the fuel system can cause a rough-running condition. This condition can be discovered by the installation of a clear plastic tube in the fuel return line from the injection nozzles (Fig. 18–18). If there is the presence of bubbles in the flow of fuel in this tube, there is air in the system. To remove the air, the fuel system should be bled.

The bleeding of the system simply involves loosening the fuel lines at the injectors (Fig. 18–19) one at a time. With the engine running, carefully loosen the line nut to allow some fuel to escape. If air is in that line, it will be pushed out with the fuel and will be noticeable. As soon as all the air is out of the line, the nut should

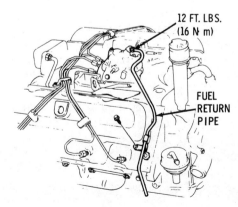

12 FT. LBS. (16 N·m)

FUEL RETURN PIPE

Figure 18–18 To bleed the fuel system, a clear tube should be installed between the outlet at the injector pump and the fuel return line. (Courtesy of Pontiac Motor Division, General Motors Corporation)

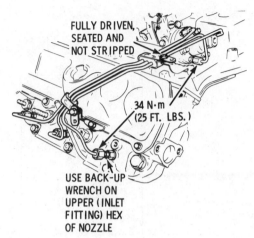

FULLY DRIVEN,
SEATED AND
NOT STRIPPED

34 N·m
(25 FT. LBS.)

USE BACK-UP
WRENCH ON
UPPER (INLET
FITTING) HEX
OF NOZZLE

Figure 18-19 After bleeding the fuel system, each injector line should be tightened to the torque spec given. (Courtesy of Pontiac Motor Division, General Motors Corporation)

be retightened to the manufacturer's specs. When only fuel is escaping from the leak caused by the loosened nut, it can be assumed that there is no air left in the line. This procedure should be repeated on each injector's line. Upon completion, the system should be free of air and all the lines should be rechecked for proper tightness.

To conduct a compression test on a diesel engine, the basic procedures for doing the test on a gasoline engine should be followed. Instead of connecting the gauge to the spark plug hole, the glow plug will be removed and the gauge installed in its hole. To disable the engine to prevent it from starting, the wire lead to the fuel injection pump solenoid should be disconnected. This will prevent any fuel from being delivered to the injectors when the engine is being cranked for the test of its compression. Because of the high compression ratios of diesel engines, you expect a reading on the compression gauge of about 300 psi. Some diesel engines will have a reading much higher than this.

Volkswagen diesels and other diesels require that you connect the compression gauge to the hole for the fuel injection nozzle. To do so, all the injectors are to be carefully removed, which requires the removal of the fuel lines that connect the injector to the injection pump. After the test is completed, the nozzles should be reinstalled and their hold-down clamps and fuel lines tightened to specs. It may also be necessary to bleed the fuel system after the engine has warmed up.

The timing of the injection pump determines the time that combustion will begin in the cylinders. This time is as critical to a diesel engine as ignition timing is to a gasoline engine. Improper injection timing can affect engine performance in many ways. There are various ways to check the injection timing of a diesel engine. Each manufacturer of diesel engines has its own procedure for the checking of timing. Before attempting to check or adjust the timing of the fuel injector pump, consult your shop manual. Included in the following discussion are the general instructions for checking and adjusting the timing on the diesel engines used in this text. Although these directions are accurate, there are variables among manufacturers, and recommended procedures should be followed for each individual engine.

The injection pump of General Motors engines has a timing mark stamped into its mounting flange and at the top of the injection pump adapter (Fig. 18-20). The two sets of lines stamped in these two places should be aligned when the injection timing is correct. To correct the timing of the pump, loosen the three retaining nuts at the flange of the pump and rotate the pump until the lines are

Figure 18-20

positioned properly. These nuts should be retightened to the recommended specs and tightened one at a time, in steps, to prevent damage to the pump's flange.

Nissan's injection pump can be quickly checked for correct timing by comparing the reference marks on the injector pump and the front end plate (Fig. 18-21). If the marks do not line up properly, the hold-down bolts should be loosened and the pump moved to align the marks. The shop manual will also include a pro-

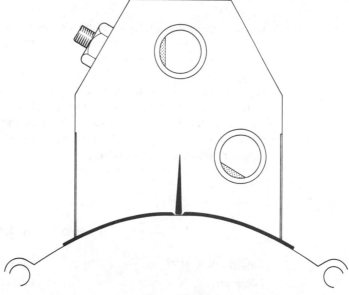

Figure 18-21

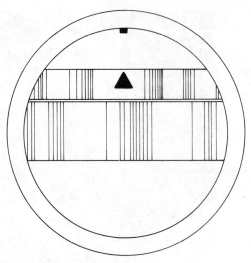

Figure 18–22

cedure for verifying the timing of the pump. If you have reason to doubt the accuracy of the marks on the pump or front end plate, follow the verification procedure given in the manual.

To set the injection timing of the Volkswagen diesel, the number 1 piston must be brought to TDC. The position of the piston is verified by the timing marks on the engine's flywheel and the bell housing (Fig. 18-22). When the piston is at TDC, the timing plug on the pump cover should be removed. This plug is provided for the checking of injection timing and expose the marks on the pump mounting plate, which should line up with the mark on the pump itself. If the timing is found to be off, the procedures, outlined in the shop manual, should be followed carefully.

SUMMARY

Diesel engines are internal combustion engines that require the correct amount of fuel to be delivered at the correct time into the correct amount of heat and air in a sealed container to cause combustion. Diagnostics of a diesel engine can be easy if you base your testing and decisions on these requirements. Detailed testing of systems and components are best done by following the manufacturer's recommendations, but often a good understanding of the engine will give you a safe and logical approach to identifying the cause of the problem. An understanding of the problem and the engine will allow you to diagnose any engine problem, diesel or gasoline, quickly and effectively.

TERMS TO KNOW

Particulate matter   Injection pump
Glow plug   Governor
Injection nozzle   Diesel timing

REVIEW QUESTIONS

1. What are the advantages of a diesel engine over a gasoline engine?
2. The primary concern during diagnostics of a diesel engine is what system?
3. What is the purpose of the glow plugs in a diesel engine?
4. How is engine speed regulated in a diesel engine?
5. Why is it necessary for diesel fuel to be delivered at high pressure?
6. What is the purpose of the swirl chamber in a VW diesel engine?
7. When are the glow plugs of a diesel engine operating?
8. Compare the control of glow plug operation in the VW, Nissan, and GM diesel engines.
9. What should be checked first if a diesel engine will not start?
10. What could be the cause of a rough-running diesel engine?
11. What could be the cause of a diesel engine that has a lack of power?
12. How can an observation of smoke help in the diagnostics of a diesel engine?
13. Why should you be familiar with the sounds of a normally operating diesel engine?
14. How can the glow plug circuit be tested?
15. When is it necessary to bleed an injector pump?
16. What is the proper procedure for bleeding an injector pump?
17. What is the proper procedure for testing the compression on a VW diesel engine?
18. What is the proper procedure for testing the compression on a GM diesel engine?
19. What is the proper procedure for testing the compression on a Nissan diesel engine?
20. How do you adjust injection timing on a diesel engine?

GLOSSARY

Accelerator pump: A mechanism that squirts fuel into the air as the throttle plates open.

Advanced timing: Ignition timing that occurs before top dead center

Air bleed: A small hole in the carburetor which allows air from the atmosphere to enter and mix with the fuel in the idle discharge passageways as an aid in the atomization of the fuel.

Air breather: Typically placed in the air cleaner assembly or in a valve cover to allow clean air into the crankcase for the PCV system.

Airflow meter: An assembly which is mounted between the air cleaner and the throttle body on fuel-injected engines, to measure the total mass of air that is entering the engine.

Airflow sensor: Located above the throttle plates, this sensor measures the volume of the air present and controls the fuel distributor which regulates the amount of fuel according to that amount of air on CIS systems.

Air pump: The main component of the air injection system, which supplies air to the exhaust port of each cylinder. This pump is driven by the engine through a belt.

Air vane meter assembly: Mounted between the air cleaner and throttle body on Multi-Point Fuel Injection systems, to measure the amount and temperature of the air that enters the engine so that the computer can measure the mass of the air entering the intake system.

Alternator: The main component of the car's charging system, composed of a rotating magnetic field which passes through a number of conductors and induces a voltage which is used to recharge the battery.

Ammeter: A meter used to measure electrical current.

Ampere: The unit of measurement for electrical current; one ampere is equal to 6.25 billion billion electrons passing through a point in one second.

Amperehour rate: A rating used by battery manufacturers which states the capability of a battery to maintain a particular current flow for a period of time.

Analog meter: A conventional-type meter used to measure a number of properties by displaying the amount measured through the movement of a needle across a fixed scale.

Antiknock value: Term referring to the ability of a gasoline to resist detonation, commonly referred to as a gasoline's octane rating.

Auxiliary air valve: Used during cold starts to allow more air to enter into the intake manifold on MPC Fuel Injection systems.

Ballast resistor: Used to ensure that a correct amount of current flows through the primary ignition circuit during all speeds and conditions.

Barometric pressure sensor: Used as an input for some computerized engine control systems and measures the barometric pressure of the atmosphere which is needed to maintain an ideal air/fuel ratio.

Bound electrons: Electrons in the valence ring of an atom which cannot be easily transferred to another atom; this is usually the case when there are five or more electrons in the valence ring.

Breaker points: Part of the primary ignition circuit and is the triggering device used in nonelectronic ignitions to open and close primary current flow.

Breakerless ignitions: Ignition systems that do not use breaker points to allow and interrupt primary current flow through the ignition coil.

Camshaft: Driven by the crankshaft at one-half of crankshaft speed, it has the purpose of controlling the opening and closing of the engine's valves.

Cannister purge, fuel vapor: During a time determined by the computer, the fuel vapors stored in the carbon cannister are pulled out of the cannister and are added to the charge of air/fuel mixture which will be burned by the engine.

Carburetor: A metering device used to mix air and fuel in the proper ratios to accommodate the engine's needs under a variety of conditions.

Catalytic converter: Used to convert poisonous gases in the exhaust of an engine to harmless gases; provides for a secondary combustion process in the exhaust by providing the presence of a catalyst which causes oxidation to occur, thereby reducing the amount of pollutants in the exhaust.

Cetane: A term used to rate the ease with which a diesel fuel will ignite.

Centrifugal advance: That mechanism of the distributor assembly which causes ignition timing to advance according to increases in engine speed.

Check engine light: A light used on GM cars equipped with the CCC system to warn the driver of a problem as well as to display to an automotive technician the results of the computer's self-diagnostics.

Choke piston: (*See* Choke pull-off).

Choke pull-off: Used to open the choke when the engine starts allowing more air to enter the engine, typically governed by engine vacuum.

Choke unloader: Provides a way to open the choke if the engine is flooded during starting attempts or if the engine is quickly accelerated during cold operation with the choke normally closed.

Closed loop: A mode of computer operation when the engine and its systems are monitored and controlled by the computer.

Coil reserve: Refers to the amount of energy that remains in the ignition coil after the firing of a spark plug; the coil reserve voltage plus the firing voltage equals the total output of the coil.

Coil section: Follows the spark line on a scope pattern and represents the dissipation of the remaining voltage in the coil after the firing of the spark plug.

Cold cranking amperes: A rating system used by battery manufacturers to state the capability of a battery to maintain a particular current flow for a period of time.

Cold start system: Used in the CIS system to measure the temperature and allows for a separate injector nozzle to release fuel when the engine is cold to improve cold operation and driveability.

Combustion: The process of burning; typically refers to the combination of fuel and air through the introduction of heat in an automotive engine.

Complete circuit: Provides for a path for current in one direction, from the battery through the conductor to the load and then back to the battery.

Compression: A reduction of volume caused by the piston moving toward TDC with the valves closed; this process increases the pressure of the air in the cylinder.

Compression ratio: The volume of the combustion chamber when the piston is at TDC compared to the volume of the combustion chamber when the piston is at BDC.

Condenser: An electrical device used in breaker ignition circuits to allow for the immediate collapse of the primary coil windings when the points open by preventing an arc from occurring across the open points.

Conductor: A material that has free electrons in their valence ring and allows for the flow of electricity.

Converter, dual-stage: A catalytic converter which is composed of two active catalysts, rhodium and platinum, housed separately in one container.

Converter, mini: Mounted immediately after the exhaust manifold and serves the purpose of converting the exhaust during engine warm-up conditions when the normal converter is not warmed up enough to be effective.

Converter, three-way: A catalytic converter that reduces the amounts of NO_x, HC, and CO in the exhaust.

Converter, warm-up: (*See* Converter, mini).

Cross-fire injection system: An injection system used on some GM engines which incorporates two throttle body injection units which individually feed half of the engine's cylinders.

Current: Controlled movement of electrons from atom to atom.

Current draw: The amount of current required to operate a device based on the operating resistance of that device.

Delay valve: A valve used to slow vacuum advance by delaying the vacuum signal to the distributor.

Detonation: The colliding of two flame fronts in the combustion chamber; one flame front is the result of normal ignition, the other results from the heat of compression. Sometimes this occurrence is referred to as preignition or pinging.

Dielectric paper: A paper that is used as an insulator, typically in a condenser and ignition coil.

Dieseling: A condition that results from too high an idle speed, in which the engine continues to run after it is turned off.

Diode: A semiconductor used to prevent current flow in one direction and to allow it in the other, most commonly used to rectify ac into dc in the engine's alternator.

Displacement: The total volume of a cylinder. An engine's size or displacement is determined by finding the displacement of one cylinder and multiplying this by the total number of cylinders.

Display pattern: An optional pattern on the scope which displays the trace of each cylinder, side by side, across the screen, used primarily for the comparison of firing voltages.

Diverter valve: Located in a common passageway from the air pump to the air injection rail, this valve prevents backfiring through the exhaust during deceleration by not allowing the air from the air pump to reach the exhaust ports.

D-jetronic: Same as MPC injection system.

Driveability: The ability of a car to start properly, run smoothly, and have immediate throttle response.

Dwell: The length of time that the points are closed or that there is primary current flow.

EGR valve: A valve typically regulated by engine vacuum which controls the amount of exhaust gas that is recirculated through the engine as part of the combustion process and part of the emission control devices on an engine.

Electrical power: Amount of voltage used to push a certain amount of current through some resistance or load, often referred to as wattage.

Electrical resistance: Any opposition to current flow, measured in ohms.

Electricity: The exchange of electrons and the movement of the electrons among atoms.

Electrolyte: The active solution used in batteries to cause the chemical reactions necessary to form and store electricity.

Electromagnet: A magnet formed by the presence of a current carrying conductor around a soft-iron core.

Electromagnetic switch: A switch that uses the principles of magnetism to open and close a circuit, rather than some mechanical means.

Filament: A fine conductor used in light bulbs which glows as current flows through it.

Firing line: A vertical line in a secondary ignition trace on a scope which represents the amount of voltage that was necessary to overcome the resistance in the secondary, to complete the path for current in that circuit.

Firing order: The order in which the cylinders are on the power stroke and when the spark plugs should fire.

Flashover: A leakage of voltage from the secondary circuit (spark plug wires) to the engine block.

Four strokes: The four movements of the piston and the subsequent events that complete one cycle of an engine; the four strokes are: intake, compression, power, and exhaust.

Free electrons: Those electrons that are present in the valence ring and are easily transferred or moved to another atom; these are usually in rings composed of three or less electrons.

Free-running current draw: The amount of current used by a motor when there is no load applied to it and it is not performing any work.

Fuel delivery volume: The amount of fuel delivered by the fuel pump and lines to the fuel induction system. Measurement of this is part of diagnostics of the fuel system.

Fuel distributor: Regulates the amount of fuel delivered to each cylinder, used in mechanical fuel injection systems.

Gas cap, nonvented: Prevents the escape of gasoline vapors from the gas tank to the atmosphere.

Glow plug: A heater-type plug used to warm the combustion chamber of a diesel engine in an attempt to increase the ease of cold weather starting and of cold driveability.

Governor: A component of a diesel injector pump which meters the amount of fuel necessary to maintain a particular engine speed.

Ground: A metal portion of a car used as a common negative post of the battery. Electrical circuits on the automobile have their circuits completed through this common ground.

Ground electrode: One of the electrodes of a spark plug which extend into the combustion chamber of an engine. This electrode is attached to the metal shell of the spark plug and becomes part of the common ground of the engine once it is installed.

Hall-effect switches: Uses the influence of magnetic fields on a semiconductor to cause the circuit to be complete or be open.

Heat range, spark plug: The heat transfer characteristic of a spark plug, determined by the length and shape of the spark plug's insulator tip.

Idle discharge hole: Allows fuel to be drawn off the float bowl to allow the engine to run during idle conditions when the throttle plates are closed or slightly open.

Ignition coil: Steps up battery voltage to thousands of volts which are necessary to complete the secondary circuit.

Impedance: Operating resistance of a component; the actual amount takes into consideration such factors as temperature, counter EMF, and pressure.

Induced voltage: Voltage that is produced by the passing of a conductor through a magnetic field. This is the principle used in the generation of electricity in an alternator.

Induction system: The system responsible for the delivery of fuel and air to the individual cylinders.

Inductive cross-firing: An occurrence causing the induction of high voltage in a spark plug wire due to the presence of a high-voltage and current-carrying conductor, another spark plug.

Inertia switch: A switch used in some fuel injection systems which shuts off the flow of current to the fuel pump in the event of a collision to prevent fires.

Initialization mode: A mode of operation of a computer after it advances from the open-loop mode and is in the process of preparing the system to advance into the closed-loop mode.

Injector pulse width: The length of time that a fuel injector is open to spray fuel into the engine. This time is controlled by a computer or by fuel pressure.

In-line engine: An arrangement of the cylinders of an engine where all the cylinders are in a straight line.

Inputs: Information received by the computer through the changing of resistances and the opening and closing of circuits.

Insulator: A material composed of atoms that have many bound electrons which tend to prevent the transfer of electricity.

Insulator tip: That part of the insulator of the spark plug which surrounds the center electrode and extends into the combustion chamber. Attention is paid to this part of the spark plug because it will portray the effects of the combustion process in its color and condition.

Internal combustion engine: An engine in which the combustion takes place in its sealed containers, within the engine itself.

Ion: A negatively or positively charged atom.

Kilovolts: One thousand (1000) volts; this measurement is typically used to measure secondary voltages.

Knock sensor: A sensor used to limit detonation through the sensing of inaudible abnormal engine sounds which are sent to the computer, which then makes the necessary adjustments to the ignition timing or to the air/fuel mixture in the attempt to prevent detonation.

Lambda sensor: An oxygen sensor used in the Bosch fuel injection systems.

Lean mixture: A mixture of fuel and air which contains more than 15 parts of air to each 1 part of fuel.

Magnetic field: A field of force which forms around a magnet as the opposite poles of the magnet attract each other.

Magnetism: An alignment of electrons in a metal that causes a concentration of force fields; it is also a form of energy that can cause current flow in a conductor.

Manifold vacuum: That amount of vacuum which is present in the intake manifold below the throttle plates. The amount of this vacuum is dependent on the operation of the engine and the ability of the outside air to enter the engine.

Metering rod: A rod used to increase the size of the hole of the main jet in the main metering system of some carburetors, allowing more fuel to flow through.

Microprocessor: A small computer that processes information.

Monolith: A single structure, as in the use of a monolith structure in a catalytic converter, which is a solid piece of catalyst cut to take on a honeycomb-like structure.

Needle valve: Located at the fuel inlet of the carburetor and moves in and out of the seat, controlling the inlet of the fuel.

Octane rating: (*See* Antiknock value)

Ohms: Unit of measurement for electrical resistance.

Ohm's law: A law that serves as a functional theory of electricity which states that one volt of electrical pressure is needed to push one ampere of electrical current through one ohm of electrical resistance.

Open: A condition of an electrical circuit where the path for electrons is not complete; this condition can be caused by a fault in the system or by the operation of a switch which is used to turn off the circuit.

Open loop: A operating mode of a computer when the computer is not making decisions and the engine is operating under preprogrammed information. This mode usually is the mode of operation during engine warm-up.

Oxidation: A chemical process in which the fuel and air delivered to the engine combine and become harmless compounds during the process of combustion.

Oxygen sensor: A sensor that measures the amount of oxygen in the exhaust and sends information about the oxygen content to the computer.

Palladium: A catalyst used in a catalytic converter for converting HC and CO into water and carbon dioxide.

Particulate matter: A pollutant released by diesel engines as a result of the impurities of diesel fuel.

Platinum: A catalyst used in a catalytic converter for converting HC and CO into water and carbon dioxide.

Port injection: A fuel injection system in which the injectors are placed in the individual cylinder's intake ports.

Power valve: A valve in the power circuit of a carburetor which increases the amount of fuel that enters the incoming air as the load on the engine increases.

Pressure differential: A condition in which two different pressures are exposed to each other, such as the introduction of the high pressure of the atmosphere to the low pressure in the intake manifold.

Pulse air system: A system that uses the pulses of the exhaust to draw in fresh air into the exhaust ports in order to reduce the amounts of HC present in the exhaust. This system replaces the air pump system, which has the same purpose.

Reactance: Causes a temporary resistance to current flow as current is applied to a winding of wire. It is the result of the magnetic field caused by the flow of current opposing the continued flow of current through the winding.

Rhodium: A catalyst used in conjunction with palladium and platinum which reduces the amount of NO_x in the exhaust.

Rich mixture: A air-and-fuel mixture in which there is less than 15 parts of air mixed with 1 part of fuel.

Saturation: A condition that exists within a coil of wire when maximum current is flowing within it.

Semiconductors: A material that neither conducts nor insulates well against electricity. Usually composed of elements that have four electrons in their valence ring.

Sensors: (*See* Inputs)

Series circuit: An electrical circuit in which current has only one path to take, which is continuous, starting at the positive terminal of the battery and ending at the negative side of the battery.

Short: An electrical fault which allows current to take an alternative path to ground, causing the circuit to experience increased current flow due to the decreased resistance.

Solenoid: An electromagnetic switch which uses electrical energy to perform a mechanical task.

Suppression wires: A type of spark plug wire which allows for clearer reception of FM radio waves.

Sweep rate: The speed of the movement of the horizontal trace of a scope.

Tachometer: A meter used to measure engine speed.

Tetraethyl lead: The compound that had been added to automotive gasoline to raise the octane rating of the fuel. This compound will be removed from all fuels because of its pollutant qualities.

Thermo-time switch: A switch that connects the starting system with the cold start valve, allowing for the delivery of current for a short period of time while the engine is being started.

Transistors: A semiconductor that is used as an electrical switch or amplifier. The development of these components has led to expanded use of solid-state components and computers on the automobile.

Vacuum: Any air pressure that is lower than its surrounding atmospheric pressure. The vacuum produces in the engine is the result of the intake stroke, which increases the volume of the cylinder, causing the pressure to decrease.

Vacuum advance: A component used to alter ignition timing according to engine load.

Vacuum amplifier: A device used to allow a weaker vacuum signal to control a much stronger one in order to perform work.

Vacuum leak: A condition in which air enters and occupies a vacuum that was formed previously, therefore reducing the amount of vacuum present.

Valence ring: The outermost shell of electrons around an atom, which can combine with other atoms to share or exchange electrons.

Valve springs: Springs mounted over the valve which have the purpose of closing the valves as soon as the camshaft permits.

Venturi: A streamlined restriction in the bore of a carburetor which causes an increase in air velocity and lowers the pressure of the air passing through it.

Venturi vacuum: The vacuum present directly after the venturi.

Volatility: A term that expresses the ease with which a liquid becomes a vapor.

Voltage: Electrical pressure, measured in volts.

Voltage drop: The amount of voltage given off or changed to heat energy as an electrical load or resistance has current flowing through it.

Wattage: The amount of voltage used to push a certain amount of current through some resistance or load, measured in watts.

ACRONYMS

AC: Alternating Current

ACC: Air Conditioning Clutch Compressor

A/CL BiMet: Air Cleaner Bimetallic Sensor

ACT: Air Charge Temperature Sensor

ACV: Air Control Valve

AFC: Air-Flow-Controlled Fuel Injection

AIR: Air Injector Reactor

AIS: Air Injection System

ALCL: Assembly Line Communication Link

ALDL: Assembly Line Data Link

AMgV: Air Management Valve

AMP: Ampere

Anti BFV: Anti-backfire Valve

ASV: AIR Switching Valve

ATDC: After Top Dead Center

ATS: Air Temperature Sensor

BARO: Barometric Pressure

BDC: Bottom Dead Center

BHS: Bimetal Heat Sensor

BMAP: Barometric and Manifold Absolute Pressure Sensor

BOB: Breakout Box

BP: Barometric Pressure Sensor

B/P: Integral Back Pressure EGR Valve

BPEGR: Back Pressure EGR

BPV: Bypass Valve

BSV: Backfire Suppressor Valve

BTDC: Before Top Dead Center

BV: Bowl Vent Port

C-4: Computer Controlled Catalytic Converter

CAN: Charcoal Cannister

CanCV: Cannister Control Valve

CANP: Cannister Purge Solenoid

CAP: Clean Air Package

CCC: Computer Command Control

CCP: Cannister Purge

CCS: Controlled Combustion System

CFI: Cross Fire Injection (GM)

CFI: Central Fuel Injection

CIS: Continuous Injection System

CO: Carbon Monoxide

COC: Conventional Oxidation Catalyst

CO2: Carbon Dioxide

C-EMF: Counter Electromotive Force

CP: Crankshaft Position Sensor

CPRV: Cannister Purge Regulator Valve

CRT: Cathode Ray Tube

CSC: Coolant Spark Control

CSSA: Cold Start Spark Advance System

CSSH: Cold Start Spark Hold System

CTAV: Cold Temperature Actuated Vacuum Switch

CTO: Coolant Temperature Override

CTS: Coolant Temperature Switch

CV: Control Valve

CWM: Cold Weather Modulator

DC: Direct Current

DFI: Digital Fuel Injection

DFS: Decel Fuel Shutoff

Dist-O-Vac: Distributor Modulator System

DMS: Distributor Modulator System

DOHC: Dual Overhead Camshaft

DRCV: Distributor Retard Control Valve

DVAC: Distributor Vacuum Advance Control Unit

DV/DV: Differential Valve/Delay Valve

DVOM: Digital Volt-Ohmmeter

DV-TW: Delay Valve—Two-Way

DVVV: Distributor Vacuum Vent Valve

ECA: Electronic Control Assembly

ECM: Electronic Control Module

ECT: Engine Coolant Temperature Sensor

ECU: Electronic Control Module

EDM: Electronic Distributor Modulator

EEC: Electronic Engine Control

EECS: Evaporative Emission Control System

EESS: Evaporative Emission Shed System

EFC: Electronic Fuel Control

EFE: Early Fuel Evaporation

EFI: Electronic Fuel Injection

EGC: Exhaust Gas Check Valve

EGO: Exhaust Gas Oxygen Sensor

EGOR: EGO Signal Return

EGR: Exhaust Gas Recirculation

EGRC: EGR Control Solenoid

EGRV: EGR Vent Solenoid

EMF: Electromotive Force

EPR: Exhaust Pressure Regulator

ESC: Electronic Spark Control

EST: Electronic Spark Timing

EST/ESC: Electronic Spark Timing with Electronic Spark Control

EVP: Exhaust Valve Position Switch

FBC: Feedback Carburetor

FBCA: Feedback Carburetor Actuator

FCS: Fuel Control Solenoid

FDV: Fuel Decel Valve

HALL: Hall-Effect Sensor

HC: Hydrocarbon

HCV: Exhaust Heat Control Valve

HEI: High Energy Ignition

HG: Mercury

HIC: Hot Idle Compensator

IAC: Idle Air Control

IAT: Intake Air Temperature Sensor

IMCO: Improved Combustion

IMS: Ignition Module Signal

IN.HG: Inches of Mercury

ISC: Idle Speed Control

ITS: Idle Tracking Switch

ITVS: Ignition Timing Vacuum Switch

IVV: Idle Vacuum Valve

KAM: Keep Alive Memory

KILO: (1000)

KNK: Knock Sensor

LED: Light-Emitting Diode

LOS: Limited Operation Strategy

MAF: Mass Air Flow Meter

MAP: Manifold Absolute Pressure

M/C Solenoid: Mixture Control Solenoid

MCT: Manifold Charge Temperature Sensor

MCU: Microprocessor Control Unit

MCV: Manifold Control Valve

MEGA: 1 million

MILLI: One-thousandth (1/1000)

MPC: Manifold Pressure Controlled Injection

MPI: Multi-Point Injection

MVPC: Motor Vehicle Pollution Control

NOX: Oxides of Nitrogen

OSAC: Orifice Spark Advance Control

PCV: Positive Crankcase Ventilation

PIP: Profile Ignition Pickup

PPM: Parts per Million

PPS: Ported Pressure Switch

PROM: Programmable Read-Only Memory

PSOV: Cannister Purge Shutoff Valve

PTC: Positive Temperature Coefficient Heater

PVS: Ported Vacuum Switch

RV: Rollover Valve

SA-FV: Separator Assembly-Fuel Vacuum

SCC: Stepped Speed Control

SDV: Spark Delay Valve

SFI: Sequential Fuel Injection

SHED: Sealed Housing Evaporative Determination System

SIL: Shift Indicator Light

SILN: Silencer Exhaust Air Supply

SOHC: Single Overhead Camshaft

SOL V: EGR Solenoid Vacuum Valve Assembly

SPOUT: Spark Output

SRTN: Signal Return Line

SSI: Solid-State Ignition

SV-CBV: Solenoid Valve Carburetor Bowl Vent

SVV: Solenoid Vent Valve

TAB: Thermactor Air Bypass Solenoid

TAC: Thermostatic Air Cleaner

TAD: Thermactor Air Diverter Solenoid

TAV: Temperature-Actuated Vacuum

TBI: Throttle Body Injection

TCP: Temperature-Compensated (Accelerator) Pump

TDC: Top Dead Center

TFI: Thick Film Ignition

TIV: Thermactor Idle Vacuum Valve

TK: Throttle Kicker Actuator

TKS: Throttle Kicker Solenoid

TPS: Throttle Position Sensor

TRS: Transmission Regulated Spark Control System

TSP: Throttle Solenoid Positioner

TSP-VOTM: TSP Vacuum-Operated Throttle Modulator

TVS: Thermal Vacuum Switch

TVV: Thermal Vent Valve

VACVV-D: Distributor Vacuum and Vent Control Valve

VACVV-T: Thermactor Vacuum and Vent Control Valve

VACW-D: Vent Valve Vacuum Delay

VAF: Vane Air Flow Meter

VAT: Vane Air Temperature

VCTS: Vacuum Control Temperature-Sensing Valve

VCV: Vacuum Check Valve

VDV: Vacuum Differential Valve

VIN: Vehicle Identification Number

VM: Vane Meter

VRDV: Vacuum Retard Delay Valve

VREF: Reference Voltage

VRV: Vacuum Regulator Valve

VSS: Vehicle Speed Sensor

VVA: Venturi Vacuum Amplifier

VVV: Vacuum Vent Valve

WOT: Wide-Open Throttle Valve

INDEX